Overseas Basing of U.S. Military Forces

An Assessment of Relative Costs and Strategic Benefits

Michael J. Lostumbo, Michael J. McNerney,
Eric Peltz, Derek Eaton, David R. Frelinger,
Victoria A. Greenfield, John Halliday, Patrick Mills,
Bruce R. Nardulli, Stacie L. Pettyjohn,
Jerry M. Sollinger, Stephen M. Worman

Prepared for the Office of the Secretary of Defense

The research described in this report was prepared for the Office of the Secretary of Defense (OSD). The research was conducted within the RAND National Defense Research Institute, a federally funded research and development center sponsored by OSD, the Joint Staff, the Unified Combatant Commands, the Navy, the Marine Corps, the defense agencies, and the defense Intelligence Community under Contract W91WAW-12-C-0030.

Library of Congress Cataloging-in-Publication Data

Lostumbo, Michael.
Overseas basing of U.S. military forces : an assessment of relative costs and strategic benefits / Michael J. Lostumbo, Michael J. McNerney, Eric Peltz, Derek Eaton, David R. Frelinger, Victoria A. Greenfield, John Halliday, Patrick Mills, Bruce R. Nardulli, Stacie L. Pettyjohn, Jerry M. Sollinger, Stephen M. Worman.
pages cm
Includes bibliographical references.
ISBN 978-0-8330-7914-5 (pbk. : alk. paper)
1. Military bases, American—Costs. 2. United States—Armed Forces—Foreign service. 3. United States—Defenses. 4. United States—Military policy. I. Title.

UA26.A2L67 2013
355.7068'1—dc23

2013013168

Published 2013 by the RAND Corporation
1776 Main Street, P.O. Box 2138, Santa Monica, CA 90407-2138
1200 South Hayes Street, Arlington, VA 22202-5050
4570 Fifth Avenue, Suite 600, Pittsburgh, PA 15213-2665
RAND URL: http://www.rand.org/
To order RAND documents or to obtain additional information, contact
Distribution Services: Telephone: (310) 451-7002;
Fax: (310) 451-6915; Email: order@rand.org

Preface

The United States has largely withdrawn its forces from Iraq, and it has identified the end of 2014 as the date when most forces will be out of Afghanistan. As U.S. forces from those conflicts return home, the Department of Defense (DoD) is reviewing its global basing structure to determine how it should be reconfigured to meet the strategic needs of the country. Congress has also turned its attention to future basing, and the conference report for the National Defense Authorization Act for Fiscal Year (FY) 2012 contained the following language:

> **SEC. 347. STUDY ON OVERSEAS BASING PRESENCE OF UNITED STATES FORCES.**
>
> (a) INDEPENDENT ASSESSMENT. The Secretary of Defense shall commission an independent assessment of the overseas basing presence of United States forces.
>
> (b) CONDUCT OF ASSESSMENT. The assessment required by subsection (a) may, at the election of the Secretary, be conducted by (1) a Federally-funded research and development center (FFRDC); or (2) an independent, non-governmental institute which is described in section 501(c)(3) of the Internal Revenue Code of 1986 and exempt from tax under section 501(a) of such Code, and has recognized credentials and expertise in national security and military affairs appropriate for the assessment.[1]

DoD asked RAND's National Defense Research Institute (NDRI) to conduct the requested independent assessment, and this report constitutes NDRI's response to that request.

This research was sponsored by the Office of the Under Secretary of Defense for Policy and conducted within the International Security and Defense Policy Center of the RAND National Defense Research Institute, a federally funded research and development center sponsored by the Office of the Secretary of Defense, the Joint Staff,

[1] U.S. House of Representatives, *Conference Report on H.R 1540, National Defense Authorization Act for Fiscal Year 2012*, Report 112-239, Washington, D.C.: U.S. Government Printing Office, December 12, 2011.

the Unified Combatant Commands, the Navy, the Marine Corps, the defense agencies, and the defense Intelligence Community.

For more information on the RAND International Security and Defense Policy Center, see http://www.rand.org/nsrd/ndri/centers/isdp.html or contact the director (contact information is provided on the web page).

Contents

Figures

Tables

Summary

The United States is at an inflection point in its defense planning due to a number of factors: the end of the Iraq War, the planned end of U.S. combat operations in Afghanistan in 2014, increased emphasis on security commitments and threats in the Pacific, and fiscal constraints. The 2012 Defense Strategic Guidance sets the course for this shift and has significant implications for overseas military posture, which needs to be designed to effectively and efficiently support the strategy as an integral component of overall defense capabilities. To that end, the National Defense Authorization Act for Fiscal Year (FY) 2012 directed the Department of Defense (DoD) to commission an independent assessment of the overseas basing presence of U.S. forces. The legislation specifically asked for an assessment of the location and number of forces needed overseas to execute the national military strategy, the advisability of changes to overseas basing in light of potential fiscal constraints and the changing strategic environment, and the cost of maintaining overseas presence. DoD asked the RAND National Defense Research Institute to carry out that independent analysis.[1]

Overseas posture should be designed as part of an integrated set of capabilities to execute the U.S. defense strategy. The starting point for this analysis was the strategy contained in the 2012 Defense Strategic Guidance and the development of an understanding of the capabilities that posture brings to bear. These capabilities—the benefits produced by overseas presence—include improving operational responsiveness to contingencies, deterring adversaries and assuring allies, and facilitating security cooperation with partner militaries. Posture also incurs risks associated with overseas facilities, including uncertainty of access in time of need and the vulnerability of such bases to attack from hostile states and nonstate actors, and costs. Basing U.S. forces abroad increases costs even in countries that provide financial and other support, with the amount varying by region and military service. To inform the assessment of overseas forces, we examined how overseas posture translates to benefits, the risks it poses, the cost of maintaining it, and how these costs would likely change were U.S. overseas presence to be modified in different ways, for example, by changing from permanent to rotational presence.

[1] The complete list of specific tasks Congress requested is provided in Chapter One.

This examination revealed some aspects of U.S. posture that are fundamental to carrying out the U.S. national security strategy. It also indicated that, beyond these enduring posture needs, there are posture changes involving both increases and reductions in overseas presence that could be advisable to consider, and these are identified in this report. Additionally, by identifying the benefits, risks, and costs associated with overseas posture, this report should inform more general deliberations about the U.S. posture now and in the future.

Strategic Benefits of Overseas Posture

Contingency Responsiveness

In-place forces provide the immediate capabilities needed to counter major acts of aggression by countries that the United States has identified as posing a substantial military threat to U.S. interests. Forward-based U.S. forces should be configured to provide the initial response necessary to prevent quick defeat while awaiting the arrival of aerial, maritime, and ground reinforcements—the last of which travel mostly by sealift. Initial response forces could be ground forces, such as those stationed in South Korea, or air or maritime forces. However, if ground forces must deploy even for short distances, the advantage gained from forward basing erodes or disappears if such forces do not have collocated, dedicated lift. This is especially true for heavy forces, which cannot deploy rapidly by air. In critical situations, lighter ground forces can deploy by air from the United States almost as quickly as they can from within a region. Additional aircraft can self-deploy, assuming they have access, and their support equipment can be airlifted or prepositioned in the region. Only when equipment has been prepositioned can heavy forces provide rapid reinforcement.

For smaller-scale contingencies, the starting location of lighter ground forces does not meaningfully influence deployment responsiveness, provided en route air bases with adequate throughput capacity are available. Overall response time, however, often hinges on the throughput capacity of the destination airfield, especially in more austere areas. Exceptions would be when multiple simultaneous events occur or other ongoing operations limit aircraft availability for a new mission. Over the long term, purchasing large fleets of intertheater cargo aircraft and forward basing overseas present alternative paths for enabling rapid deployment in small-scale contingency situations. That is, large lift fleets sized for major wars can support rapid response to globally distributed smaller-scale contingencies. Maritime forces that establish presence in new areas where events threaten U.S. interests can provide additional flexibility. These maritime forces also complement land-based presence in regions of enduring concern, when tensions rise. Forward, land-based presence does make a difference, however, for special operations forces performing missions in which mere hours can make a difference.

The U.S. strategy calls for global capabilities, so posture decisions should maintain an effective global en route infrastructure—facilities, access agreements, fuel storage, and other assets. This infrastructure must include multiple routes to key regions to ensure resiliency, to overcome the risks of natural and man-made disruptions, and to increase overall capacity. The United States can maintain global expeditionary capabilities and relatively rapid response capabilities as long as this infrastructure and a robust fleet of strategic lift assets are maintained. Broadly distributed maritime presence also strongly contributes to flexible rapid-response capabilities. A strategy that calls for protection against identified threats that could lead to major, high-intensity conflict must maintain some forces in place, supported by prepositioned equipment. In general, after the initial phase of operations to stabilize or even resolve a situation, the response by the U.S. military to a contingency of any substantial size will come primarily from forces deployed from bases in the United States.

Deterrence and Assurance

While the U.S. overseas posture does contribute to deterring potential adversaries and assuring friends and allies, it does not mean that all overseas facilities and forward capabilities can be justified on this basis; they are not all equally important in this regard. Deterrence relies on perceptions of the will of a nation and its abilities relevant to a particular conflict. The overseas posture contributes to both these aspects. The presence of U.S. forces in a region shows a commitment and U.S. interest in the security of the area, which speaks to the willingness of the United States to become involved in future conflicts to stabilize situations, secure U.S. interests, and protect the global commons. The forces there also help by providing relevant capabilities. In our assessment, the most important capability in this regard is an ability to prevent a quick victory by an adversary that could change the security situation on the ground.

The U.S. military presence in a region also helps to assure allies. It is a physical symbol of U.S. commitment to the security of a region, and in that sense, it can become a factor in the strategic calculations of allies. Without that assurance, they might make different choices that could influence a wide range of their strategic decisions: security policy choices, including formal and informal alliances; diplomatic positions; force structure choices; and budgetary decisions. While countries are no longer faced with the binary choice of the Cold War—between aligning for or against the Soviet Union—the United States still has an interest in harmonizing the security outlook and choices of allies. A U.S. military presence in or near an ally's territory can be an important factor in building and sustaining alliance relationships.

Certain types of capabilities are more likely to contribute to deterrence than others, particularly forces that can respond to prevent a quick victory and missile-defense capabilities to defend allies from coercive attacks. In some areas, like South Korea, this leads the United States to maintain continuous presence. In other areas, the United States may not have a permanent presence but does seek to maintain an ability to respond in times of crisis. Forces that can respond quickly include several different types of units—carrier strike groups (CSGs) and amphibious ready groups (ARGs)/ marine expeditionary units (MEUs)—give the United States presence in a number of potentially unstable regions, a variety of combat aircraft can quickly deploy to areas at risk, and Army airborne and some other units are configured to deploy quickly. For some crises, these quickly deployable forces will be sufficient. In others, they will play a role as an initial entry force, to be supported by larger deployments that take more time to deploy. In either case, the en route mobility infrastructure allows the United States to project substantial combat power around the globe, contributing to deterrence and assurance.

Security Cooperation

Forces based overseas benefit from the interoperability and adaptability skills and the greater cultural awareness gained from more frequent training with foreign partners. These skills are also important for U.S.-based forces to develop through rotational and temporary deployments. Security cooperation benefits the participating U.S. forces by training them to operate with foreign forces, both technically and culturally. To understand how military force can be used to build coalitions in support of U.S. interests and to influence adversaries takes considerable understanding of the customs and value systems of the foreign forces involved. Living and working on foreign soil offers opportunities for U.S. forces to experience these differences in depth and incorporate them into their skill set.

While the incremental costs of security cooperation activities are lower with U.S. forces based overseas, the savings are not close to sufficient to offset the higher costs of basing forces overseas. But, more important, security cooperation activities comprise a very small fraction of the operating costs of U.S. forces based overseas, in part because they can be combined with basic unit training needs or other activities. This low marginal cost leads to much greater frequency of security cooperation than would otherwise occur. In short, having overseas presence significantly increases the frequency and range of security cooperation activities.

While U.S.-based forces are capable of building partner-nation security capabilities, overseas basing is especially beneficial when conducting security cooperation activities with more advanced militaries, for example, those in Europe and South Korea. Forward basing helps strengthen personal and unit relationships, which are especially important for coalition interoperability. Most important, it provides frequent opportunities for intensive bilateral and multinational training, including specialized military capabilities. For other types of training in many parts of the world (e.g., foreign internal defense, peacekeeping, counterterrorism), use of rotational or temporary deployments is likely to be more cost-effective.

Given that forward-based forces appear to get the greatest security cooperation benefit from large-scale, multinational training, maintaining training facilities in

Europe and enhancing those in the Pacific could be valuable. In Europe, while rotational forces are planned to play a greater role in security cooperation, some level of forward-based forces and facilities to house, support, and train rotational units are important enablers. Substituting rotational forces for permanently stationed forces could increase flexibility to conduct security cooperation globally and provide opportunities for the benefits of security cooperation to accrue to a broader range of U.S. and foreign forces. On the other hand, it would risk reducing the depth of relationships and expertise that develop from more frequent security cooperation interactions engendered by close, continuous proximity. The Army's recent implementation of regionally focused units could help reduce some of the disadvantage in this area.

Risks of Overseas Posture

Political Risks

While the U.S. forward presence provides strategic benefits, it also carries with it a number of risks. U.S. peacetime military presence on foreign soil comes only with the acquiescence of the host nation. Therefore, if a host nation revokes U.S. access, DoD may be evicted from or prohibited from using bases where it has made significant investments. During a crisis, for example, the host nation might restrict the use of its facilities and territory. Access in a crisis should not be considered as binary (i.e., either providing full access or nothing at all). In practice, it tends to be granted by degrees. Some access limitations can be quite restrictive—for instance, limiting cooperation to overflight rights or limiting the number of landings allowed. Others may allow for some types of combat operations but not others, such as combat strike missions. Such restrictions can have operational effects, hindering the effectiveness of U.S. operations.

Political access cannot be guaranteed in advance, even when formal agreements exist, but there are factors that are likely to influence access decisions, such as the level of overlapping threat perception and interest, host-nation domestic public opinion about the conflict and the U.S. role in the conflict, and the perceived likelihood of reprisals. Moreover, some of these negative factors are more likely to influence the decisionmaking of unstable host nations. For example, if a host government faces significant internal instability, this could lead to a politically constrained view of acceptable U.S. access. While these access risks will endure, the United States can hedge against them by having diversity in its global presence. Relationships and facilities in several countries can provide alternatives if any one country chooses not to provide access during a future crisis. Still, this diversity of access locations comes at a cost, so carefully selecting the partners and the investments the United States makes in those partner nations will be an important part of a successful implementation strategy. This cost can be limited while mitigating some risk through the pursuit of access bases in some regions. Investments in these minimally manned access sites to enable future U.S. operations could be thought of as a form of war reserve.

Operational Risks

In recent years, the advent of long-range precision-guided weapons has put at risk a number of U.S. forces and facilities that previously enjoyed sanctuary, with further increases in the accuracy of such threats on the horizon. Some adversaries will have capabilities to inflict substantial damage on forward bases and forward-deployed forces, such as CSGs. Several current U.S. overseas facilities already face a substantial threat from these weapons—for instance, the accuracy and number of precision-guided weapons China currently fields are highly advanced. As a result, of particular concern is the large percentage of U.S. facilities that sit within high-threat areas, with over 90 percent of U.S. air facilities in Northeast Asia within heavy-threat areas from systems that China currently fields. While their capabilities are not currently as numerous or accurate, Iran and North Korea are investing in building such capabilities, and others could follow suit.

The impact of these weapons could be profound, potentially necessitating changes to U.S. military concepts of operations and force structure, as well as adjustments to basing and forward presence practices. If the United States is going to operate military forces within range of large numbers of such systems, it may need to employ a diverse strategy of active defenses, passive defenses, and either hardening or, when feasible, dispersal to reduce the effectiveness of such weapons. Essentially, a strategy would be for the United States to take away the easy and highly efficient use of such weapons, especially considering their limited supply.

Violent Extremism Risks

The U.S. military has suffered attacks from a number of different violent extremist groups. In considering risks to forward-deployed forces from violent extremist groups, the security of the facilities is not the only consideration. In many cases, U.S. military personnel will be most at risk when they are traveling outside of their work facility. In many instances, assessments of previous violent extremist activity in the area will be quite informative; however, such an assessment may miss the wider reach of some groups that have a capability to conduct operations far away from their traditional base of support.

Costs of Overseas Posture

In considering future posture changes, the condition of current facilities could influence those decisions, if conditions are poor enough that closure avoids large, future infrastructure reinvestments. Although the data on installation conditions are weak, when analyzed in combination with other qualitative evidence from U.S. military representatives overseas, they suggest that installation conditions overseas are at least as good as those in the United States and U.S. territories—possibly better. This implies that, given the small differences in average conditions, restoration and modernization needs in the United States would be about the same in relative terms for existing overseas facilities.

Despite substantial host-nation financial and in-kind support, we found that stationing forces and maintaining bases overseas does entail measurably higher direct financial costs to DoD. Host-nation support—substantial from Japan and South Korea (in terms of both in-kind support and cash payments) and from NATO allies (mostly through indirect in-kind support)—offsets some, but not all, of the higher costs of overseas basing, as well as the higher costs of having a more distributed basing structure. If the U.S. overseas posture were to shift toward less-developed areas of the world where resources are less plentiful, U.S. contributions could increase and those of host nations could decline, although the lower cost of living in some such areas could have a countervailing effect.

We found that there are annual recurring fixed costs to having a base open, ranging from an estimated $50 million to about $200 million per year, depending on service and region, with additional variable recurring costs depending on base size. This is important because it means that, if forces were to be consolidated on fewer, larger bases, whether in the United States or overseas, the fixed-cost portions of the closed bases would be saved. There are efficiencies to be gained from using fewer, larger bases rather than a more distributed posture. This effect, by itself, would be a significant contributor to cost reductions were forces realigned from overseas or inactivated in place. The fixed costs per base do not appear to be systematically higher overseas, with the exception of the Air Force bases, compared with facilities in the United States.

In contrast, the recurring variable costs per person are systematically higher overseas in Europe and the Asia-Pacific region due to higher allowances related to the cost of living, higher permanent-change-of-station move costs, and the need to provide schools more comprehensively, with the incremental overseas cost per person varying widely from about $10,000 to close to $40,000 per year. The variation depends on service and location, with factors such as dependent ratios, local cost of living, and housing type driving these differences. Thus, the cost effects of posture changes depend greatly on the service and region under consideration.

Combining analysis of variable costs with the fixed cost findings indicates that consolidating forces at fewer bases would provide more savings when the forces move to the United States and the overseas base closes, compared with consolidating two overseas facilities. The fixed costs would be saved whether consolidating in the United States or overseas, but closing an overseas base and consolidating in the United States also reduces variable costs due to the incremental overseas personnel-related costs. However, the United States cannot repurpose overseas bases like it can in the United

States, and repurposing U.S. bases could produce non-DoD economic benefits. We did not examine the benefits gained by the broader economy as the result of U.S. base closures.

The One-Time Costs from Closing Bases and Restationing Forces

By far, the largest one-time transition cost for closing bases and restationing forces to the United States would be the cost of construction when capacity limitations in the United States result in the need for new facilities. These costs are not incurred, however, if the units at the closed bases are inactivated as part of force reductions. These costs are also avoided if the units returning to the United States can use facilities that are vacated due to the inactivations of other units. Since the Army is planning to reduce its overall force by 80,000 troops, and posture options contemplate only a fraction of that number being realigned to the United States, we estimate lower and upper bounds for the construction costs of realigning Army units to the United States. While the Marine Corps is downsizing as well, the information we garnered indicates that any realignments would require new facilities. We assume that the Air Force and Navy would also have to expand their U.S. facilities if forces were realigned to the United States.

The Costs of Rotational Presence

As pressure has risen to consider reducing the permanent stationing of U.S. forces overseas, rotational presence is being increasingly considered to provide for some of the same benefits, because it is believed to be more efficient or at least less expensive. Our analysis indicates that whether this hypothesis is correct and the degree of the cost difference depends heavily on the rotational design (frequency and duration of rotations) and the type of permanent presence change.

Generally, we found that the savings produced by only realigning forces from an installation while keeping it open is not sufficient to offset the cost of providing full presence through rotational deployments. In most cases, realignments of permanent forces can underwrite only partial-year rotational presence in the same location. If an installation is closed as well, this will usually provide some net savings, albeit limited in some cases, even if the realigned unit is replaced with full-year rotational presence to the region. The net savings depend greatly on the service, unit type, location, and rotational design; for ground forces, sealift to move equipment or available equipment for prepositioning is necessary for savings. Furthermore, if a base were to be closed and its forces realigned, another permanent base in that country or region must be maintained to support the rotating forces, or a host nation must agree to provide access to one of its bases.

Note that our cost assessments of rotational presence include only the costs associated with supporting and moving units and people, assuming no additions in force structure would be needed to enable the rotations. Sustaining rotational presence in a location requires a "rotation" base in the force structure to enable personnel tempo goals, such as time at home between deployments, to be met. This report does not examine the associated constraints on increasing rotational presence by unit type and service. If additional units had to be added to the force structure to support rotations, this would add substantially to the rotational presence costs presented in this report and would likely make rotational presence more expensive than permanent basing in a location when the latter is an option.

Opportunities for Efficiencies and Reducing Costs

These cost considerations do suggest opportunities for efficiency, through two paths. The first is increased centralization, which is already being implemented in South Korea and Europe. The second is achieving presence through one or two long rotations per year to a location, accompanied by base closures while retaining at least one base with operating support in a region. Both of these should be considered in light of any negative effects and other objectives. In particular, more distributed forces can provide strategic advantages, long rotations could negatively affect quality of life, and consolidation could be detrimental in areas under high threat of precision-guided missiles.

Foundational Elements of Overseas Posture

The examination of strategic benefits made it clear that there are several elements of overseas posture that are vital for successful execution of the strategic guidance. A robust global en route infrastructure, in conjunction with substantial lift fleets and other global enablers, such as communications capabilities, provides the foundation for a global response capability that can leverage the entire force. This is complemented by the Navy and Marine Corps' at-sea deployments. In-place forces where major attacks are considered possible threats to U.S. security interests or allies are essential to deter high-end threats and prevent quick defeats in the event of aggression. The United States has also made commitments to some key allies to provide them with air and missile defense, necessitating forward ground and maritime forces to provide these capabilities. The combination of mandates to uphold commitments, preserve relationships with allies, and be able to counter major threats to national and global security requires at least some forces in Europe, the Middle East, and Asia—but how much in each is less clear and, thus, the subject of our discussion of options.

Analysis of Illustrative Postures for Insights on the Trade-offs Among Strategic Benefits, Risks, and Costs

To understand the consequences of changing the United States' current overseas posture, we developed three illustrative postures and applied our qualitative findings and quantitative models to determine how they would affect strategic benefits, risks, and cost. All of the postures share some foundational elements. Beyond these common aspects, each posture emphasizes a different goal—cost-reduction, global responsiveness and engagement, or major contingencies (see Table S.1). These alternative postures are not meant to be policy options, but rather are analytic tools that enabled us to evaluate the range of strategic benefits and costs that follow from revising U.S. overseas military presence. Because each illustrative posture prioritizes a particular objective, the analyses allow us to estimate the scope and type of effects that pursuing each objective would have in its purest form.

The illustrative cost-reduction posture (CRP) aims to minimize the cost of the U.S. global posture while simultaneously maintaining enough forward presence to achieve national security goals, including enabling global power projection and protecting the global commons. This posture rests on the notion that closing bases and bringing forces back to the United States would yield significant cost savings and on the assumption that the United States could meet its national security objectives with a smaller overseas presence in selected regions and through new means for maintaining alliances and pursuing security cooperation. This posture closes/realigns a substantial portion, but still a minority, of overseas facilities and forces. It represents the minimum forward military presence that the United States would need to remain a globally responsive military power.

The global responsiveness and engagement posture (GREP) aims to create an overseas military presence that maximizes the United States' ability to rapidly respond to smaller-scale contingencies and, to increase military burden sharing, to build the

Table S.1
Illustrative Postures

Illustrative Posture Type	Priority	Characteristics
Cost-reduction	Save money but retain ability to project power globally	• Fewer bases and forces overseas • Larger bases • Preserve key mobility infrastructure, expansible bases, multi-purpose facilities
Global responsiveness and engagement	Maximize U.S. ability to respond rapidly to small-scale contingencies and enhance partner capacity	• A hub with a number of access sites (spokes) in each region • Mixture of forces, especially those that are versatile • Distributed forces—permanent and rotational
Major contingency	Secure access to bases and position forces to deter and, if necessary, respond to Iran, North Korea, and China	• Additional primary bases with combat forces • Large number of dispersed expansible bases that forces frequently rotate to • Hardened bases • Concentrated in high-threat regions • Dispersal across threat rings • Increased rotations to reinforce high-threat zones

military capabilities of allies and partners and their willingness to participate in global security efforts. U.S. force posture would resemble a regional hub-and-spoke network in which permanently stationed U.S. forces consolidate at regional hubs (i.e., one or more primary bases) that can support rotational forces that periodically deploy to the spokes (i.e., access bases) for operations or exercises. The United States would station a mixture of forces at each hub to provide a wide range of capabilities for rapid response and engagement activities.

The illustrative major contingency posture (MCP) positions U.S. forces overseas so that they would be situated to deter or engage in large-scale operations against specific potential adversaries: Iran, North Korea, and China. The United States would place greater forces forward capable of conducting major operations against these potential adversaries. Conversely, the United States would divest itself of overseas bases and forces that would not be useful against one of these three adversaries. Consequently, the United States would retain only those bases in Africa and Europe that provide critical enabling capabilities for intertheater operations or that could be used for operations in the Middle East.

The following summarizes the analysis of the postures:

- The CRP is the only illustrative option that would reduce overall costs, illustrating a rough limit from posture changes of about $3 billion per year in savings, with a majority coming from Europe, after an initial investment with a 1.5- to 3-year payback. This would come at the expense of reduced levels of security cooperation activities and potentially assurance of allies.
- The GREP would expand security cooperation opportunities and create the potential for more robust access to bases for broadly distributed contingencies. Annual recurring costs would not change, but there would be meaningful transition costs to realign a small number of forces to provide the recurring savings to reinvest in rotations in new areas.
- The MCP would provide the highest level of deterrence and assurance of allies and partners for the three principal state-based security threats of concern. This would come at the expense of reduced security cooperation in Europe, where assurance of allies and partners could also decline. The MCP also risks increased exposure of forward-stationed forces to anti-access threats, and it would add annual recurring costs as well as require significant investment.

Analysis of the illustrative postures led to several insights. Only by substantially reducing forces and bases in one or more regions and limiting the level of replacement by rotations would posture changes yield meaningful savings. This would force one or more trade-offs in strategic benefits. Conversely, it appears to be infeasible to increase engagement substantially with new partners while also significantly reducing overall costs. Realigning forces from one region to the United States to produce operating cost savings to be reinvested elsewhere, whether for rotational or permanent presence, for operating-cost neutrality is likely to require some investment. Similarly, increasing presence for specific major threats could require substantial investments. The contrasts between the CRP and the other two postures suggest that implications for security cooperation, deterrence, and assurance are likely to be greater than for global responsiveness and access risk when considering posture options, as long as the options protect global en route infrastructure, emphasize maintenance of geographically distributed access to bases, and maintain maritime capabilities.

Posture Options to Consider Depending Upon Strategic Judgments

Europe

Further posture reduction could be considered in Europe, but this could have negative repercussions for alliance cohesion, assurance of partners, and interoperability. Europe has long hosted the bulk of U.S. overseas forces, but that presence has been reduced substantially over the last 20 years. The forces that remain in Europe focus particularly on security cooperation, so further reductions would limit those activities, with air bases also enabling direct operational support around the European periphery. Further reductions could be made as part of overall defense-resource trade-offs to reduce costs or to meet needs in other regions, but may be detrimental to the NATO alliance.

If substantial reductions were made, limiting continuing presence to the maintenance of capabilities for global power projection, bases for operations around the periphery, and forces for formal commitments, the United States could save up to $2 billion per year. This would diminish security cooperation activities, with impact greatest in three categories of security cooperation: (1) multinational training capacity, for example, through the closure of the Joint Multinational Training Command (JMTC); (2) forces that focus on the strategic and operational level of engagement, such as headquarters units; and (3) enabling units that build specialized capabilities, such as logistics, medical, air-ground operations, and intelligence units.

Some of the negative effects of reductions in permanent presence might be mitigated by using rotational forces and more specialized capabilities (e.g., Special Operations Forces, missile defense) to replace some of the lost presence. If training with European allies remains a priority, then JMTC would likely need to be retained. However, retaining JMTC and achieving the same level of tactically oriented security cooperation from the United States through rotations would consume about half the potential savings. Higher, strategic-level engagement could be hindered without keeping major commands in Europe with high-ranking flag officers and their staffs. This could be done without high levels of assigned forces, though major force reductions could result in the loss of U.S.-held senior leadership positions within NATO.

Pacific

In Asia, the United States faces competing demands. The United States aims to deter North Korea and other major conflicts in Northeast Asia, but the concentration of U.S. forces in South Korea and Japan keeps those forces under threat from numerous precision-guided missiles. In the meantime, the United States has an interest in increasing security engagement with partners in South and Southeast Asia, where the United States has a much smaller presence consisting of rotational forces.

The majority of the U.S. facilities in South Korea and Japan sit in the heaviest threat zones and face potential long-range threats from China's precision-guided weapons. If the United States wishes to maintain a forward presence at these locations to deter and assure, there are divergent options that could reduce vulnerabilities to attack. These facilities could be hardened and protected with missile defenses, or the number and mix of aircraft and ships could be reduced and restationed elsewhere in the Pacific—basing availability permitting—or in the United States. Such protections would not make these bases invulnerable to attack, but they can be valuable if they take away fairly easy and efficient ways to disable bases and destroy forces.

Through rotational presence in Singapore, the Philippines, and Australia, the United States is trying to increase the level and sophistication of activities with those countries and other regional partners. Whether policymakers view this as sufficient could have implications for the overseas posture. Currently, no partner nations in South or Southeast Asia have offered access to their territory for the permanent presence of U.S. forces. Given this reluctance, if greater security cooperation is desired, the Navy or Marine Corps presence in the region may be the first option, because they do not rely on significant levels of host-nation hospitality. Alternatively, additional countries may agree to host rotational presence of U.S. forces to facilitate expanded interactions.

Related to this rotational presence and broader efforts at engagement, the Marine Corps posture in the Pacific is in transition. In accordance with agreements with Japan and Australia, the Marine Corps plans to reduce some forces in Okinawa, maintain a rotational presence in northern Australia, establish a presence in Guam, and increase forces in Hawaii. However, if Marine Corps forces distributed in the Pacific do not gain the dedicated lift that would enable them to take advantage of their positioning, it may be advisable to consider shifting some of them to the continental United States, given the lower costs there compared with Pacific island locations. For humanitarian response and security cooperation, the 31st MEU, with a collocated ARG in the Pacific, provides unique capabilities. The absence of dedicated lift for the other ground and logistics forces in Okinawa or rotary-wing aviation in other parts of the Pacific makes their forward position less of an advantage. Depending on how decisionmakers assess the benefit that additional Marine Corps forces beyond the 31st MEU contribute by being based in Okinawa or elsewhere in the Pacific with respect to assurance, security cooperation, or responsiveness, keeping them there merits weighing against the somewhat higher costs, the potentially limited mobility advantage, the potential threats to Okinawa from China, and the opposition in some quarters in Okinawa to a continued U.S. presence there. Among these considerations, the biggest is likely to be how a reduction of forces in the region would affect Japanese and other nations' perceptions of U.S. commitments to the region. The broader decision to keep these forces in the Pacific also merits linkage to Navy force structure and positioning considerations with respect to amphibious ships.

Overall, depending on how decisionmakers judge the likely effect of modest force reductions in Asia on regional perceptions of the U.S. commitment to the region, how critical they believe large in-place forces are to deterrence, and the degree to which forces should be kept in higher-threat zones, modest reductions in the Asia-Pacific region, including some of the Marine Corps forces and an Air Force base and wing, could produce some savings—contributing roughly equal amounts of up to $450 million per year—while preserving in-place forces in South Korea and some additional capabilities in Japan for broader regional security. This would reflect the call for pursuing new approaches to defense in the face of resource constraints. Any of these steps, though, might appear incompatible with the U.S. government's stated intention to rebalance toward Asia, even if alternative approaches could provide similar capabilities. Concerted efforts to explain to allies how security could still be provided would have to be made, with some risk of not fully assuring key U.S. allies in the region.

Alternatively, emphasizing different aspects of the 2012 Defense Strategic Guidance could lead to increased presence in Asia and the Pacific. If increased security cooperation in South and Southeast Asia is highly valued and increases in rotational presence are pursued, this would increase costs. However, if done in combination with modest reductions in Northeast Asia, costs in the greater region might be held relatively steady. If such rotations were added while maintaining or increasing presence oriented toward meeting perceived needs to increase assurance and deterrence, then annual recurring costs in the region would increase, potentially substantially. Any costs for hardening of facilities or additional force structure to support rotations would be in addition to the cost estimates in this report. The region presents a complex set of judgments and trade-offs regarding assurance, deterrence, security cooperation, and risks, with a range of options corresponding to different judgments on how different posture choices are likely to affect these factors.

The Middle East

As a result of the wars in Iraq and Afghanistan, the United States currently has substantial forces in the Persian Gulf, but the number and composition of any remaining forces after the drawdown in Afghanistan remains undetermined. As noted in the 2012 Defense Strategic Guidance, the United States intends to "continue to place a premium on" presence in the region. In addition to maintaining capabilities to counter violent extremists and uphold commitments to partner states in the region, the United States has an interest in preventing Iran from disrupting commerce, seeking to politically pressure or destabilize neighboring states, or developing the capability to threaten regional states through nuclear coercion.

The United States currently has a network of air bases, significant maritime presence, and prepositioned equipment in the Persian Gulf region, with plans for ground-force rotations. U.S. military presence in the region is predominantly rotational, as most host nations prefer that to permanent presence, and infrastructure to host these rotations is maintained. Whether further increases to this presence would improve deterrence and regional stability or would be needed to effectively respond to aggression—were deterrence to fail—is not clear.

Foreign military presence has long been a sensitive political issue for many Middle Eastern countries. Hard to gauge are the potential political risks of increasing forces in the region, the willingness of regional leaders to accept this presence, or problems that such a sustained, significant presence could pose to partner nations. This could come from domestic sources, where such presence could spur opposition to the regimes in the host nations, from other states, or nonstate actors in the region. If a host-nation government faces the prospect of significant internal unrest, decisionmakers may want to weigh carefully whether they continue to make investments to military facilities in that nation. Political instability could well result in diminished or lost American access, as well as new security concerns. On the other hand, presence could facilitate improvement in partner capabilities and strengthened relationships, in addition to contributing to deterrence of potential adversaries and assurance of partners.

Thus, the central posture question in the region is how responsiveness and deterrence needs in the Persian Gulf should be weighed against the potential for political tensions and risks. Depending on the weight given to these two competing sets of factors, decisionmakers could elect to selectively reduce rotations in the region, maintain the status quo, or seek to increase rotations to the region across the services. To give some sense of the costs that could be avoided or how they would increase as a result of these choices, annual armored brigade combat team (ABCT) rotations to Kuwait would cost roughly $200 million per year, maintaining a composite air expeditionary wing through continuous rotations costs about $300 million per year, and quarterly fighter-squadron rotations would be $50–100 million per year, depending on the aircraft types and how the rotations are executed and not accounting for the possibility of any needed increases in force structure to provide a sufficient rotation base.

Posture Choices

Tables S.2 and S.3 highlight a few of the major posture choices that emerge from a consideration of the strategic benefits, risks, and costs of posture changes. They consider both potential reductions in current posture as well as potential additions. In both

Table S.2
Current Elements of Overseas Posture That Could Be Evaluated for Realignment/Closure

Shift in Priority or Evaluation of Needs	Potential Realignment/Closure
Less security cooperation in Europe	• Most Army units and bases in Europe • Some Air Force units in and bases in Europe (some need to be retained for global mobility and bases from which to execute operations)
High anti-access/area-denial missile threat in Asia	• Some reduction in air units and bases in Japan or South Korea • III MEF HQ and ground forces in the West Pacific (retain MEU)
Limited assurance and deterrence value	• III MEF HQ and ground forces in the West Pacific (retain MEU)
Limited deterrence benefit in the Middle East	• Reduced rotations in the Middle East

Table S.3
New Elements of Overseas Posture That Could Be Considered

Shift in Priority or Evaluation of Needs	Potential Addition
More security cooperation emphasis with new partners	• Increased rotations to Southeast Asia, Africa, and Eastern Europe • Additional ARG in the West Pacific
Increased risk of Iranian aggression	• Increased rotations to the Middle East—all services • Increased air and missile defense assets • Increased armor prepositioning
High anti-access/area-denial missile threat in Asia	• Hardening of bases • Increased access to partner bases across the Asia-Pacific region
Increased need for assurance of Asian partners	• Increased air and naval presence

cases, the purpose is to highlight how changing concerns and priorities might translate into potential posture changes.

Significant savings would require choosing from what we find to be a relatively small set of options. The only substantial ones would be Army and Air Force units and bases in Europe. Smaller opportunities would be some of the Marine Corps and Air Force forces in the Pacific and rotational forces in the Middle East. Reductions in Asia are likely to create more deterrence and assurance risk than reductions in Europe, while reductions in the Middle East would have mixed effects. Reductions in Europe would likely affect security cooperation.

Potential increases revolve around three considerations: the value of increasing security cooperation with new partners to build capacity, posturing to deter and respond to potential Iranian aggression, and pivoting to Asia for increased deterrence and assurance of allies. Pursuing the types of options in Table S.3 would increase recurring costs and would involve additional investments in some cases. They could be pursued independently or in conjunction with some of the reduction options (to reflect shifting priorities) to reduce the cost impacts.

There are some clear limits to how far consolidation in the United States could be pursued, beyond which achieving national security goals and executing the 2012 Defense Strategic Guidance would become untenable. There is a minimum threshold of foreign posture that the United States must retain. Beyond that, there is additional posture that is almost certainly advisable to retain or even add. But there are a number of choices in each region for which different judgments could lead to differing calculations of the advisability of reductions, additions, or changes in the nature of posture. These posture options for potential consideration represent policy choices that do not have any one empirical "answer"—only the cost side of the equation can be determined with some degree of certainty. Instead, decisions will reflect judgments based on the perceived values assigned to the competing goals—i.e., how they are prioritized—and the degree to which overseas posture is perceived to advance strategic goals.

Acknowledgments

We deeply appreciate the contributions of several of our RAND colleagues. Caroline Baxter provided invaluable assistance in creating and applying Army and Marine Corps rotational cost models, which provided quantitative analysis that informed the report's conclusions. Abby Doll was instrumental in helping prepare the chapter on Analysis of Illustrative Postures, including compiling data, write-ups, integration of material from several other chapters, and cross-checking of information. Scot Hickey programmed and applied the deployment response and lift requirements model. Colleagues Alan Vick, Wasif Syed, Mike McMahon, Tony Atler, Scott Boston, Richard Darilek, and Brad Martin assisted with the development of the illustrative postures. Mike McMahon and Brad Martin helped shape the discussions of Navy and Marine Corps presence and posture considerations. Arthur Bullock conducted queries on the RPAD database for installation conditions. Pat Boren provided second destination transportation data, and Jennifer Moroney, Jessica Yeats, Brian Gordon, MAJ Christopher Springer, and MAJ Stoney Trent provided valuable insights on security cooperation. We also deeply appreciate the members of the Illustrative Postures Red Team: Lynn Davis, Paula Thornhill, Tom Szayna, John Yurchak, Todd Nichols, and Anthony Rosello. Our colleague Bryan Hallmark and Army Fellow MAJ Michael Baim deepened our understanding of training issues. Mark Totten conducted queries on the Work Experience and monthly Active Duty Pay files to generate the personnel allowance estimates. Daniel Sommerhauser assisted with the development of the statistical models for the cost analysis. We also acknowledge the invaluable leadership and guidance of James Dobbins, director of the International Security and Defense Policy Center.

We also recognize the invaluable assistance provided by our sponsors and other members of the Department of Defense and the military services who gave unstintingly of their time and facilitated our access to information. Deputy Assistant Secretary of Defense for Plans Robert Scher and his staff ensured that we had the support across the department to get the information required for the study. Within his staff, our daily contact, Nina Wagner, was exceptional in requesting and following up on needed data and coordinating with others across the department. Within Europe, the commanders of U.S. Air Forces in Europe and U.S. Army, Europe, and the deputy commander of U.S. European Command—Gen Philip Breedlove, LTG Mark Hertling,

VADM Charles Martoglio—gave us their valuable perspectives and ensured their staffs went the extra mile in providing us with information and discussing key issues. We owe a special thanks for LCDR Todd Phillips for organizing the visit and traveling with us and Al Viana for overall coordination. The people in their organizations as well the other combatant commands, the services' headquarters, and offices across the Office of the Secretary of Defense were critical to enabling the study. They include, among others:

Rob Presler, RDML Rick Snyder, CDR Scott Starkey, John Hamilton, Millie Waters, Jeffrey Hall, Col Kevin Fox, Col Scott Shapiro, Maj Benjamin Oakes, COL John Morehead, LTC Ivan Udell, Tim Rose, LCDR Mike Williams, CDR Matt Konopka, Gregg Nishimura, MG James Boozer, COL Andrew Heppelmann, LTC Robert Murphy, MAJ Garrett Trott, MAJ Thomas Switajewski, COL Stuart Bradin, CDR Marvin McGuire, Michael Fitzgerald, Michael Ramirez, Maj Benjamin Dainty, Lt Col Kevin Schiller, SMSgt Leonard, Lt Col James Damato, John Faulkner, Lt Col Trevor Matsuo, Michael Koning, Steve Billington, Maj Julie Gaulin, Maj Matthew Ashton, Maj Kevin Lord, Saxon Yandell, Maj Kevin Lee, BG Kevin McNeely, CAPT Brian Hoyt, Col Aaron Maynard, LCDR Doug Peterson, Don Timm, Col Phil Everitte, CDR Mike Riordan, Tom Roberts, LCDR Dude Underwood, Bob Chadwick, Allison Sands, Patricia Bushway, Robert Lange, Robert Coffman, Marianne Petty, CAPT David Berchtold, Larry Klapper, Richard Snow, Maria Probst, Lee Alloway, Chris MacPherson, Tony Cusimano, Dave Meyer, Clifford Ivery, William Lively, John Wicinski, Hans Filip, Brian Ondrick, and Robert Brady.

Personnel at U.S. Pacific Command who were generous with their time and assistance include MG Anthony G. Crutchfield, Col Jeffrey D. White, CDR Homer Denius, Daniel Garcia, Lt Col Jeffrey T. Kronewitter, William J. Wesley, COL Boccardi, Maj Todd Larsen, D. Edward Durham, and LTC Christopher S. Martin.

We also acknowledge the invaluable contributions of our reviewers. These include Michael O'Hanlon from the Brookings Institution and our RAND colleagues James Hosek, Jim Quinlivan, and Daniel Norton. Their careful reviews substantially improved the report. We also gratefully acknowledge the work of our editor, Bryce Schoenborn, our graphic artists, Mary Wrazen and Sandra Petitjean, and our production editor, Matthew Byrd. Their diligent and timely work was instrumental in meeting our very demanding deadlines.

Finally, we recognize the indispensable administrative support provided by Angelina Becerra, Jocqueline Johnson, and Erin Smith.

Abbreviations

AAMDC	Army Air and Missile Defense Command
ABCT	armored brigade combat team
ABW	air base wing
ACF	area cost factor
AEG	air expeditionary group
AEW	air expeditionary wing
AFPAM	Air Force Pamphlet
AFRICOM	U.S. Africa Command
AFTOC	Air Force Total Operating Cost
AMC	Air Mobility Command
AMS	avionics maintenance squadron
AOC	air operations center
AOR	area of responsibility
APOE (D)	aerial port of embarkation (debarkation)
ARG	amphibious ready group
AT&L	Under Secretary of Defense for Acquisition, Technology and Logistics
AWACS	Airborne Warning and Control System
BAH	basic allowance for housing
BCT	brigade combat team
BEAR	basic expeditionary airfield resources
BMD	ballistic missile defense

BRAC	Base Realignment and Closure
CAB	combat aviation brigade
CAOC	combined air and space operations center
CENTCOM	U.S. Central Command
CIVPERS	civilian personnel
CJTF-HOA	Combined Joint Task Force–Horn of Africa
COCOM	combatant command
COLA	cost of living allowance
CONUS	continental United States
CRP	cost-reduction posture
CSG	carrier strike group
CSL	cooperative security locations
CVN	carrier
DDESS	Domestic Dependent Elementary and Secondary Schools
DDG	destroyer
DLA	Defense Logistics Agency
DMDC	Defense Manpower Data Center
DoD	U.S. Department of Defense
DoDDS	Department of Defense Dependents Schools
DoDEA	DoD Education Agency
EAD	echelon above division
EPAA	European Phased Adaptive Approach
EUCOM	U.S. European Command
FCI	facility conditions index
FCM	FORCES Cost Model
FDD	forward distribution depot
FDNF	Forward Deployed Naval Forces
FID	foreign internal defense

FIP	Facilities Improvement Program
FORCES	Force and Organization Cost Estimating System
FOS	forward operating site
FRP	fleet response plan
FW	fighter wing
FY	fiscal year
GCC	Gulf Cooperation Council
GDPR	Global Defense Posture Review
GREP	global responsiveness and engagement posture
HNS	host-nation support
HUMRO	humanitarian relief
IBCT	infantry brigade combat team
ISR	intelligence, surveillance, and reconnaissance
ISR-S	Installation Status Report for Services
JCS	Joint Chiefs of Staff
JMTC	Joint Multinational Training Command
LCS	littoral combat ship
LMSR	large, medium-speed, roll-on/roll-off ship
MAGTF	marine air-ground task force
MARFORAF	Marine Corps Forces, Africa
MARFOREUR	Marine Corps Forces, Europe
MARFORPAC	Marine Corps Forces, Pacific
MAW	marine aircraft wing
MCAS	Marine Corps Air Station
MCM	mine countermeasures
MCO	major combat operation
MCP	major contingency posture
MCPP-N	Marine Corps Prepositioning Program–Norway

MDS	Mission Design Series
MEB	marine expeditionary brigade
MEF	marine expeditionary force
MEU	marine expeditionary unit
MILCON	military construction
MK	Mihail Kogalniceanu (air base in Romania)
MOB	main operating base
MOD	Ministry of Defense
MOFA	Ministry of Foreign Affairs
MOG	maximum on ground
MPSRON	maritime prepositioning ships squadron
MSC	Military Sealift Command
MTON	measurement ton
NATO	North Atlantic Treaty Organization
NDAA	National Defense Authorization Act
nm	nautical mile
NORTHCOM	U.S. Northern Command
NRF	NATO Response Force
NSA	naval support activity
NSIP	NATO Security Investment Program
OCO	overseas contingency operation
OCONUS	cutside the continental United States
OEF	Operation Enduring Freedom
OHA	overseas housing allowance
OIF	Operation Iraqi Freedom
OOT	over the ocean transportation
OSD	Office of the Secretary of Defense
PACAF	Pacific Air Forces

PACOM	U.S. Pacific Command
PCS	permanent change of staion
PE	program element
PERSTEMPO	personnel tempo
PHIBRON	Amphibious Squadron
PKK	Kurdistan Workers' Party
PKO	peacekeeping operations
prepo	prepositioning
PRV	plant replacement value
QDR	Quadrennial Defense Review
R&M	restoration and modernization
RAF	Royal Air Force
RPAD	real property database
SACO	Special Action Committee in Okinawa
SBCT	Stryker brigade combat team
SDT	second destination transportation
SF	Special Forces
SMA	special measures agreement
SOFA	status of forces agreement
SOUTHCOM	U.S. Southern Command
SSN	attack submarine, nuclear
THAAD	Terminal High-Altitude Area Defense
TRANSCOM	U.S. Transportation Command
TSC	theater sustainment command
UAE	United Arab Emirates
UAV	unmanned aerial vehicle
UDP	Unit Deployment Program
UIC	Unit Identification Code

UK	United Kingdom
USAF	U.S. Air Force
USAFE	U.S. Air Forces in Europe
USAG	U.S. Army garrison
USAREUR	U.S. Army, Europe
USARPAC	U.S. Army, Pacific
USNAVEUR	U.S. Naval Forces Europe
USPACFLT	U.S. Pacific Fleet
U.S.C.	U.S. Code
USFJ	United States Forces, Japan
USFK	United States Forces, Korea
USMC	U.S. Marine Corps
USN	U.S. Navy
WMD	weapons of mass destruction
WRM	war reserve materiel

CHAPTER ONE

Introduction

The U.S. military posture offers broad scope and geographic reach to contribute to the nation's security objectives. It is a physical expression of the enduring global interests of the United States. The presence of U.S. forces and access to bases in so many countries provides flexibility to address those security objectives. It allows U.S. forces to respond quickly to a variety of situations, such as natural disasters and countering piracy. It allows U.S. forces to train more often with partners and, of course, to fight the nation's wars.

The United States not only responds to world events but also seeks to shape them, and the U.S. military serves as an important instrument in this effort. A forward presence allows for more opportunities to engage allies and partners, to enhance capabilities for collective security, and to build coalitions. It also influences the behavior of those who might disrupt the international order.

These benefits were encapsulated in the 2010 Quadrennial Defense Review (QDR) Report, which asserts that U.S. military personnel forward-stationed or rotationally deployed "help sustain U.S. capacity for global reach and power projection."[1] The QDR Report notes that overseas basing deters adversaries, assures allies and partners, supports partnership capacity-building efforts, and supports efforts to respond to contingencies.[2]

While the U.S. global defense posture provides certain unique advantages, it comes with associated costs and risks. Maintaining forces overseas or deploying them temporarily increases costs over stationing and training them within the United States. The nations who host U.S. forces sometimes restrict their activities, and in some cases the forward locations are in areas of heightened risks, either because of adversary military capabilities or greater exposure to violent extremist groups.

The U.S. military has maintained a substantial overseas presence since World War II. While there has been continuity in the countries that host major U.S. military facilities, the number and types of forces in those facilities have changed substantially.

---

[1] Department of Defense (DoD), *Quadrennial Defense Review Report*, Washington, D.C., February 2010a, p. 62.

[2] DoD, 2010a, pp. 62–64.

Today, the war in Iraq has ended, and a transition of responsibility for security to Afghan forces has begun. With the end of these conflicts, the national security focus is shifting to the broader range of challenges that confront the United States. These changes and U.S. fiscal pressures also create an appropriate point to consider the U.S. military overseas posture that will serve the interests of the American people in the coming years. This is clearly recognized in the 2012 DoD Strategic Guidance.

The 2012 Defense Strategic Guidance includes the same posture objectives as the 2010 QDR, while emphasizing the needs to "rebalance toward the Asia-Pacific region," continue to "place a premium" on presence in the Middle East, and evolve global posture in line with the changing security environment—most notably in Europe while maintaining North Atlantic Treaty Organization (NATO) commitments. "Building partnership capacity" is emphasized throughout while calling for a rethinking of how this can be efficiently achieved, such as with rotations and advisory capabilities and the need to "make thoughtful choices" in light of the strategic shifts and constrained resources. The 2012 Defense Strategic Guidance seeks to maintain global presence, the ability to respond globally, and the wherewithal to protect global "freedom of access."[3]

Purpose

The National Defense Authorization Act for Fiscal Year 2012 (Public Law 112-81) directed DoD to commission an independent assessment of the requirements for and the costs of basing forces outside the United States. DoD asked RAND's National Defense Research Institute, a federally funded research and development center, to carry out the independent analysis. The legislation indicated that the assessment should include but not be limited to the following five tasks:

1. An assessment of the location and number of United States forces required to be forward based outside the United States to meet the National Military Strategy, 2010; the QDR; and the engagement strategies and operational plans of the combatant commands.[4]
2. An assessment of the following aspects of the defense posture:
    a. the current condition and capacity of the available military facilities and training ranges of the United States overseas for all permanent stations and deployed locations, including land and improvements at such facilities and ranges and the availability of additional land, if required, for such facilities and ranges

[3] DoD, *Sustaining U.S. Global Leadership: Priorities for 21st Century Defense*, Washington D.C., January 2012a.

[4] *Sustaining U.S. Global Leadership: Priorities for 21st Century Defense* was released after Public Law 112-81 was enacted; hence it was not referenced. We assume in this report that the assessment should also consider this new strategy document.

b. the cost of maintaining such infrastructure.

3. A determination of the amounts received by the United States, whether in direct payments, in-kind contributions, or otherwise, from foreign countries by reason of military facilities of the United States overseas.
4. A determination of the amounts paid by the United States in direct payments to foreign countries for the use of facilities, ranges, and lands.
5. An assessment of the advisability of the retention, closure, or realignment of military facilities of the United States overseas, or of the establishment of new military facilities of the United States overseas, in light of potential fiscal constraints on DoD and emerging national security requirements in coming years.

The 2012 Defense Strategic Guidance articulates several goals relevant to posture: being able to respond to aggression and crises globally, retaining a stabilizing international presence, protecting a free and open international economic environment, deterring malevolent actors, and assuring and building the capacity of current and future partners. However, there are costs to employing an overseas posture to accomplish these goals, and the guidance emphasizes the need to make judicious choices in an era of resource constraints. It is valuable to make these costs clear so that the benefits and costs of overseas posture can be considered within overall defense planning and the broad set of resource and budget trade-offs.

To accomplish the spectrum of tasks, this report provides an assessment of the critical underpinning elements of U.S. overseas posture, the benefits it provides, and what it costs in light of facility conditions, host-nation support, and unique financial factors. We analyze the strategic benefits, risks, and costs of overseas posture, and we assess the degree to which different posture alternatives would affect these benefits, risks, and costs. This informs our assessment of the advisability of changes to U.S. overseas posture and also enables this report to serve as a decision-support aid for national security planners and decisionmakers in overseas posture option development and decisionmaking deliberations.

Approach

To address task 1, we sought to determine the minimum essential elements of posture required to execute the U.S. national security strategy and meet formal commitments. In some cases, we identify specific locations and unit types that are fundamental to meeting the strategy. An example of what we mean by an *element* of posture is the global mobility infrastructure that supports deployments and sustainment from the United States or even from one region to another. Within this element, there may be some bases that are essential, or there may be others that could be equally valuable alternatives, as long as a region is sufficiently covered.

Beyond determining the minimum essential needs in general terms with some key facilities highlighted, three factors preclude a precise, comprehensive, location-by-location, unit-by-unit response to task 1. First, as will be discussed in this report, many of the strategic benefits of overseas posture cannot be definitively measured and quantified, so it is not possible to define precisely how an individual facility or unit presence achieves national security goals. There is some band of uncertainty producing a range of postures that could be considered as fulfilling the intent of the strategy, based on different judgments regarding the benefits and risks. Second, in some areas, there are multiple locations and forms of presence that can play similar roles or provide similar benefits in pursuit of achieving national security goals. Third, the engagement strategies and operational plans of the combatant commands are based in part on their assigned forces and other resources available to them. The combatant commands should incorporate available resources into plans and will naturally do so. However, there may be other means for accomplishing objectives, and there may be cases where thresholds for achieving an objective are exceeded, opening room for discussions of rebalancing in the face of constraints as implied by task 5. Given these factors, the report recognizes that, beyond what we can clearly identify as needed minimums, other aspects of posture are likely advisable and cost-effective to maintain or even add—in conjunction with possibilities for realignment and closure.

Overseas posture provides several strategic benefits that can be valued only qualitatively through expert judgment, informed by the available theory, evidence, and experience. These include responsiveness to contingencies, assurance, deterrence, and the advancement of security cooperation. There are also risks associated with overseas posture that can hinder its value in achieving these benefits. As a starting point, we assessed how overseas posture contributes to these strategic goals and what factors generate associated risks. Then we examined how changes in postures would affect pursuit of these goals and the risks. The benefits and risks should be weighed against the costs, with the weights of the strategic benefits dependent upon the judgments of decision-makers. To enable this, we developed a detailed, comprehensive cost model to estimate the cost effects of changes in overseas posture to include unit stationing, facilities, and rotational presence costs. Other portions of the analysis directly address tasks 2, 3 and 4 to the extent feasible based on available data, with the results supporting the development and evaluation of posture options.

To inform decisions about "the advisability of the retention, closure, or realignment" of overseas military facilities called for in task 5, we developed three illustrative postures that emphasized different goals for overseas posture and costs to help us understand how possible changes could affect strategic benefits, risks, and costs. This was intended to develop an understanding of the trade-offs encountered when considering alternative overseas postures—and on what information and judgments decisions might turn. The evaluations in the report are intended to serve as decision-support aides, including the results of our cost assessment of each element of overseas

States will continue to place a premium on U.S. and allied military presence in – and support of – partner nations in and around this region.

Most European countries are now producers of security rather than consumers of it. Combined with the drawdown in Iraq and Afghanistan, this has created a strategic opportunity to rebalance the U.S. military investment in Europe, moving from a focus on current conflicts toward a focus on future capabilities. *In keeping with this evolving strategic landscape, our posture in Europe must also evolve.* As this occurs, the United States will maintain our Article 5 commitments to allied security and promote enhanced capacity and interoperability for coalition operations. In this resource-constrained era, we will also work with NATO allies to develop a "Smart Defense" approach to pool, share, and specialize capabilities as needed to meet 21st century challenges.

Building partnership capacity elsewhere in the world also remains important for sharing the costs and responsibilities of global leadership.... *Whenever possible, we will develop innovative, low-cost, and small-footprint approaches to achieve our security objectives, relying on exercises, rotational presence, and advisory capabilities.*

To enable economic growth and commerce, America, working in conjunction with allies and partners around the world, will seek to protect freedom of access throughout the global commons– those areas beyond national jurisdiction that constitute the vital connective tissue of the international system.

Provide a Stabilizing Presence. U.S. forces will conduct a sustainable pace of presence operations abroad, including rotational deployments and bilateral and multilateral training exercises. These activities reinforce deterrence, help to build the capacity and competence of U.S., allied, and partner forces for internal and external defense, strengthen alliance cohesion, and increase U.S. influence. A reduction in resources will require innovative and creative solutions to maintain our support for allied and partner interoperability and building partner capacity. *However, with reduced resources, thoughtful choices will need to be made regarding the location and frequency of these operations.*

...it will be necessary to examine how this strategy will influence existing campaign and contingency plans so that more limited resources may be better tuned to their requirements. This will include a renewed emphasis on the need for a globally networked approach to deterrence and warfare.[44]

We conduct our analysis of posture requirements and the advisability of changes through the lens of these statements.

[44] DoD, 2012a.

The Current U.S. Global Posture

In 2012, the bulk of the U.S. overseas military presence is located in EUCOM, CENTCOM, and PACOM, with fewer forces stationed in U.S. Africa Command (AFRICOM) and U.S. Southern Command (SOUTHCOM). See Table 1.1 for the current levels of assigned U.S. forces in each theater.

European Command

In EUCOM, there are permanent U.S. garrisons in Germany, Belgium, the UK, Italy, Spain, Greece, Turkey, and Portugal, as well as smaller facilities in Hungary, Romania, Poland, Bulgaria, and Norway. In short, the United States has nearly 80,000 active-duty personnel stationed at 39 bases in 15 countries in EUCOM. Figure 1.8 shows a map of major EUCOM installations, while Table 1.2 shows the number of major U.S. military installations in EUCOM, by type, for each service, corresponding to the installation locations shown in Figure 1.8.

Table 1.1
U.S. Military Personnel Assignments, by Combatant Command

Region	Army	Navy	Marine Corps	Air Force
CONUS	448,827	269,397	166,712	260,521
AFRICOM	65	6	131	17
CENTCOM	1,834	4,103	596	518
EUCOM	41,933	5,847	1,083	30,900
NORTHCOM	101	26	10	122
PACOM[a]	58,402	36,976	23,109	35,330
SOUTHCOM	745	587	38	229
Unknown[b]	55	215	4,300	1118
Total	551,962	317,157	195,979	328,755

SOURCE: Personnel data provided to RAND by DMDC on May 31, 2012.

NOTE: NORTHCOM = U.S. Northern Command. Does not include deployed locations (either for contingencies or rotational deployments). To develop the picture of the current U.S. military overseas posture—its major forces and bases presented here—we integrated a range of data and information sources. The data in Table 1.1 show personnel based on home station assignment and do not reflect those currently deployed in support of contingency operations. The numbers of personnel supporting contingency operations is often in flux, and this cost analysis focuses mostly on forces that are permanently stationed or conducting training or presence-oriented rotational operations, as opposed to ongoing contingency operations. In the rest of this section, we show further detail for each overseas region, including numbers of personnel and installations.

[a] Includes forces based in Alaska and Hawaii.

[b] Refers to unknown locations OCONUS.

given set of maintenance policies, along with transit time from homeports, one can directly translate naval force structure into a level of presence potential.

Most CONUS-based ships are operated on 27–32 month fleet response plan (FRP) cycles built around periodic extended depot maintenance phases[56] in which major shipyard or depot-level repairs and upgrade and modernization work is performed. During these maintenance periods, a ship is unavailable for tasking for deployment or training. Upon completion of maintenance, the crew must then go through basic unit-level and integrated training phases to be certified for major combat operations (MCOs) and thus available for deployment and employment for the full spectrum of missions.[57] From this point until the start of the next maintenance phase, ships are available for deployed operations. As with all of the services, the Navy has personnel tempo (PERSTEMPO) goals to protect the quality of life of personnel and to preserve the effectiveness of the all-volunteer force. The Navy's policies call for at least 50 percent of sailors' time in homeport during each FRP cycle, and the time at home between deployments should be at least as long as the previous deployment length, which puts a limit on steady-state overseas presence during the time available for deployment (the ship and crew are available throughout this period for surges to emergent contingency response needs).[58] Thus, the limit on steady-state presence for ships under these policies is a function of the number of ships in the fleet, the length of maintenance cycles for a given class of ships, and operational funding to keep the ships underway and to sustain training.

To support rotational or surge deployments, the Navy maintains overseas naval stations, support activities, bases, and support facilities for resupply and maintenance and command and control, along with access to ports of call. Examples include Naval Support Activities in Italy and Bahrain, Naval Support Facility Diego Garcia, and Naval Base Guam. These facilities support fleet operations and thus "presence."

The Navy also forward stations some forces in homeports overseas to increase naval presence in a region when critical long-term needs exceed the supply of presence that can be provided through FRP cycles. These forward-stationed forces are designated Forward Deployed Naval Forces (FDNF). Beyond increasing presence by elimi-

---

[56] Called "maintenance availabilities."

[57] When the basic phase is complete, "units may be tasked with independent operations in support of Phase 0 (Shaping/Deterrence), Homeland Security, Humanitarian Assistance/Disaster Relief, or other specific, focused operations." At the end of integrated training, ships/crews are ready for MCO-surge operations, with advanced integrated training then necessary to become fully MCO-ready. See Department of the Navy, *Fleet Response Plan*, OPNAV Instruction 3000.15, August 31, 2006; and Department of the Navy, *Surface Force Training Manual*, COMNAVSURFOR Instruction 3502.1D, July 1, 2007b. Ships also undergo extended mid-life maintenance/upgrades that affect total availability. These are combined with refueling for nuclear-powered ships.

[58] In 1985, the Navy established a PERSTEMPO program to balance pursuit of national security objectives with maintaining a reasonable quality of life for its sailors. The goals cited here are specified in Department of the Navy, *Personnel Tempo of Operations Program*, OPNAV Instruction 3000.13C, January 16, 2007a.

nating transit time, FDNF ships and crews operate with different policies that enable continuous availability for employment in the homeported region (either deployed at sea or from a port where continuous maintenance and training policies enable response within 30 days), increasing the amount of presence achieved for a given level of force structure. Instead of operating with FRP cycles, the ships do not enter into extended depot maintenance periods that require a training stand-down, but rather undergo depot-level maintenance through more frequent, shorter maintenance events,[59] which keep them ready for relatively short-notice taskings, with this mirrored by the crews, which stay trained and ready. Ships generally are kept in theater for ten years before being "hull-swapped" with similar ships, at which point they enter a shipyard for more extended maintenance, to possibly include deferred "deep" maintenance. Additionally, crews are able to maintain up to 100 percent presence in the homeported region because they can provide trained and ready overseas presence without being at sea away from their homeport, removing the impact of PERSTEMPO limits on presence. In effect, to gain the full presence potential of FDNF, the Navy employs different maintenance, training, and PERSTEMPO policies.

FDNF improves the Navy's ability to keep forces on station in specific areas, thereby increasing ability to engage, provide deterrence, and be responsive during contingencies. However, having forces in the FDNF also does not guarantee constant working presence or continuous readiness for all missions. FDNF ships still need maintenance, and crews cannot be ready at all times for all missions, particularly complex ones. In addition, total maintenance needs can be affected by the high operating tempo and deferment of some depot maintenance activities, such as structural preservation tasks. The lack of an FRP cycle with periodic stand-downs requires more intensive PERSTEMPO management of crews and affects long-term PERSTEMPO.

The Navy could try to increase forward presence by extending FDNF maintenance and training cycle policies to the total force. However, given PERSTEMPO limits and transit time effects, the forward presence increase would remain somewhat limited. To overcome this limit, the number of FDNF ships could be increased. However, the lack of stand-down periods and continuous readiness would begin affecting the PERSTEMPO of a larger portion of personnel, with unclear effects on recruiting and retention. Either way, it would fundamentally change the Navy's maintenance paradigm with unclear effects, likely including increased operating costs and potentially shorter ship life cycles, as they would not be able to rotate off of the high FDNF-like operating tempo. Such changes could drive imbalances in shipyards and other maintenance providers that would drive inefficiencies. If either of these two directions were to be considered—broadly applying FDNF policies or increasing the size of FDNF—a more in-depth examination of the broader impacts on ship life, personnel, maintenance, and total costs would need to be undertaken. We considered this beyond the

[59] Called "continuous maintenance availabilities."

scope of this study. In another FDNF alternative, crews could be flown into an overseas port in a crew-swap paradigm without formally homeporting the ship and assigning the personnel overseas. This would eliminate the loss in presence from sailing to and from the United States and would eliminate the impact of the dwell limit, but it would require increased personnel in the force. Given the unknown impacts of broad changes in maintenance, training, and personnel policies, combined with the study's assumption of force structure neutrality, we limit changes in naval force structure considered in this report to those achievable through very modest changes in FDNF.

In summary, the Navy primarily provides forward presence through rotational forces based in the United States. These forces go through maintenance periods, train, and are certified both as individual deploying units and as strike groups and then deploy, with the typical deployment running to 6–7 months. The biggest factors influencing availability for presence are the size of the fleet and the policies for maintenance, training and readiness, and PERSTEMPO. Although forward stationing of assets can improve availability to an extent, naval presence is more influenced by the size of the fleet than the level of overseas basing.

It should also be noted that were other services, particularly the Army and the Air Force, to switch from an emphasis on permanent basing to one focused primarily on rotational deployments to achieve overseas presence (all services currently use some combination of the two), they would ultimately hit the same limits, with force-structure size becoming the determinant of presence limits. This is reflected in steady-state deployment limits recognized through applications of PERSTEMPO limits with the Army Force Generation (ARFORGEN) model and the Air Force's Air and Space Expeditionary Force Construct. It was also reflected in the personnel stress on the force that occurred at the peak of OIF and OEF, when dwell-to-deployment ratios did not always meet goals for the force.

How This Report Is Organized

This report has three main parts, each with a different focus. The first part describes the strategic benefits and considerations that should be taken into account when making decisions about overseas posture, along with the risks that might jeopardize those benefits. The second part deals with the relative costs of overseas posture, and the third part describes and assesses illustrative postures with respect to the strategic considerations and how they would affect costs.

Chapters Two, Three, and Four take up the issues of strategic benefits, with Chapter Two describing how U.S. forces deploy to contingencies, the role of posture in those deployments, and the role of overseas basing in enabling direct operational execution. Chapter Three considers assurance and deterrence. Chapter Four discusses the relationship between security cooperation and basing. Chapter Five describes the risks to those benefits.

The chapters of the report that deal with cost issues are Chapters Six, Seven, and Eight. Chapter Six discusses the current condition of overseas installations and how this compares with U.S. installations. These costs would include any need either to modernize an installation or its restore facilities and capabilities. Chapter Seven examines the host-nation support that DoD receives when it stations forces in a foreign country. Chapter Eight describes our approach to building a cost model that enables us to determine the incremental costs beyond U.S. stationing and maintaining overseas bases and forces, as well the difference in permanent and rotational presence options.

Chapter Nine describes three illustrative postures as a way of illustrating the issues that policymakers must consider as it contemplates structural change driven by the influences described above. Chapter Ten assesses the postures, describing how they affect strategic benefits, risks, and costs. Finally, Chapter Eleven presents the study's conclusions. The report also has nine appendixes with additional analysis details and information. Two related appendixes, not available to the general public, were distributed separately, one dealing with mobility and another with EUCOM operational plans.

CHAPTER TWO

Strategic Considerations: Benefits of Overseas Posture to Contingency Response

An important strategic benefit often attributed to forward military presence is its contribution to contingency response by enabling military forces to respond quickly to a wide range of situations and geographic regions. Indeed, U.S. overseas posture has its roots in contingency responsiveness, particularly where there have been threats of major wars by strong adversaries, epitomized by positioning large forces in Europe and Northeast Asia, and this is where attention often first turns when posture changes are considered. While this role still remains, the nature of threats has evolved considerably, U.S. combined arms response capabilities have changed dramatically, and partner capabilities have improved significantly. As a result of these changes, assumptions about how forward posture should be configured to contribute to responsiveness may need to evolve.

In warfighting campaigns that involve a large U.S. deployment, the vast majority of the forces will come from the United States, with joint capabilities and limited ground forces preventing defeat in the face of any larger attacks. Those forces deploy to contingencies using a sophisticated network of air and sea bases and fleets of transport aircraft and ships. These fleets and en route infrastructure, along with at-sea presence, also enable flexibility to respond rapidly across much of the globe to a broad range of unpredictable events. Additionally, geographically distributed forward bases and airfield access buttress response capabilities by enabling air assets to quickly support widely ranging operations. Forward presence, then, is backed by the full military capacity of the nation, which largely resides in the United States.

Thus, for contingency responsiveness posture plays several roles, providing:

1. in-place forces enabling response to high-consequence, low-probability major events, in conjunction with longer reach assets
2. global infrastructure to enable high-volume force flows for major wars and rapid response to smaller contingencies in unpredictable locations
3. seaborne forces to also respond quickly to globally dispersed, unpredictable, small-scale contingencies and major events

4. basing access to enable air; intelligence, surveillance, and reconnaissance (ISR); and logistics support from nearby safe havens.

For major threats from highly capable adversaries, forward posture underpins the ability to prevent initial setbacks or defeat until sufficient reinforcements arrive to achieve U.S. goals. Large force deployments require large numbers of ships, with delivery potentially constrained by reception port capacity. If a large force is needed quickly, it must be in place, as the U.S. stationed large ground forces backed up by high levels of prepositioned equipment in Germany in the face of the perceived Soviet threat during the Cold War. But this type of situation no longer dominates overseas presence considerations. With the changes in threats and U.S. weapon capabilities, the needed posture capabilities for initial major contingency response has changed. Forward and long-range offensive strike capabilities from air and maritime forces, air and missile defense assets, and limited initial ground forces can respond in a few days or even hours to a contingency. This force can be reinforced by a buildup of ground and more significant air forces as needed for the situation and mission.

The implication for U.S. posture is that heavy forces needed early in an operation must either be in place or their heavy equipment must be prepositioned in place or at sea in the region. Air assets can deploy rapidly as long as there are bases available with enabling assets ready to receive and support them. Rapid reinforcement by limited numbers of light ground forces by air and by sea, in the case of deployed marines, and by maritime forces in the area is also feasible. Large-scale reinforcement will be by sea from the United States, given the physical characteristics of such forces. Additional forward force posture, in a region near but not directly in the location of a contingency, will have a relatively small effect on force closure.

The threat environment and the wide spectrum of security interests of the United States also require flexibility, so a robust, globally distributed infrastructure, supported by air and sealift fleets, is critical. This runs counter to positioning large forces in one or two overseas areas, which would reduce agility of the total force. Heavy forward presence in a region helps in just one location, while robust lift capability and en route infrastructure helps in broadly distributed locations in an environment of high uncertainty as to contingency locations. CONUS-based lift forces and global infrastructure allow relatively effective deployment response that can draw on the full set of capabilities of the U.S. military. Such reach and agility would be difficult to achieve through forward force presence alone with the current size of the U.S. military.

Whether forces also need to be positioned forward to support broadly distributed contingency operations is less clear. Close or far, heavy forces cannot be deployed quickly by air. This leaves rapid response to air and lighter ground forces, and to units in a region with equipment already aboard ships. But even for light ground forces, given current U.S. airlift fleets, deployment of air-deployable brigade-sized forces can be almost as quick from the United States as from a forward base. The forward-based air deployment advantage is limited to two sets of conditions: (1) when very small units, such as special forces or small infantry units (e.g., a battalion or smaller), need to be deployed in situations where hours can make a difference and (2) situations in which airlift is constrained, such as when there are multiple ongoing airlift-intensive events. Forward maritime presence, both from Navy systems and from marines aboard naval ships, also provides capabilities to respond quickly to contingencies in areas where the United States does not have a permanent presence, or when the response requires more than is immediately provided by land-based forces in the area. Air forces can deploy with support equipment light enough for relatively rapid contingency response from afar. In addition, air forces do not necessarily have to deploy directly to the point of conflict. They just have to be within operational range, creating more flexibility and creating value in having forward bases that have the infrastructure and support equipment to enable the base to be used for wide-ranging operations, even if the bases do not continuously host forces.

This chapter explains and provides support for these conclusions and has three parts: (1) the role of overseas posture in deployment responsiveness for major contingencies, (2) the role of posture for smaller contingencies, and (3) the role of overseas posture in providing bases from which direct operational support can be provided. The deployment of military assets involves consideration of what forces are needed, the infrastructure-based routes available to those assets, the location of those forces, what their mission or missions are, where they are to be sent, the availability of transportation assets, and the interaction among these considerations. This chapter connects these considerations. It starts with a description of the implications of different types of forces for deployment. It then turns to the value of en route infrastructure needs, which also explains several of the factors that affect the deployment analysis that follows. The initial part of the deployment analysis discusses the breakpoints in terms of time and force size at which intertheater sealift becomes a better deployment option than air-based deployment, which is largely what limits the value of forward regional presence from a deployment standpoint to early deployment of lighter forces, in-place presence, and prepositioned equipment. This has implications for the conclusions—both with respect to major wars and smaller-scale contingencies—because it also illustrates why, beyond the necessary in-place forces, forward ground-force presence within a region has limited effect on force closure for major operations. The chapter then examines in more detail how much responsiveness or resource value overseas basing has for the deployment of rapid response forces. It then concludes with a discussion of how posture affects direct operational support responsiveness that enables immediate support from bases from which assets such as fighter aircraft, intratheater airlift for sustainment, and ISR can operate.

Force Types and Implications for Deployment and Presence

While posture generally is about where units are stationed, those decisions are influenced considerably by how those units move and are thus not the driving factor for USN and sea-based forces. We start with an overview of different unit and asset types and relevant considerations for deployment responsiveness.

During the Cold War, approximately one-third of the U.S. Army was stationed in Europe, with an Army headquarters, two corps with four divisions each, two armored cavalry regiments, a support command, and supporting units in place, augmented by prepositioned equipment sets for several divisions. In addition, one division and significant echelon-above-division (EAD) forces, including a support command, were stationed in Korea. These were both indicative of the recognition that if large forces were needed to defend quickly against a major threat, they or least their equipment must be in place. Today, that permanent presence is being reduced to two BCTs and five enabling brigades in Europe and has been reduced to a division headquarters with one brigade and a reduced EAD presence in Korea. These reductions came about as the Cold War threat in Europe ended and as the situation in Korea has changed, especially regarding improved South Korean capabilities.

The Army today is both much smaller and much more U.S.-based than the Army of the Cold War era. While it is feasible to deploy light or even medium-weight Army units by air, many Army units are very heavy and generally require sealift for deployment. This requires time and usually railroad transport to get to a seaport, days to load ships, followed by some nine to 16 days at sea to arrive in the area of operations.[1] In such cases, the time for deployment completion for units based in the United States does not differ greatly from those based abroad—and could be no different depending upon the coordination of sealift—with prepositioned equipment having an advantage. For example, Figure 2.1 compares approximate sealift deployment time to Kuwait for an ABCT from Germany, from the United States, and with using prepositioned equipment aboard ships located at Diego Garcia.[2] This assumes no warning time or order to begin moving the prepositioned ships early.

Air Force units are generally regarded as self-deploying. That is true with respect to the aircraft themselves. They are the fastest-deploying assets but must have access

[1] Busan, South Korea, is about 5,230 nm from Long Beach, California, and Ad Damman, Saudi Arabia, is about 8,420 nm from Norfolk, Virginia. A large, medium-speed, roll-on/roll-off (LMSR) ship travelling at 24 knots can cover these distances in 9.1 and 15.3 days, respectively, with 16 hours of additional time added to the latter to account for travel through the Suez Canal.

[2] The time from Europe in Figure 2.1 assumes that LMSRs are available in Europe by the time the ABCT reaches the port. Currently, an LMSR would need to travel from the East Coast to Europe first, and this could increase the overall timeline. Some of the delay might be mitigated by advance warning, and some might be in parallel with the time to prepare and move the ground forces, but it would still likely increase the total time. Units in CONUS may be able to begin the sealift portion of their deployment earlier than those in Germany, potentially reducing or eliminating the difference in closure time seen in Figure 2.1.

in a plan that both the Secretary of Defense and the relevant combatant commander sign as being executable with the current U.S. posture. Substantial posture changes may require more detailed examinations of operational impact, including evaluation of changes to contingency plans. While we did not review or critique these plans, we did analyze the role of posture in deploying forces to both large-scale conflicts and smaller-scale contingencies to inform long-term posture deliberations.

Small changes in forward presence, on the order of a brigade or regiment or two, do not affect force closure timelines for MCOs. Instead, such decisions are important with respect to the size of the immediate response force for MCOs considered necessary to be in place for initial response to prevent defeat or other objectives. In any MCO with the current posture and force structure, these forward forces will only represent a small fraction of responding forces. MCO deployments are large and require extensive deployments by CONUS-based forces. For example, in 1990s Korea-conflict simulations, RAND used a ground force of five Army divisions, two armored cavalry regiments, four separate field artillery brigades, three Army combat aviation brigades, and six Marine Corps brigades or regiments.[5] This force is the equivalent of some 23 combat brigades. During the same time period, DoD and Army planning documents and RAND analysis associated with the Bottom-Up Review posited major conflicts requiring between 14 and 27 Army brigade equivalents.[6] During Operation Desert Storm, the Army deployed a force equivalent to 23 brigades.[7] The invasion of Iraq in 2003 was conducted with approximately 19 brigade equivalents (13 U.S. Army, four USMC, and 2 British Army).[8] The question for defense planners with regard to MCOs is whether there are enough ground forces and prepositioned equipment in place at the start of a contingency to hold and prevent defeat, whether there are enough land and sea-based air assets within operating range at the start of a contingency or soon thereafter, whether sufficient infrastructure is in place to deploy required reinforcements, and whether there are sufficient lift assets. With respect to airlift, the latter two requirements also provide the ability to rapidly deploy small, additional forces required to ensure a successful initial defense.[9]

This reinforcement capability and limited insensitivity of large-scale force closure to having forces regionally based or CONUS-based does rely on adequate sea and airlift, which come at substantial cost. In 2013 dollars, LMSRs—used for large-scale sea-

---

[5] Unpublished researched by Barry Wilson, Bruce W. Bennett, and Carl Jones of RAND.

[6] Ronald E. Sortor, *Army Active/Reserve Mix: Force Planning for Major Regional Contingencies*, Santa Monica, Calif.: RAND Corporation, MR-545-A, 1995, pp. 20–21, 41, 56–57, 62–63.

[7] Sortor, 1995, p. 41.

[8] Gregory Fontenot, E. J. Degen, and David Tohn, *On Point: The United States Army in Operation Iraqi Freedom*, Annapolis, Md.: Naval Institute Press, 2004, pp. 76–80, 448–473.

[9] Further discussion of these findings is in an appendix distributed separately (not available to the general public).

borne deployments—are estimated as costing about $560 million and C-17s as about $240 million per aircraft, with annual operating costs of $8–21 million and $14 million, respectively.[10] Without these lift forces, larger in-place forces in key threat areas might be required. But this is not a simple trade-off of like capabilities (i.e., forward-based forces for lift). Without substantial lift, forward-based forces would be somewhat "stranded" in place, with limited agility. In contrast, the current levels of lift capacity provide flexibility to deploy large forces to a broad range of locations, where infrastructure permits.

En Route Infrastructure and Capabilities

The foundation of the global reach of the U.S. military comes from access to a broad network of capable en route bases that support evolving needs from changes in the strategic environment. On a continuing basis, TRANSCOM identifies routes, air bases and sea ports, and/or multimodal locations to provide desired global capabilities. As part of this overall plan, the Air Mobility Command (AMC) Global En Route Strategy calls for a system of mutually supporting routes centered on two regions, Europe/Eurasia/Middle East and the Pacific, with multiple routes in each region for resiliency and higher capacity. The en route system, as it stood in 2010 and which was used in the deployment analysis discussed later, is illustrated in Figure 2.2. While the system has evolved some since 2010, this gives a sense of its global coverage.[11]

Resiliency in the en route system is an important consideration because it helps mitigate risk to deployment operations. Having alternate bases available helps ensure that if one base or route becomes unavailable or degraded due to such factors as weather, political decisions, accidents, airport limitations, or attacks, other options are readily available. An example is illustrated in Figure 2.3, which shows two alternate routes from Dover AFB to Kuwait.[12] The northern route travels through Royal Air Force

---

[10] Based on Congressional Budget Office, *Options for Strategic Military Transportation Systems*, September 2005, with cost estimates converted to 2013 dollars. An LMSR can carry 359 to 513 C-17 loads depending upon the equipment composition.

[11] The AMC en route system was analyzed in the Global Access and Infrastructure Assessment, completed in December 2010. It began with country-level assessments aggregated to 19 COCOM-defined regions such as the Levant, the Arabian Peninsula, and the Central and South Asian States. For each region, TRANSCOM assessed the likelihood of an event requiring the movement of forces and the current adequacy of access. This led to a detailed analysis of ten regions with a high probability of a mobility event or access limitations, or both. The final phase identified critical nodes that support multiple regions. The resulting strategy highlights the interdependencies of the routes and base capabilities and identifies required capabilities at each location. The most recent DoD analysis produced a network of proposed air bases to be established by 2025. Airfields that are part of en route infrastructure are categorized according to a four tier-tier system—from most capable (tier I) to more austere contingency locations (tier IV) based on the operations, refueling, passenger and cargo handling, and command and control capabilities at each location. See Air Mobility Command, *Air Mobility Command Global En Route Strategy White Paper*, version 7.2.1, July 14, 2010, pp. 31–33.

[12] These routes are based on a 3,200 nm C-17 critical leg and assume that fuel is available in Kuwait. The northern route is 5,665 nm long and the southern one is 5,962 nm long.

Army, USMC, and USN unit cargo will be primarily by AMC aircraft. The workhorses of AMC for strategic movements are its fleets of 222 C-17s and 54 C-5s. While some passengers do move with the materiel, generally most personnel are transported on commercial chartered aircraft or aboard Civil Reserve Air Fleet aircraft.[18]

Surface movement is provided by Military Sealift Command (MSC) ships, as well as commercial vessels under programs analogous to Civil Reserve Air Fleet. Of the 115 noncombatant, civilian-crewed ships MSC operates worldwide, the most modern are the nine LMSRs, which were procured after difficulties encountered in the first Gulf War demonstrated the need for large, relatively fast vessels. Another category is the 32 Maritime Prepositioning Program ships, which include ten LMSRs and 12 container and roll-on/roll-off ships. Once ships used for equipment prepositioning have been offloaded, they can enter the common user sealift pool.[19] The preponderance of unit cargo in any large deployment operation will be carried by roll-on/roll-off capable vessels, which are capable of rapid loading and unloading operations and, especially in the case of the nominally 24-knot LMSRs, relatively fast transits. This fleet of ships is crucial for large deployments, particularly for MCOs.[20]

It might seem obvious that deployment by air is faster than by sea, but airlift may not be faster than sealift, depending upon the amount of materiel to be moved, the distance, and throughput capacity of facilities. Only for very small unit movements is air deployment consistently and significantly faster than by sea, and even here an exception is being within about 1,000 nm. Both modes, though, depend on reception capacity at ports of debarkation and, in the case of aircraft, the MOG at en route locations. The MOGs of some of the en route system bases are shown in Table 2.1. Sealift requires ports with adequate draft, berths with lengths capable of handling large roll-on/roll-off ships, and nearby unit marshaling areas for reception, staging, and onward movement.

---

[18] The Civil Reserve Air Fleet is a DoD program to contract for the services of specific U.S.-owned aircraft during national emergencies or other defense-oriented situations when civil augmentation of the existing military airlift fleet is required. See Joint Publication 1-02, *Department of Defense Dictionary of Military and Associated Terms*, Joint Chiefs of Staff, November 15, 2012, p. 47.

[19] The other prepositioning ships include three container ships, two aviation logistics support ships, two dry cargo/ammunition ships, one tanker, and one offshore petroleum distribution system ship, and the HSV-2 Swift. There are also two joint high-speed vessels (USNS *Spearhead* and USNS *Choctaw County*) and two high-speed vessels (MV *Huakai* and MV *Westpac Express*) designated for intratheater sealift, which are fast but limited in range and capacity. The remaining MSC ships serve various missions, such as ammunition transport, replenishment oilers, oceanographic survey, and ocean surveillance. For the current inventory of MSC ships, see Military Sealift Command, "MSC Ship Inventory," U.S. Navy, undated a. See also Military Sealift Command, *2012 Ships of the U.S. Navy's Military Sealift Command*, U.S. Navy, undated c.

[20] TRANSCOM and the Cost Assessment and Program Evaluation Office discuss the mobility requirements for MCOs in the Mobility Capabilities and Requirements Study 2016. The posture-related results of this study are described in an appendix distributed separately (not available to the general public).

Table 2.1
Representative En Route System MOGs

Base	Parking MOG	Working MOG (Mechanical)
Ramstein AB, Germany	11	2 C-5/4 C-17
Naval Station Rota	17	3
Incirlik AB, Turkey	6	6
Bagram AB, Afghanistan	3	2 Wide Body/4 Narrow Body
Cairo West, Egypt	3	Maintenance Recovery Team
Yokota AB, Japan	6	3
Osan AB, South Korea	3	3
Clark AB, Philippines	4	N/A
Ambouli IAP, Djibouti	1	0
U-Tapao, Thailand	7	N/A
Peya Lebar AB, Singapore	3	1
Richmond, Australia	1	1

SOURCE: Air Mobility Command, 2010.

NOTE: Parking MOG is based on the number of wide body spots available. IAP = international airport; AB = Air Base.

For airlift, aerial port of debarkation (APOD) MOGs of 2 or 3 are not unusual. For simplicity, assume two hours are needed to unload a C-17 and launch it again. A MOG of 2 would then permit 24 aircraft landings to deliver materiel every 24 hours. Therefore, in ten days, due to MOG restrictions, which could actually result from constraints at an en route stop due to limited fuel capabilities, for example, airlift could deliver 240 C-17 loads. Depending upon the unit type and shipload configuration, an LMSR can deliver between 359 and 513 C-17 loads at once.[21] Thus, for any destination within ten days' sailing of a seaport, approximately 2,000 nm, a single LMSR could potentially deliver considerably more materiel than C-17s constrained by a MOG of 2. Of course, airlift has the advantage of delivering a stream of materiel over time, whereas the LMSR arrives with all the materiel at once.

[21] According to the Surface Distribution and Deployment Command Transportation Engineering Agency, it takes 287 C-17 sorties to deploy an IBCT and 575 sorties to deploy an ABCT (based on a 3,200 nm critical leg). The ratio between a ship's capacity and the actual amount of cargo that can be loaded is known as the stow factor. A stow factor of 0.65 means that 65 percent of the available space can be used to load cargo. Using a stow factor of 0.65, it requires 0.8 LMSRs to transport an IBCT; if the stow factor is 0.75, the number of LMSRs required is 0.69. Similar figures for an ABCT are 1.1 and 1.3 LMSRs.

Table 2.2
Army Force Packages

Force Package	Passengers	Weight (short tons)	Vehicles	C-17 Sorties	CRAF Sorties
Port opening (Army)	237	1,848	141	40	0.9
Theater opening (Army)	959	3,907	529	115	1.1
Deterrent task force	6,358	26,741	2,632	696	11
FID/PK task force	5,757	20,858	2,209	575	9.7
HUMRO task force	3,012	14,868	1,683	483	5.6
Accompanying supplies	—	265–868	—	4.2–14.7	—

SOURCE: RAND analysis.
NOTES: CRAF = Civil Reserve Air Fleet.

Table 2.3
Aircraft Force Packages

Force Package	Passengers	Weight (short tons)	C-17 Sorties
F-16 squadron (bare base)	1,241	2,854	56
F-16 squadron (host-nation infrastructure)	998	978	25
F-15 squadron (bare base)	1,293	2,951	58
F-15 squadron (host-nation infrastructure)	1,064	1,063	27
C-130 squadron (bare base)	1,100	2,181	44
C-130 squadron (host-nation infrastructure)	918	887	23

SOURCE: RAND analysis.
NOTES: F-16 and F-15 squadrons have 24 aircraft. C-130 squadrons have 14 aircraft.

cations.[36] This also shows the limits of deployment responsiveness given en route and destination infrastructure. When this is the case, closure times vary little. For example,

[36] The estimation of the differences between deployment of forward-deployed forces and CONUS-based forces by air was accomplished through the development and employment of an Excel model based on Air Force Pamphlet 10-1403, 2011. As the pamphlet states, the factors are intended for "gross estimates." Detailed mission planning would require other tools and the tailoring of assumptions to the specific situation. For our purpose of comparing performance from overseas bases versus from U.S. bases, the estimates are sufficiently accurate and internally consistent for determining whether there are meaningful differences. Key inputs that drive the model's results are the number of aircraft available; the working MOG at the APOE, each en route stop, and the APOD(s); the distances between each location; the number of C-17 loads required to move the force package; the refueling and unload times; and the aircraft utilization rate. We used the expedited unload time of one hour and forty five minutes, and working MOGs were initially derived from TRANSCOM sources and then varied for

Table 2.4
Marine Corps Force Packages

Force Package	Passengers	C-17 Sorties	CRAF Sorties	Notes
Marine expeditionary brigade	15,230	263	53	4 MPSRON ships
Naval construction force	929	7	3	Part of MPSRON
Naval support element	—	15	9	Part of MPSRON
Marine expeditionary unit	2,216	—	—	Amphibious ready group or 1 MPSRON ships

SOURCE: RAND analysis; U.S. Marine Corps, *Prepositioning Programs Handbook*, 2nd edition, January 2009.

NOTES: CRAF = Civil Reserve Air Fleet; MPSRON = maritime prepositioning ships squadron.

given aircraft are in place, the first aircraft arrival and unload time starting at Royal Australian Air Force Base Tindal in Australia and going to Polonia, Indonesia, a distance of 2,280 miles, would be approximately 11 hours with no en route stops, and it would require 21 C-17s to fill the air bridge. From Hickam AFB, Hawaii, at 6,103 miles, it would be approximately 25 hours for first aircraft arrival, including two en route stops for refueling, and it requires 57 C-17s to fill the air bridge. The time from first aircraft arrival to force closure will then be the same for both cases, leaving just a 14-hour deployment time difference compared to the overall 16-day timeline. In terms of a military advantage, a 14-hour time difference is not apt to be significant, except for very time-critical missions, such as responding to a surprise attack on U.S. personnel that does not end abruptly to protect or rescue them. This would likely be the purview of special operations forces or potentially small, light conventional Marine Corps or Army units, not the type of force packages modeled here. Figure 2.11 shows the relatively similar times for deployment for the FID force package as distance increases when the air bridge is kept fully utilized.

What varies instead of time is the number of aircraft needed to achieve these response times. Figure 2.12 shows the number of airlift assets needed to achieve the fastest possible response time (left y-axis) for the FID force package given en route and destination throughput constraints for the different scenarios considered. The number grows in a linear fashion with distance (right y-axis) when the lowest working MOG at the APOD or one of the en route bases is the deployment constraint. Working MOG at a CONUS APOE or a major overseas strategic mobility hub, such as Ramstein AFB, is typically not the constraint. The aircraft requirement advantage would be the potential advantage of forward basing for brigade-sized deployments. Results for the HUMRO scenarios are similar to the more sortie-intense FID cases, but the aircraft

exploratory purposes. Available aircraft were also treated as a variable. Key outputs of the model are closure time and the number of aircraft required to fill the air bridge.

For smaller-scale contingencies requiring initial response from force packages built around light- and medium-weight brigade-sized units, there is limited response time advantage for regionally based forces compared with those in CONUS. This rests upon the assumption that when relatively rapid response is truly critical, with days making a difference, airlift allocation will likely be high. In contrast, heavy brigades such as ABCTs can often be deployed more quickly by sealift, particularly when moving prepositioned afloat assets to a port close to the point of operations, since sealift is faster than airlift for deploying such units. So, if they might be needed quickly, they do need to be in place or have prepositioned equipment. But this will tend to be critical for higher-intensity, higher-threat situations more closely related to MCO planning.

An exception to the limited-advantage conclusion for smaller-scale contingency deployment is that forward-deployed forces require fewer aircraft to deploy to contingency locations in the same region. Positioning an overseas garrison a short distance from a potential deployment location, especially if within unrefueled range of a C-17, can dramatically reduce the number of aircraft required to execute the movement, freeing up lift to move other assets to meet up with prepositioned equipment or aviation units. This could be important if there are critical, time-sensitive operations occurring in parallel that limit aircraft availability. This would be more important if airlift fleets are ever reduced, although choices between locations would then also likely have to made, reducing the overall level of deployment response flexibility.

Available basing also affects the ability to support operations effectively with assets that typically need to operate from safe havens or must have a more robust support infrastructure. These include combat aircraft, ISR platforms, and sustainment assets. This chapter provides rough approximations of the operating ranges for exemplar assets in these categories and how resource efficiency changes with operating range. This can be used to assess whether a range of contingencies in different regions can be supported effectively from different sets of bases.

CHAPTER THREE

Strategic Considerations: Benefits of Overseas Posture for Deterrence and Assurance

This chapter begins by describing the role that the overseas defense posture plays in deterring potential adversaries. It then discusses how a forward presence might actually detract from U.S. interests. The chapter concludes with a discussion of the complement of deterrence: assurance.

The Role of Foreign Posture in Deterring Potential Foes

One of the expected benefits of foreign presence is its contribution to deterrence. A 2004 DoD report to Congress on the global U.S. posture opens with this rationale: "Together with our overall military force structure, our global defense posture enables the United States to assure allies, dissuade potential challengers, deter our enemies, and defeat aggression if necessary."[1] The 2006 Deterrent Operations Joint Operating Concept provided a more explicit link between foreign posture and deterrent goals. More recently, the 2010 QDR and the 2012 Defense Strategic Guidance linked U.S. forward presence to deterrent goals and to strengthening alliances. The latter stated:

> U.S. forces will conduct a sustainable pace of presence operations abroad, including rotational deployments and bilateral and multilateral training exercises. These activities reinforce deterrence, help to build the capacity and competence of U.S., allied, and partner forces for internal and external defense, strengthen alliance cohesion, and increase U.S. influence.[2]

In short, the U.S. global presence is intended to contribute to long-standing strategic goals of assuring partners and deterring potential adversaries.

---

[1] DoD, 2004a, p. 4

[2] DoD, 2012a, p. 5.

Deterrence Ideas and Practice

While much of the conventional deterrent potential of the United States comes from the perceived dominance of the U.S. military, it is not enough simply to have a powerful military. The United States must also be seen as willing to use force in the face of aggression it has tried to deter. Deterrence involves perceptions about military might but also political will. To deter a specific act of aggression, a country needs both the capacity and the will to use military force to prevent that aggression from being successful. Foreign bases and force presence contribute to both of these ingredients of deterrence. They indicate a willingness of the United States to become involved in conflicts abroad, and they shape perceptions about the effectiveness of the U.S. military to project power quickly and sustain it over time.

Fundamentally, deterrence seeks to manipulate the cost-benefit calculations of another to shape behavior to better advantage. Ultimately, deterrence requires an appreciation of the values, motivations, beliefs, and fears of an adversary.[3] The challenge of deterrence is to find ways to manipulate behaviors in situations where the stability of a government regime could be at stake and when colored by the strong fears and emotions that influence decisions of war.[4]

The credibility of a deterrent threat rests on the perceived willingness and capacity of the country to implement it, particularly when trying to protect a third party. Going to war is both costly and risky. Going to war to help defend another country when you have not been directly attacked may not always be considered an automatic decision by the aggressor, degrading deterrence. Deterrent commitments need to be reinforced through a variety of pronouncements and visible displays of support to establish credibility and convince others of a willingness to risk lives and suffer losses. Since Vietnam, the U.S. willingness to sustain losses in a distant theater has often been called into question, often erroneously, but when trying to deter a future action, perceptions and expectations factor into these decisions. If ties to a country or region are weak, it is more likely that adversaries will discount deterrent moves as a bluff.[5]

The second element of a credible deterrent threat is the ability to carry it out. Can a state back up its deterrent threat with appropriate military capacity and action? Most relevant to this calculation is not the overall net assessment of national military power among the protagonists, but an assessment of the forces likely to be devoted to the conflict. For instance, during the Vietnam War, the United States kept a substantial

---

[3] See Robert Jervis, Richard Ned Lebow, and Janice Gross Stein, *Psychology and Deterrence*, Baltimore, Md.: Johns Hopkins University Press, 1985. Prospect theory focuses on the cognitive aspects of deterrence. For an essay applying prospect theory to deterrence, see Jeffrey D. Berejikian, "A Cognitive Theory of Deterrence," *Journal of Peace Research*, Vol. 39, No. 2, 2002, pp. 165–183.

[4] See Keith B. Payne, "The Fallacies of Cold War Deterrence and a New Direction," *Comparative Strategy*, Vol. 22, No. 5, 2003, pp. 411–448.

[5] Kenneth Watman, Dean Wilkening, Brian Nichiporuk, and John Arquilla, *U.S. Regional Deterrence Strategies*, Santa Monica, Calif.: RAND Corporation, MR-490-A/AF, 1995, pp. x–xi.

fraction of its forces devoted to deterring Soviet aggression in Europe. Furthermore, a weaker state still may not be deterred, even if the local balance of forces is unfavorable, if it cares more about winning the conflict and is prepared to take greater losses.

This chapter addresses how the United States can convincingly show its willingness and capacity to deter aggression of many different potential actors in many different regions of the world, as well as the role of forward presence in these efforts. In 2006, DoD offered some specifics about how posture might contribute to deterrent strategy. As the Deterrent Operations Joint Operating Concept asserted:

> US capabilities resident in forward-stationed and forward-deployed multi-purpose combat and expeditionary forces enhance deterrence by improving our ability to act rapidly around the globe. Forward presence strengthens the role of partners and expands joint and multinational capabilities. Our presence conveys a credible message that the United States remains committed to preventing conflict and demonstrates commitment to the defense of US and allied vital interests.[6]

Building on this, we can assess whether overseas posture contributes to deterrence in the following ways:

1. Showing a costly commitment to a country or region
2. Maintaining capabilities to prevent a quick victory
3. Improving capabilities of allies and friends, particularly through security cooperation
4. Improving understanding of regional dynamics.

Costly Commitments: To the extent that deterrence involves calculations of military power and intent, the forces deployed abroad demonstrate the investment of both the United States and the host nation in the security of that region. The costs of stationing those forces forward are a strong indication of a willingness to become involved in stabilizing and engaging the region and are more convincing than statements alone.[7] In situations where putting those forces forward puts them at increased risk, it becomes a clear statement to adversaries and allies alike that, should an attack be launched against the host, the United States will become immediately involved in the host nation's defense. Security treaties, like those the United States has with NATO allies, Japan, and Korea, can also play a role in emphasizing to both allies and potential foes the willingness of the United States to tie its security to that of other countries. To the extent that these treaties contain strong mutual defense clauses, they can contribute both to deterring potential adversaries and assuring partners.

---

[6] DoD, "Deterrence Operations Joint Operating Concept" Version 2.0, December 2006, pp. 33–34.

[7] See Paul K. Huth, "Deterrence and International Conflict," *Annual Review of Political Science*, Vol. 2, 1999.

Preventing a Quick Victory: Preventing an adversary from achieving a quick victory, particularly by using military force to change facts on the ground, is important to deterrence strategy.[8] The ability of U.S. forces stationed abroad to respond immediately or more quickly to a crisis than those based in the United States strengthens the overall deterrent value of the force. Several examinations of historical crises suggest that, particularly with authoritarian adversaries, having the means to prevent such leaders from achieving a quick victory is important.[9] Deterring something that has yet to happen is considered easier than taking action to reverse something that has already happened.[10] Having a capability to quickly meet an act of aggression carries this advantage. U.S. forces based at the point of friction, or near enough to the point of friction that they can arrive in time to prevent a quick victory, reduces these temptations.[11] Having forces in the region as tensions rise could also be preferable in circumstances where introducing new forces to a region might be considered destabilizing. The ability of the United States to reinforce the early deployments with a more sustained and much larger force through visible global deployment capabilities strengthens the deterrent value of these early-response capabilities.

Improve Partner Capabilities: In trying to improve our ability to deter attacks against an ally or friend, the extent to which the partner nation is seen as militarily competent and able to make a substantial contribution to its own defense is relevant. In this way, efforts to train and exercise with partners, plan and organize joint commands, and procure compatible equipment all contribute to enhancing the capabilities of the host nation and their combined operating ability with U.S. force.[12]

Understanding Regional Dynamics: The presence of U.S. forces in the region also provides the time and contacts for the U.S. military to better understand regional dynamics. This can contribute to deterrence in that devising successful deterrent strategies requires a nuanced understanding of the values and interests of a potential adver-

---

[8] Paul K. Huth, "Extended Deterrence and the Outbreak of War," *American Political Science Review*, Vol. 82, No. 2, June 1988.

[9] John Mearsheimer, *Conventional Deterrence*, Ithaca, N.Y.: Cornell University Press, 1983. Analogous to the temptation of an individual, when you leave an ambitious country or leader unchecked and easily able to take by force objects of desire, you are essentially inviting such behavior. Thus, keeping enough forces to prevent an adversary from achieving gains easily has support from several studies. As Watman et al. found, "When they resort to force, regional adversaries typically seek short, cheap wars. Therefore, those U.S. military forces that can credibly deny a quick victory will be most impressive to the opponent. In other words, *it is those forces that are in the region, or that can deploy to the region on short notice, that will have the greatest deterrent effect*." Watman et al., 1995, p. xii.

[10] Thomas Schelling, *Arms and Influence*, New Haven, Conn.: Yale University Press, 1966.

[11] This is immediate deterrence, which seeks to prevent a particular action from happening in the near future, as opposed to general deterrence, which is more an assessment of the relative power balance. See Richard Ned Lebow, "Deterrence: Critical Analysis and Research Questions," in Myriam Dunn and Victor Mauer, eds., *The Routledge Companion to Security Studies*, London: Routledge, August 2011, p. 3.

[12] The next chapter of the report is devoted to the security cooperation aspects of forward presence.

sary, as well as those of partners and allies. It is much easier to develop that understanding through repeated contact with relevant officials in the area. Forward presence provides more opportunities for such contact and greater depth of understanding of regional political factors.

Deterrence must be tailored toward a specific threat. The 2012 Defense Strategic Guidance is specific about trying to deter North Korea and terrorist groups (in particular al-Qa'ida and Hezbollah), but is more cautious about naming other potential adversaries. Instead, it tends to focus on threatening methods of attack, such as anti-access capabilities, in which it names China and Iran; nuclear, chemical, and biological weapons; and even cyber security, in addition to general "aggression."[13] Considering the global interests of the United States, its forces must be flexible and prepared for a wide variety of conflicts located in many different environments. Still, looking at some of the specifics about how U.S. posture might contribute to deterring some of these identified or implied threats or strategic concerns may be instructive. These different strategic concerns are addressed in turn below.

U.S. forces based in South Korea are intended to show North Korea that the United States remains committed and retains the capacity to assist tangibly in the defense of South Korea.[14] This threat could take may forms: invasion, acts of military coercion, terror attacks from ballistic missiles, or chemical weapons. The South Korean military has the ability to confront many of these attacks on its own, but the United States remains prepared to assist. The kinds of capabilities that might be useful are often specialized, such as missile defense capabilities, expertise in chemical weapons, and sensors.

The sustained presence of U.S. forces in South Korea has many operational benefits that in turn augment the deterrent value of both U.S. and South Korean forces. That sustained presence allows for the development of bilateral military plans, an organizational structure that can command and control the application of coalition military power, regular opportunities for coalition training and exercises, and a greater incentive to consider interoperability in making acquisition decisions. Finally, the U.S. forces gain in-depth knowledge of the local conditions that could be invaluable in a contingency. This would include many factors, from geography and weather to political and cultural organization and values.

The South Korean military has now developed capabilities to deal with many kinds of North Korean attacks on its own. It has invested in advanced capabilities, conducts sophisticated training, and constantly maintains a state of high readiness. For most situations, the presence of the United States demonstrates the U.S. commit-

---

[13] DoD, 2012a, pp. 4–5.

[14] Robert F. Willard, USN, Commander, U.S. Pacific Command, "Testimony Before the Senate Armed Services Committee on U.S. Pacific Command Posture," February 28, 2012, p. 4.

ment to defending Korea, coupled with the capabilities of the South Korean forces for immediate defense.

Like North Korea, Iran's pursuit of nuclear weapons is opposed by the United States. The United States also opposes Iran's support of Hezbollah and its threats to its neighbors. A number of U.S. capabilities in the region could be said to be contributing to deterring Iran. U.S. combat aircraft in the Arabian Peninsula, along with naval warships and MEU/ARG presence in the Arabian Sea could contribute to deterrence of a number of regional conflicts, including threats from Iran to use land-based cruise missiles to close the Strait of Hormuz.

U.S. forces in the region could also be used to try to hinder Iran's nuclear weapons program, though the decision to use military force would need to take into account a variety of competing interests and potential outcomes. Such an action would likely involve long-range strike platforms, such as bombers or sea-launched cruise missiles. Some strike missions can be conducted from distant bomber bases, some of which are in the United States, without assistance from forward presence. Others may require assistance from forward facilities, particularly if the mission needs support from off-board sensors. Depending on the air defense characteristics, it may also require fighter assistance, electronic warfare, combat search and rescue, and tanker support. As the Defense Strategic Guidance specifically links Iran with an anti-access strategy, the U.S. presence in close proximity to Iran has to be critically assessed not only for the combat power it can deliver against Iran, but also for its likely resilience to anti-access actions. The presence of U.S. forces in the region contributes to deterrence by maintaining an ability to respond quickly to a number of different potential Iranian provocations and by demonstrating a willingness of the United States to remain involved in regional security.

The United States has an extensive relationship with China that is qualitatively different from those with North Korea and Iran, which are both under trade sanctions. With ongoing territorial disputes in the South and East China seas, and with the long-standing impasse over Taiwan's status, there are many issues to work though. Friction points have led each side to try to use military capabilities for influence. For instance, in 1995 and 1996, China sought to influence the behavior of Taiwan by launching ballistic missiles that landed near Taiwan. The United States responded with warnings and the deployment of an aircraft carrier to the area. From the U.S. perspective, the United States sought to deter China from attacking Taiwan through the positioning of military forces.[15] The key capability development that could threaten perceptions about the U.S. ability to effectively counter some Chinese acts of aggression is the invest-

[15] For more details about the deterrent aspects of this crisis, an example of the application of immediate, extended deterrence, see Robert S. Ross, "The 1995–96 Taiwan Strait Confrontation" *International Security*, Vol. 25, No. 2, 2000, pp. 87–123.

ments China has made in long-range precision weapons, which are explored in more detail in Chapter Five. Still, there are many reasons for the two sides to avert conflict.

Countering terrorist groups remains an important national security priority. While initially there was considerable skepticism about the possibility of deterring terrorist groups, there is a growing body of literature devoted to assessments of influencing terrorist groups.[16] To the extent that it can identify and locate terrorist leadership and operational cells, the United States will want to maintain capabilities that it can employ to act on such information to capture and, if necessary, kill those involved. However, there are other ways to hinder terrorist operations. While individual terrorists may not be deterrable, terrorist groups must operate like other organizations: They must raise funds, recruit and train personnel, and design and carry out operations. These organizational necessities present opportunities to hinder operations and weaken terrorist organizations.[17]

Currently, the United States is drawing from forward forces to hinder terrorist groups in several areas. In addition to the large effort to disrupt terrorist groups in Pakistan, the United States also has active counterterror initiatives in Yemen and Somalia. For both, the United States benefits from having access to facilities nearby in Djibouti.[18] UAVs are primary tools for U.S. counterterrorism operations, both as sensors to locate and monitor terrorist groups and as platforms to launch strikes.

To assert that foreign bases contribute to deterrence is not to imply they are the only way to achieve deterrent effects. If the United States pulled its combat aircraft out of range of the Persian Gulf and reduced or stopped its maritime patrols in the region, the United States might still be able to deter Iran by maintaining both a clear diplomatic strategy to convince friends and Iran of the continued U.S. interest in the Gulf and by maintaining an ability to deploy forces to the region; however, that response would be less prompt because initial forces would not be in place in the region for immediate action. Some of that delay could be mitigated by prepositioned equipment, which would allow the United States to fly forces to the region quickly to fall in on their equipment. Other risks to removing presence from a region where the United States seeks to deter conflict could come from loss of access to facilities that make up the en route infrastructure. Follow-on forces could also be slower if the en route infrastructure were degraded. In fact, they could be very slow if regional partners did not provide prompt or sufficient access to bases in the area, or if those facilities needed improvements before they were suitable for U.S. combat operations.[19] Furthermore,

---

[16] See Andrew R. Morral and Brian A. Jackson, *Understanding the Role of Deterrence in Counterterrorism Security*, Santa Monica, Calif.: RAND, OP-281-RC, 2009.

[17] Morral and Jackson, 2009.

[18] See Craig Whitlock, "Remote U.S. Base at Core of Secret Operations," *Washington Post*, October 25, 2012.

[19] Improvements required could be to the physical plant, such as lengthening or resurfacing runways, or they could be equipment upgrades necessary to maintain and operate U.S. warplanes.

while a diplomatic strategy may be effective in certain circumstances, in others it will be necessary to couple it with clear demonstrations of relevant military capabilities.

The Adversary Gets a Vote

There are ways in which forward presence could detract from U.S. interests. Adding military capability to a region can be destabilizing. Steps that one side takes to increase its security by deterring a potential foe may often be viewed by that adversary as threatening to its own security, particularly when the move strengthens the position of a world power in another region. As Robert Jervis said, "A world power cannot help but have the ability to harm many others that is out of proportion to the others' ability to harm it."[20] Just as the Soviet Union's preparations to deploy nuclear weapons to Cuba in 1962 evoked a crisis, an action that a country initially intends to increase its security can lead to an unstable situation and evoke a strong counter-action.

Forward forces may be exposed, leaving them more vulnerable to an early attack. In efforts to prevent a future conflict, states often signal resolve through the deployment of military forces. There is a danger that, in crafting such signals, a state can lose sight of the consequences of such deployments if deterrence fails. Exposed forces might create incentives for adversaries to preemptively strike in hopes of catching U.S. forces off guard and vulnerable.[21] Such inherent risks of forward presence should be integral to considerations about the size and location of U.S. forces and periodically reassessed as new threats develop.

In 1941, the United States was surprised by the Japanese attack on Pearl Harbor, but such an attack by a weaker state on a stronger one was not a unique event in history, and weaker states have often tried to deter stronger states.[22] When seeking to deter another state, the vulnerabilities of one's own forces cannot be ignored. Sometimes weaker states attack stronger ones.[23] Historically, there are three situations in which such attacks have occurred: when weak states had high motivations, when weak states suffered from misperceptions, and when the stronger state had military vulnerabilities.[24]

---

[20] Robert Jervis, "Cooperation Under the Security Dilemma," *World Politics*, Vol. 30, No. 2, January 1978, p. 185.

[21] For an assessment of the conditions in which preventative attacks might be attractive, see Karl P. Mueller, Jasen J. Castillo, Forrest E. Morgan, Negeen Pegahi, and Brian Rosen, *Striking First: Preemptive and Preventive Attack in U.S. National Security Policy*, Santa Monica, Calif.: RAND Corporation, MG-403-AF, 2006.

[22] For a pre-invasion assessment of the deterrent interaction between the United States and Saddam Hussein in 2003, see Robert L. Jervis, "The Confrontation Between Iraq and the US: Implications for the Theory and Practice of Deterrence," *European Journal of International Relations*, Vol. 9, No. 2, June 2003, pp. 315–337.

[23] Bruce Bueno de Mesquita, *The War Trap*, New Haven, Conn.: Yale University Press, 1981, pp. 141–142, found that weaker nations initiated one-third of 20th century conflicts, as of 1981.

[24] Unpublished research by Barry Wolf of RAND.

Not all deterrent threats will work, even when the overall balance of forces is favorable. In posturing military forces to convey deterrent threats, leaving them vulnerable to devastating attacks could be quite risky. Douglas MacArthur's poor defensive preparations in the Philippines at a time of heightened tensions left the U.S. Pacific bomber force vulnerable to attack and gave the Japanese an opportunity to destroy a major retaliatory capability of the U.S. military after the Pearl Harbor attack. This suggests that U.S. forward presence should be of sufficient size to absorb attack and still present credible denial of quick victory to the adversary.

A forward posture can also be detrimental to U.S. interests to the extent that it limits flexibility. The agreements and current practices that allow access for U.S. forces to a country could, in some cases, actually end up tying down military assets and capabilities that could better be used elsewhere.[25] In fact, this problem was addressed in the 2010 QDR, when part of the DoD posture strategy sought flexibility and continuous adaptation in the U.S. posture.[26] This can be an issue of friction even with some of the United States' oldest and closest allies that host U.S. forces, particularly if their governments or populations oppose the U.S. action in another country. In such situations, it may be difficult for U.S. forces to operate from allied facilities. Even when the host country has more neutral feelings about the U.S. action, it may be concerned that taking U.S. capabilities elsewhere increases its own vulnerabilities.[27]

Forward-deployed forces provide the United States with many capabilities and options that could not be achieved if those same forces were stationed permanently in the United States, but these forces do have limitations. While forward-deployed forces may provide some measure of deterrence against some major conflicts, the forces that are forward-deployed are not sufficient of themselves to address conflicts of every scope. U.S. forward forces can be the leading edge of a U.S. response, but they rely on augmentation forces from elsewhere.

In many cases, this reliance has become less of a concern because allies have developed fairly substantial military capabilities in their own right. NATO allies, South Korea, and to some extent Japan, though it has policy constraints, can meet many potential threats on their own. In those cases, the U.S. presence is probably more about the level of commitment shown by the United States. In most areas, given the strong capabilities of the U.S. military compared with other militaries, potential adversaries are more likely to question the U.S. commitment, rather than capability. A substantial

---

[25] For example, see the description of how a proposed U.S. plan to remove four fighter aircraft based in Iceland threatened the U.S.-Icelandic Defense Agreement in Valur Ingimundarson, "Relations on Ice Over U.S. Jets," *New York Times*, July 12, 2003. The four aircraft were eventually removed, but the incident shows how aligning the interests of allies can be challenging and that it can lead to situations where the United States keeps forces in a country even when the security rationale is questionable.

[26] DoD, 2010a, pp. 63–64.

[27] As was the case when the United States took forces stationed in South Korea and deployed them to Afghanistan.

foreign presence is an important way to show this commitment, but it is not the only way.

Assuring Allies

The United States seeks not only to deter foes, but also to assure allies and friends of mutual security commitments. Like deterrence, the concept of assurance gained prominence in the nuclear age. In particular, the idea originates from a commitment by one party to extend nuclear guarantees to protect an ally. Today, the reliance on nuclear weapons in U.S. security policy is diminished, but the dominant capabilities of conventional U.S. military forces make conventional assurance a relevant concept.

During the Cold War, the United States frequently had to reassure key allies—despite the Article 5 assurance of the NATO Treaty that an attack against one would be treated as an attack against all—that, should the Soviet Union threaten them with nuclear weapons or even with only conventional capabilities, the United States was prepared to extend the deterrent and retaliatory capabilities of its nuclear force to those allies. This included taking the step of storing tactical nuclear weapons in Europe, again both to deter and assure.

Convincing allies of the authenticity of a security commitment may actually be harder than convincing potential adversaries. As a former British Defense Minister only somewhat facetiously stated, "It takes only five percent credibility of American retaliation to deter the Russians, but ninety-five percent credibility to reassure the Europeans."[28] This was most pertinent to NATO allies, and NATO documents were explicit about this. Even now, NATO's 2010 *Strategic Concept* articulates a "supreme guarantee" of security from alliance nuclear forces, "particularly those of the United States."[29] Such a policy had a number of benefits for the United States. First, it diminished incentives for countries to develop their own nuclear deterrent, which coincides with the nonproliferation goals of the United States. Second, it strengthened the alliance; the United States could hardly make a more weighty alliance commitment to bind the security of alliance members. Finally, it countered what otherwise would have been escalation dominance over allies by the Soviet Union; in any conflict with a non-nuclear ally, the Soviet Union could threaten nuclear attack, and the ally would have no ability to counter without nuclear security guarantees from another alliance member. This bargain worked. The number of NATO nuclear states remained small, while the alliance, by historical standards, became closer than ever.

---

[28] The quotation is from Denis Healy, quoted in Davis S. Yost, "Assurance and US extended deterrence in NATO," *International Affairs*, Vol. 85, No. 4, 2009, p. 756.

[29] NATO, *Active Engagement, Modern Defence*, North Atlantic Treaty Organization Strategic Concept, 2010, p. 14.

These extended nuclear protections were not only for NATO members. The United States also offered similar guarantees to Japan and Korea.[30] These countries have the economic and technical means to pursue nuclear weapons, but have refrained from acquiring them, despite North Korea's pursuit of nuclear weapons and the long-standing nuclear forces in nearby China and Russia. The formal treaties with Japan and Korea, the presence of U.S. troops, and U.S. policies of extended deterrence all likely influenced the nuclear weapons policies of these allies.

Today, when the United States announces that assuring allies is part of its strategy, certainly this historic nuclear assurance is part of what it means; however, the concept has expanded to include a conventional aspect as well and a specific link to the U.S. posture. The 2010 QDR states, "The long-term presence of U.S. forces abroad reassures allies and partners of our commitment to mutual security relationships." In the same way that nuclear weapons were offered previously, here the dominance of U.S. conventional military capabilities are offered to help deter threats to allies and host nations to meet broader security goals. The United States wants to signal that it will come to the aid of close allies and nations hosting U.S. bases that are threatened by a variety of potential attacks. This can be codified in treaty alliances, bilateral agreements, or sometimes informally.

The 2001 QDR elevated the concept of assuring allies to one of four top security goals. It stated:

> The presence of American forces overseas is one of the most profound symbols of the U.S. commitment to allies and friends. The U.S. military plays a critical role in assuring allies and friends that the Nation will honor its obligations and will be a reliable security partner. Through its willingness to use force in its own defense and that of others and to advance common goals, the United States demonstrates its resolve and steadiness of purpose and the credibility of the U.S. military to meet the Nation's commitments and responsibilities.[31]

Assuring friends and allies with conventional capabilities has similarities with nuclear guarantees but also some key differences. As was the case with extending nuclear protections, by extending to allies protection through conventional forces, the United States can also meet its nonproliferation objectives and strengthen its security alliances. The nature of the threat is very different, though, and that leads to different needs for the United States and different needs and incentives for host nations.

---

[30] David J. Trachtenberg, "US Extended Deterrence: How Much Strategic Force is Too Little?" *Strategic Studies Quarterly*, Vol. 6, No. 2, Summer 2012, pp. 265–298.

[31] DoD, "Quadrennial Defense Review Report," September 30, 2001, p. 11.

Aligning Interests

While overlapping interests create the conditions for establishing U.S. presence in a host nation, overlapping interests do not mean identical interests or imply identical threat perceptions and goals. There are two reasons why the interests of the United States may not be congruent with those of alliance partners and host nations. First, even if they agree on a common enemy, they will not have identical views about that adversary and how it should be handled. Second, the United States takes a global view of security interests and commitments in a way that a local partner is unlikely to share.

Understanding and overcoming these different threat perceptions and relationships with potential adversaries is a key ingredient to the assurance that the United States seeks to provide. The hosts live in the region, and their strategic outlook is as much shaped by their history as it is shaped by a range of present-day political, economic, and human interactions. These factors combine to produce different perceptions of interests and different cost-benefit calculations, especially when considering the intentions of neighbors and the threats of war. Allies and host nations may not simply view their interests as the inverse of their adversaries' being deterred, and, because the consequences of conflict are likely to be disproportionately borne by them, they are likely to be much more cautious.

Another difficulty in aligning interests comes from the fact that the United States has global interests and tends to utilize assets globally to pursue its interests. While other countries have global interests too, the geographic scope and scale of U.S. activities to shape events is unparalleled. In many cases, this means that while the United States seeks access to a country to pursue global interests, the host nation typically takes the local perspective of domestic politics and the threats it faces before it agrees to the continued presence of the U.S. military.

The 2010 QDR, reinforced by the 2012 Defense Strategic Guidance, links U.S. posture to the goals of having flexible and adaptive forces and a robust deployment architecture that allows them to deploy globally.[32] This goal stems from a view of the global threats with which the United States is concerned, as well as the widely distributed operations in which the U.S. military now commonly engages. A base in Country A may be used to conduct surveillance of Country B, which might accrue more to the benefit of Country C than to that of the "host" for U.S. forces, Country A. In most host countries, spelling this out may prove inconvenient. The fact is that when the United States conducts military operations today, it does so utilizing capabilities from its global force. This operational reality probably could be better communicated to host populations. The Defense Strategic Guidance describes the broad global agenda of the United States in the following way:

---

[32] To summarize, it identifies five goals: reassuring allies, flexibility for United States to respond globally, robust lines of communication, acceptance by host nations, and continuous adaptation. See DoD, 2010a, pp. 63–64.

> Across the globe we will seek to be the security partner of choice, pursuing new partnerships with a growing number of nations—including those in Africa and Latin America—whose interests and viewpoints are merging into a common vision of freedom, stability, and prosperity.[33]

Aligning interests can help the United States maintain the flexibility that it seeks from its posture. One potential drawback of a foreign presence is the extent to which it ties U.S. forces to a particular threat. To use Korea as an example, if the United States had to operate in Korea today, it would require access and cooperation from Japan, as was the case during the Korean War. What is different now is that U.S. operations would need not only the cooperation of Japan but probably also that of several other countries to utilize fully all the conventional assets it might want to employ to help defend South Korea. UAV pilots might be located thousands of miles away while they direct their aircraft over Korea. Similarly, satellite ground stations and intelligence analysis operations are not in Korea. The forces operating in Korea would rely on a global logistics network of air bases and seaports.

In seeking flexibility, the United States may have to work to assure host nations. The host nation may legitimately be concerned that when U.S. forces are taken away for other operations, it might embolden the adversary they seek to deter. Flexibility is quite important to the United States to balance its commitments around the world and maintain the capability to respond dynamically. Assuring allies thus requires an ongoing dialogue about interests, threat perceptions, and intentions.

Implications for Posture

Regional forces, particularly those directly in place where the threat is greatest, that are seen to prevent quick victory by adversaries can contribute to deterrence, as can missile defense capabilities that prevent coercion by states with substantial missile arsenals. Forward forces that can provide such capabilities are CSGs, expeditionary strike groups, light Army units, and various combat and sensor aircraft. In addition, missile defenses, either land-based or maritime, can provide the capability to parry the coercive use of long-range missiles. To be able to respond within a number of hours, it is not enough to be forward, but to be forward in the right region. Of these forces, missile defenses and Army ground forces need to be located in the affected area. Even deployed CSGs and expeditionary strike groups may take several days to reach a crisis area if they are not already close by. In some instances, the presence of U.S. forces becomes important to emphasize not just the military capability to respond to a threat, but also the interest and political will to do so. This is very important to assure allies of the U.S. commitment to their security.

---

[33] DoD, 2012a.

While forward forces often can contribute to deterrence, there is no accurate way to estimate how much of a forward presence is required to meet deterrent goals. Since there is significant cost to keeping a forward military presence, careful evaluation of the necessity of each forward unit is warranted. Much of the deterrent influence of the U.S. military comes from its ability to generate highly capable forces and deploy and sustain them anywhere in the world, which relies upon a network of air and sea transportation hubs described in Chapter Two. As such, when considering the needs for forward-deployed forces to meet deterrent goals, it should be in the context of maintaining capabilities to thwart quick victories and to assure allies of U.S. commitments.

CHAPTER FOUR

Strategic Considerations: Benefits of Overseas Posture for Security Cooperation

U.S. government strategic guidance documents, such as the 2012 Defense Strategic Guidance, 2010 QDR, and DoD defense posture reports to Congress, assert that overseas posture provides significant strategic benefits, including improved security cooperation between the United States and partner countries.

DoD doctrine defines security cooperation as "activities undertaken by the Department of Defense to encourage and enable international partners to work with the U.S. to achieve strategic objectives."[1] Examples of security cooperation include programs that train and equip foreign partners, provide professional military education, conduct military exercises, and exchange information. Security cooperation is a means to several important U.S. national security goals. It helps build the capacity and professionalism of partner security forces (sometimes called "building partner capacity"), which can reduce instability, assure partners, and deter adversaries. It may also strengthen partners' willingness to use their security forces in support of U.S. foreign policy goals. Finally, it can improve U.S. military capabilities, particularly those required for effective coalition operations.

This chapter provides a qualitative assessment of the ways overseas posture might benefit U.S. security cooperation efforts. Specifically, we analyzed whether U.S. forces based overseas might provide advantages in terms of cost, partner willingness to deploy forces, partner capability development, and U.S. training. We also analyzed whether rotational forces (i.e., forces that deploy overseas regularly) provide advantages similar to those of forces permanently based overseas.

RAND has produced extensive analyses of security cooperation and nation building, but there has been very little research about whether forces based overseas provide security cooperation advantages over U.S.-based forces.[2] We focused our research for

---

[1] Deputy Secretary of Defense Gordon England, *DoD Policy and Responsibilities Relating to Security Cooperation*, Department of Defense Directive Number 5132.03, October 24, 2008.

[2] Jennifer D. P. Moroney, Beth Grill, Joe Hogler, Lianne Kennedy-Boudali, and Christopher Paul, *How Successful Are U.S. Efforts to Build Capacity in Developing Countries?* Santa Monica, Calif.: RAND Corporation, TR-1121-OSD, 2011; Jennifer D. P. Moroney, Patrick Mills, David T. Orletsky, and David E. Thaler, *Working*

this component of the study on a number of snapshots that might inform a broader qualitative analysis. First, we focus primarily on larger-scale security cooperation conducted overseas, since these activities are more likely to influence decisions about overseas basing or allocations of rotational forces. Second, while we looked at the conduct of security cooperation by all four military services, we focused our analysis primarily on the Army and Air Force. We look specifically at the Marine Corps in terms of how rotational forces compare with forces permanently stationed overseas and at the Navy in terms of conducting security cooperation with sea-based forces. Third, while we address security cooperation conducted around the world, we focused especially on EUCOM, because of its emphasis on security cooperation and relatively high level of overseas forces.

Overall, we conclude that overseas basing provides some advantages in each of these areas but only in certain circumstances, as described below. Although CONUS-based forces have also conducted extensive security cooperation to positive effect, we also conclude that overseas basing both improves and increases security cooperation.

Does Overseas Basing Provide Cost Advantages for Security Cooperation?

The data generally supported our hypothesis that forces based overseas can conduct security cooperation activities with relatively low marginal cost and at a lower cost than CONUS-based forces, particularly for large-scale activities. However, the cost differential is much smaller than the overall incremental cost to base forces overseas, so the decision to base forces abroad should depend upon other factors. This section looks at the marginal cost of conducting security cooperation, given that forces are already based overseas, and the cost differential of security cooperation conducted by forward compared with CONUS-based forces. Chapter Eight assesses the cost differential between U.S. and foreign basing as a whole. Security cooperation cost efficiencies are simply one additional factor, but one that could slightly offset higher overseas basing costs.

---

with Allies and Partners: A Cost-Based Analysis for the U.S. Air Forces in Europe, Santa Monica, Calif.: RAND Corporation, TR-1241-AF, 2012; Christopher Paul, Colin P. Clarke, Beth Grill, Stephanie Young, Jennifer D. P. Moroney, Joe Hogler, and Christine Leah, *What Works Best When Building Partner Capacity and Under What Circumstances?* Santa Monica, Calif.: RAND Corporation, MG-1253/1-OSD, 2013. See also Michael McNerney and Thomas Szayna, *Assessing Security Cooperation as a Preventive Tool,* Santa Monica, Calif.: RAND Corporation, forthcoming; James Dobbins, Seth G. Jones, Keith Crane, and Beth Cole DeGrasse, *The Beginner's Guide to Nation-Building,* Santa Monica, Calif.: RAND Corporation, MG-557-SRF, 2007; James Dobbins, Seth G. Jones, Keith Crane, Andrew Rathmell, Brett Steele, Richard Teltschik, and Anga R. Timilsina, *The UN's Role in Nation-Building: From the Congo to Iraq*, Santa Monica, Calif.: RAND Corporation, MG-304-RC, 2005; and James Dobbins, John G. McGinn, Keith Crane, Seth G. Jones, Rollie Lal, Andrew Rathmell, Rachel M. Swanger, and Anga R. Timilsina, *America's Role in Nation-Building: From Germany to Iraq*, Santa Monica, Calif.: RAND Corporation, MR-1753-RC, 2003.

Security cooperation activities can range from multinational exercises with army battalions and fighter squadrons to military advisory missions to classroom education to conferences. Similarly, costs can vary from a few hundred to a few million dollars. Military planners can leverage forces based overseas to reduce costs of security cooperation by using existing assets, personnel, and U.S. training requirements. For example, RAND research has shown that USAFE security cooperation—even including high-end, more expensive activities—adds only $51 million to overall annual operating costs, roughly 1 percent of USAFE's $5 billion operating budget.[3] Air Force pilots who need to fly hours for their own training can sometimes fly those hours in concert with a security cooperation mission. Since some security cooperation costs are partially covered by fixed costs as well as annual training costs when forces are stationed in the region, more security cooperation does not result in proportionally more cost. Thus, as Air Force officials in Europe noted, as squadron numbers go down, so will security cooperation activities. Planners can mitigate this somewhat by bringing in Air National Guard or other units from the United States, but costs for security cooperation grow significantly.[4] But it is more likely that the frequency of such activities will decrease.

U.S. forces based in Europe also reduce transportation costs for security cooperation in Africa, but only by small amounts. For example, U.S. Naval Forces Africa, found that, based on an average of four security cooperation missions per year, transportation costs would be $1.2 million (or 60 percent more) for CONUS-based forces, compared with $750,000 for European-based forces. U.S. Air Forces Africa found travel costs for personnel to be about 50 percent more when using CONUS-based forces.[5] U.S. forces from Europe conduct about half of their security cooperation activities in Africa. While these examples indicate there are potential offsets to the total marginal costs of basing forces overseas, any savings would constitute a relatively small part of the overall cost factors considered in this study.

While the examples above help illustrate the potential security cooperation cost savings of overseas basing, available security cooperation data did not allow for comprehensive comparisons of the relative costs of security cooperation between forces based in the United States and those based overseas. A general review of security cooperation data was inconclusive. This could be because transporting equipment and personnel for security cooperation events within a region is not consistently less expensive than transporting them from the United States. It could also be because the data did not always reflect total costs of activities or because they did not reflect the qualitative differences between activities. More important than the relatively small potential cost savings, however, is the fact that when forces are based overseas for multiple reasons

---

[3] Moroney et al., 2012.

[4] Interviews with U.S. military leaders in Europe, August 27–31, 2012.

[5] Documents provided by U.S. Africa Command, August 27–31, 2012.

(e.g., deterrence, operational reach), the incremental cost to conduct security cooperation is low, which leads to a higher level of activity. Additional information about security cooperation costs and overseas basing is in Appendix C.

Does Overseas Basing Improve Partner Willingness to Deploy Forces?

While a partner nation's political considerations will dominate its decisions about deploying forces to support multinational operations, U.S. overseas basing and security cooperation may influence those decisions for several reasons. First, it strengthens political and military relationships that could translate into U.S. influence or at least a more general alignment of political values and goals. Second, security cooperation with U.S. forces overseas may help partners feel secure enough to deploy their forces in coalition operations rather than to focus exclusively on their own security. Third, it might support the strengthening of partner capabilities such that partners become more willing to participate in coalition operations.

Deployment decisions are often overshadowed by other factors. For example, despite hosting large numbers of U.S. forces, Japan's constitution severely constrains its ability to deploy security forces outside its borders. Australia has been the largest non-NATO contributor of troops to Afghanistan, despite an absence of U.S. bases. Although it hosts the only U.S. base in Latin America, Honduras provided no troops to Afghanistan and far fewer to Iraq than its neighbor, El Salvador. On the other hand, South Korea was the second-largest contributor of ground troops to support U.S. operations in Iraq after the UK.[6] The long-term commitment of U.S. forces to Korea may have made it politically and militarily feasible for South Korea to contribute forces to Iraq. While 20 European countries were contributing troops to the International Security Assistance Force in Afghanistan as of August 2012, it would be difficult to evaluate to what extent European deployments are due to U.S. overseas basing and security cooperation rather than shared political values derived from economic, cultural, and historical ties, as well as the integrating influence of NATO and the capabilities of the countries themselves. Table 4.1 shows the number of troops contributed by nations in four categories.

Table 4.1
International Security Assistance Force Troops, August 2012

Total	U.S.	NATO (Non-U.S.)	European (NATO Partner)	Other
128,500	90,000	34,000	2,000	2,500

SOURCE: Figures provided by U.S. Army Europe, 2012.

NOTE: Largest non-NATO contributor: Australia, with 1,550 forces.

[6] Stephen A. Carney, *Allied Participation in Operation Iraqi Freedom*, Washington, D.C.: Center of Military History, United States Army, 2011, p. 102.

Does Overseas Basing Provide Benefits for Partner Capability Development?

In some cases, basing forces overseas appeared beneficial for strengthening partner capabilities—particularly for more advanced military capabilities and interoperability—while in others it appeared that U.S.-based forces could do the job adequately. Overseas basing appears most important for security cooperation efforts in South Korea and many countries in Europe. For these countries, security cooperation is an almost constant activity based on enduring relationships and informal interactions that are not always captured in databases. Moreover, as described below, there is a qualitative difference in the type of security cooperation that is conducted by forces based in those regions. For many other countries, particularly those with less advanced militaries, security cooperation can be conducted largely through frequent but smaller-scale events using rotational or temporarily deployed forces.

U.S. military leaders in the Pacific noted that because U.S. and Korean forces operate under one of the world's most integrated command structures, security cooperation is continuous and automatic.[7] The commander of USFK also serves as commander of Combined Forces Command, which oversees both U.S. and South Korean forces. Together, these forces plan, train, collect intelligence, and respond to crises. It is likely that the constant engagement resulting from the permanent stationing of U.S. forces has played a role in the dramatic improvements in Korean military capabilities. These improvements will allow the Korean military to take wartime operational control of U.S. and Korean forces by 2015.

On the other hand, U.S. forces in Japan focus less on building the security capabilities of their host. While security cooperation certainly occurs, particularly between the U.S. and Japanese navies, the level of security cooperation depends far less on the number of U.S. forces stationed there. In Australia, USMC plans to increase training with Australian forces and allow for more training in Southeast Asia. Elsewhere in Asia, the United States builds partner capabilities primarily through rotational forces and temporary deployments rather than through permanent bases, at least partly due to partners' political limitations on establishing permanent bases in the region. Plans to increase the U.S. presence in Asia are at least as much about assurance and deterrence as strengthening partner capabilities.

In the Middle East, DoD conducts numerous exercises and other security cooperation activities with partners, using a small number of permanently assigned forces, supported by large numbers of rotational and temporarily deployed forces. The Air Force builds capabilities of partner air forces primarily through forces stationed at Ali Al Salem, Al Udeid, and Al Dhafra, with support from forces deployed from the United States. In particular, the Gulf Air Warfare Center at the UAE's Al Dhafra Air

---

[7] Interviews with U.S. military leaders in the Pacific, October 1–2, 2012.

Base strengthens interoperability among air forces in the region. The Navy uses its Fifth Fleet headquarters in Bahrain to strengthen relationships in the region, while rotating ships for exercises. The Army and USMC rely primarily on rotational and temporarily deployed forces.

In Africa, U.S. forces based at Camp Lemonnier in Djibouti build capabilities of partners in East Africa and in the Trans-Sahel region, especially for counterterrorism. While Camp Lemonnier provides benefits for conducting security cooperation in the region, it is even more important as a base for counterterrorism operations and for responding to contingencies in the Middle East and elsewhere. More important for security cooperation in Africa are U.S. forces based in Europe. For example, USAF personnel based in Europe not only conduct about 200 security cooperation activities with European partners each year, but also about 150 with African partners as well, though the latter were often smaller scale.[8] Another example is Africa Endeavor 2011, an exercise to improve communications interoperability between 35 African militaries and NATO (including U.S.) military forces and to build capability for future multinational operations.[9]

In Europe, the United States organizes about 30 regional exercises each year in locations, such as Ukraine and Romania. Regional exercises enable host nations to train larger numbers of their forces with U.S. forces. They also provide benefits to U.S. forces that train to deploy and operate in a more austere environment.[10] Although such exercises are not necessarily more effective when conducted by U.S. forces that are based overseas, the closer proximity makes them more likely to occur.

For ground-forces training, Joint Multinational Training Command's (JMTC's) training facilities conduct more multinational training than any other U.S. training facility. JMTC oversees the Grafenwoehr and Hohenfels training areas, providing military forces with live-fire and combat-maneuver training, a simulation center, and professional military education, as well as home-station training for units at 15 other locations in Europe.[11]

In 2011, JMTC trained 20,000 U.S. forces and almost 10,000 European and other forces. Since 2008, JMTC has trained 70,000 U.S. and 33,000 multinational forces from 27 countries. Most European partners would receive far less training with U.S. forces without the facilities in Europe, since most partners pay their own way for this training but may not pay to go to training sites in the United States.

Greater burden sharing with Allies in Europe is a means by which DoD could improve partner capabilities and interoperability, while also reducing costs. For example, NATO, or individual NATO members, could be encouraged to share more of the

---

[8] Interviews with U.S. military leaders in Europe, August 27–31, 2012.

[9] Interviews with U.S. military leaders in Europe, August 27–31 and October 23–24, 2012.

[10] Information provided by U.S. Army Europe, September 2012.

[11] JMTC information provided by U.S. Army Europe in September and October 2012.

costs for multinational exercises at JMTC. NATO has a clear policy that training is a responsibility for individual member countries, but multinational exercises are as much about interoperability as national training. Moreover, NATO funds Allied Command Transformation, which supports doctrine development and experimentation and oversees several subordinate commands, including the Joint Warfare Centre in Norway, the Joint Force Training Centre in Poland, and over a dozen centers of excellence. Given NATO's interest in applying its "Smart Defense" concept to improvements in interoperability and training, Allied Command Transformation could co-locate with JMTC, improve efficiencies in training across Europe, and begin to share the costs for operating JMTC.[12] NATO Common Funds currently support Allied and partner participation in JMTC exercises but do not subsidize JMTC's operating or overhead costs, despite the fact that European military forces now make up over 30 percent of the forces trained by JMTC. Although NATO has shown little appetite for taking on additional costs, future U.S. force reductions in Europe could lead to NATO having to face the choice between subsidizing (and thereby saving) JMTC and other U.S. facilities or allowing them to close. Alternatively, other NATO members, such as the UK, France, Italy, Norway, and Germany, could consider subsidizing JMTC operations costs and using it as a substitute for their own training ranges, which could potentially be reduced.

There may also be opportunities for improving security cooperation through multi national basing and other cost-sharing arrangements. For example, NATO underwrites over 60 percent of the routine operations and maintenance costs at the Chievres Airbase in Belgium, which also serves as a USAF operating site. NATO or NATO Allies also share some costs at U.S. facilities such as Ramstein Air Base, Aviano Air Base, and Sigonella Naval Air Station, as well as at co-location sites, such as Rota Naval Air Station, RAF Lakenheath, and RAF Mildenhall. Arrangements like this not only control U.S. costs but can also facilitate greater levels of security cooperation.

A 2012 RAND study for USAFE found that USAFE forces engage in security cooperation as part of their daily routine. Security cooperation permeates the everyday activities of the headquarters staff as well as the six wings that were assessed. Security cooperation is often included as a core element of USAFE training, which contributes to partner capability development, particularly capabilities for more effective coalition operations. The study found that forward basing facilitates important relationship- and capacity-building activities. A number of activities occur primarily because of forward basing, including nearly daily air refueling, frequent joint tactical air controller qualification and training performed by the Warrior Preparation Center near Ramstein Air

---

[12] NATO's Secretary General has advocated for Allies to "pool and share capabilities" in support of the Smart Defence approach for "greater security, for less money, by working together and with more flexibility." See Anders Fogh Rasmussen, "Building Security in an Age of Austerity," keynote speech delivered at Munich Security Conference, February 4, 2011.

Force Base, and some hosted events.[13] Interviews with USAREUR leaders suggested similar continuous interaction between U.S. and European ground forces.[14]

Moreover, the presence of forward-based forces facilitates coalition operations. The development and sustainment of personal and unit relationships enables smoother integration during combat operations. It also allows high revisit rates for partners who require it to increase their capacity for out-of-area operations. Years of interaction with traditional as well as newer NATO allies have born fruit for coalition building and capability. For example, European air forces have been able to operate in coalitions with the USAF in Afghanistan, Iraq, and Libya in part because they have trained and exercised together; standardized tactics, techniques, and procedures; and developed working relationships. These activities over many years have enabled the more advanced European air forces to plan and fight with the United States. Likewise, USAFE seeks to bring less-advanced partners into future coalition operations by emphasizing important contributions that do not require well-established air force capabilities or institutions. USAFE efforts to develop partner-country capacity in air controller capabilities for OEF have included allies such as Lithuania, Estonia, Slovenia, and Latvia that have no capability to deploy and operate aircraft in overseas contingencies. Security cooperation activities have also allowed countries such as Poland to execute cargo preparation operations on their own to enable deployment on U.S. airlifters to out-of-area operations.[15]

U.S. forces across Europe, not just those at training facilities, help build the capabilities of European forces. The Army and Air Force use most of their major units to conduct security cooperation to one degree or another, whether through combat units or through military police, military intelligence, medical, or air-ground operations units. U.S. forces based overseas allow for more frequent interactions between these specialized units and their foreign counterparts.

While U.S. Special Operations Forces specialize in operating in austere environments to build partner capabilities, they, too, leverage U.S. bases around the world to facilitate their security cooperation activities. For example, their ability to maintain a permanent presence in Europe, particularly through the stationing in Germany of the First Battalion, Tenth Special Forces Group, allows for a more sustained focus on improving European special operations capabilities. Through a combination of a permanent forward presence and rotational forces from the United States, Special Operations Command, Europe, conducts over 300 security cooperation events per year. The permanent presence of U.S. Special Operations Forces in Europe also facilitated the establishment of the NATO Special Operations Headquarters, which coordinates special operations education, training, and information sharing. Improved European spe-

---

[13] Moroney et al., 2012.

[14] Interviews with U.S. military leaders in Europe, August 27–31, 2012.

[15] Moroney et al., 2012.

cial operations capabilities have been especially evident in their improved performance over time in Afghanistan and Iraq.[16] Thus, while Special Operations Forces can conduct security cooperation effectively from U.S. bases, their ability to leverage overseas bases allows them to amplify their effect.

Does Overseas Basing Provide Training Advantages for U.S. Forces?

Compared with those based in the United States, forces based overseas often benefit from more frequent training with foreign partners. In particular, U.S. forces learn to operate more effectively in a coalition and adapt to foreign environments. As one military officer put it, "At U.S. training sites we train ourselves by ourselves." In Asia, Africa, Latin America, and the Middle East, however, these benefits often come at the price of the quality of the training for high-end missions. Whether because of political restrictions, limited space or infrastructure, or lack of time due to other mission requirements, training overseas is often seen as more of a constraint than a benefit. Constraints exist in Europe as well, but the interoperability benefits that come from training with relatively capable partners compensate for these constraints, which are not meaningfully different from those that exist at many bases in the United States.[17]

Along with Canada and Australia, European militaries have been the most capable and willing partners in recent coalition operations. If the United States continues to emphasize the importance of coalitions for future conflicts, interoperability with allied militaries will remain important. Many of the security cooperation activities conducted by U.S. forces in Europe provide training benefits for both European and U.S. forces. In particular, bilateral and multilateral training activities that occur across the continent—both on U.S. and partner bases—enhance U.S. military capabilities. For example, the Army base at Baumholder provides both quality training and flexibility due to the ability of U.S. and German forces in that region to use each other's ranges.

As indicated earlier, one U.S. Army overseas training center warrants particular emphasis. JMTC falls in the same category as the National Training Center at Ft. Irwin, California, and the Joint Readiness Training Center at Ft. Polk, Louisiana. JMTC operates at full capacity, with some partner countries waiting for months to use its facilities. In addition to training U.S. forces based in Europe, JMTC trains Army National Guard and Reserve units from the United States. Though focused in recent years on training relevant to Afghanistan and Iraq, in 2012 it began a transition back to full-spectrum training, hosting two Decisive Action Training Events involving

---

[16] Interviews with U.S. military leaders in Europe, August 27–31 and October 23–24, 2012.

[17] Interviews with U.S. military leaders in Europe, August 27–31 and October 23–24, 2012. Also, RAND analysis of training limitation reports provided by U.S. COCOMs.

hybrid warfare scenarios. The October 2012 event, called Saber Junction, involved the Second Cavalry Regiment and forces from 19 other countries and included force-on-force, live-fire, irregular warfare, special operations, and cyber elements. In addition to the training areas at Grafenwoehr and Hohenfels, limited activities were conducted in the German community between the two bases, creating a training area geographically as large as the National Training Center—albeit with many more restrictions on what could be done on one hand, but requirements to deal with the civilian populace on the other.

Army leaders argue that no U.S. training center can compete with JMTC in the following areas:

- Training for coalition operations at the battalion level and higher: JMTC can exercise a brigade with battalions from multiple countries. For example, in addition to its Decisive Action Training Events, two Georgian battalions trained at JMTC in 2012, and the UK was scheduled to train a battalion-plus. While feasible at U.S. centers, costs would be prohibitive for many partners and available capacity extremely limited, given requirements for U.S. training.
- Diversity of activities: JMTC executes a wide range of training and education from the individual soldier to three-star headquarters. JMTC not only supports air drop, cargo drop, and live close-air-support training like other centers, it also provides simulations, professional military education, and culturally attuned mobile training.
- Realistic interaction with local communities: The U.S. SOFA with Germany allows JMTC to work with local officials to create a unique maneuver training area that includes the community between Grafenwoehr and Hohenfels.[18]

The benefits described above come with costs. There may be ways to reduce these costs somewhat through the use of rotational forces, although—as discussed later—rotational forces bring their own advantages and disadvantages. A detailed analysis comparing the benefits and relative costs of U.S. military training centers within the United States and abroad could provide important insights for future investments.

For the Air Force in Europe, much of the training for attaining qualifications and keeping skills current is carried out in partner countries. While units based in Europe do conduct some training in the United States (e.g., at Red Flag), it would be expensive for partners to do the majority of their training events across the Atlantic. As such, USAFE elects to forge and sustain relationships with countries in Europe to gain access to their ranges and airspace and to train with them.

The training value of security cooperation events can vary depending on the type of U.S. unit and aircraft or other equipment, the goal of a training event, the level of

[18] Interviews with U.S. military leaders in Europe, August 27–31, 2012.

sophistication of the partner(s), the availability of ranges and other training environments, and a number of other factors. For example, the ability of U.S. strike or close-air-support aircraft (including F-15Es, A-10s, and F-16s) to work with both U.S. and foreign air controllers at ranges in Croatia enables those U.S. pilots both to remain current in controlled bombing missions and to encounter different levels of capability among controllers; at the same time, these events ensure qualification and currency of the foreign air controllers themselves. U.S. fighters in Europe are able to conduct dissimilar aircraft training with foreign pilots, who fly Typhoons, MiG-29s, and Tornados. This training provides U.S. pilots with adaptive skills and allows them to experience missions against pilots who employ different tactics, capabilities, and procedures, and it helps prepare U.S. pilots for both combat against adversaries and for combat integration with allies.

On the other hand, while many U.S. units and personnel derive training benefits from security cooperation activities and presence in Europe, some benefit less than others. Some training involving partners appears less beneficial to U.S. currency and readiness requirements. Partner-nation air forces have less-advanced capabilities than those of the USAF—with some trying to maintain just a basic capacity to field an air force—and U.S. tactics at times must be adjusted to account for this in ways that detract from U.S. training. Events related to fighter interoperability appear to be the most problematic. For instance, it is a challenge to find adequate air-to-air training for U.S. pilots. There are a number of reasons for this, including lack of ranges, the need to "dial back" skills and tactics during exercises to match partner capabilities and meet nondisclosure requirements, and restrictions in the AOR on important training regimens, such as launches of AIM-120 air-to-air missiles.

There is significant cultural and interoperability value in working with both highly capable and less capable partners. It appears, however, that some security cooperation activities (e.g., high-end, multinational training) are especially beneficial to U.S. forces. It also appears that U.S.-based and overseas-based forces can benefit from security cooperation activities. Finally, overseas basing may matter more for developing military-to-military relationships and interoperability at the strategic and operational level rather than the tactical level, where tasks are less complex and personnel turnover is higher. Thus, when considering future U.S. overseas force reductions, there should be particular attention paid to maintaining strategic and operational interoperability skills.

How Do Rotational Forces Change U.S. Military Strategies for Security Cooperation?

As discussed earlier, posture includes more than overseas basing. Rotational forces are a component of posture that is important for conducting security cooperation activities and they have the potential to mitigate reductions in overseas basing.

The Navy's operating practices, which provide global presence at sea, mean that much of its force conducting security cooperation is, in effect, rotational or on temporary deployment. Because the Navy deploys constantly for multiple missions (e.g., presence, deterrence), it can readily engage in security cooperation in the course of its operations, thereby maintaining a low marginal cost for these activities. The Navy certainly uses land-based forces to conduct security cooperation as well, but largely sees its bases as logistics platforms, as opposed to the focal points for security cooperation. Naval forces provide some level of flexibility as to where security cooperation takes place and where the U.S. military has bases. Under any future posture, the Navy could continue to conduct large multinational exercises or other security cooperation events in international waters. One example of how the Navy combines the advantages of forward-based and rotational forces is its use of the Fifth Fleet headquarters in Bahrain to strengthen relationships in the Persian Gulf region, while using rotational naval forces to conduct specific security cooperation activities.

The Marine Corps also conducts the majority of its security cooperation using rotational forces. With a few exceptions (e.g., marines based in Japan), marines conducting security cooperation activities are generally forward-deployed but not forward-based. When overseas, marines are often deployed on ships, conducting multiple missions as part of an MEU deployment. Because the MEU must train to conduct a range of missions from small-scale amphibious assaults to evacuations to marine interdiction to humanitarian assistance, the costs for security cooperation are relatively marginal to the overall cost of unit operations. In other cases, the Marine Corps creates special task forces for more limited missions.

The Black Sea Rotational Force temporarily deploys 200–400 marines to train partners in that region for six months per year. In 2012, the Marine Corps also began to utilize a Special Purpose Marine Air-Ground Task Force, consisting of about 180 marines on six-month rotations from Sigonella Naval Air Station in Italy, to conduct security cooperation across Africa. Also in 2012, the first of what will be regular rotations of marines arrived at an Australian base in Darwin to conduct multiple missions, including training with Australian forces and security cooperation throughout Southeast Asia.

Regardless of whether they are deployed on ships or on land, these rotational forces have the flexibility to train partners wherever the need is greatest, and to strengthen their own expeditionary capabilities. On the other hand, these forces tend to focus on tactical training and rely on the infrastructure of other services or countries. This type of rotational approach to security cooperation could sometimes prove an effective substitute for permanently stationed forces, when working with partners that are unlikely to operate with U.S. forces in future complex contingencies or in regions unlikely to host U.S. bases. For South Korea and many countries in Europe, however, the focus on tactical training and reliance on the others' infrastructure means that rotational forces could only partially mitigate the loss of security cooperation capability that would come from reductions in permanently stationed forces.

As the Army deactivates two of its remaining four BCTs in Europe, it is considering how to mitigate these reductions through rotations. The Army could participate in the NATO Response Force by rotating a portion of a "European Response Force" into the region, perhaps two times per year for two months each. If the Army were to dedicate a division to supporting these rotations as part of its "Regionally Aligned Forces" concept, it could deploy different brigade headquarters, battalion headquarters, and maneuver battalions each year to give multiple units experience in the region. A set of equipment for a combined arms battalion could also be left in Germany to reduce the transportation costs for rotational forces.

The Army could remove additional units from Europe and substitute additional rotational forces. The cost implications of additional reductions and rotations are described in Chapters Eight and Ten. Besides the potential cost savings, other benefits would include slightly increased flexibility for Army headquarters to conduct security cooperation globally and opportunities to gain multinational experience for more units. Disadvantages would include greater challenges to building deep military relationships and less expertise working with highly capable partners. Increasing use of rotational forces could also strain U.S. forces when long-duration contingency operations are also underway. The most significant disadvantage would come if there were a net reduction in multinational training for U.S. forces, particularly if the Army closed JMTC, its only multinational training center.

The Air Force could consider similar trade-offs for its forces in Europe. While inherently more mobile than the Army, transportation costs for the Air Force for aircraft support equipment and personnel can also be significant. Besides the cost implications, which are described in Chapter Eight, the advantages and disadvantages of substituting rotational forces for permanent forces are similar to those described for the Army and appear in Table 4.2.

Implications for Posture

In most cases, security cooperation will be one factor among several considerations in making posture decisions. The strongest security cooperation rationale for keeping forces in overseas facilities occurs in situations where the United States seeks to build and maintain high levels of interoperability with a regional partner. In such situa-

Table 4.2
Potential Advantages and Disadvantages of Substituting Rotational Forces for Forward-Based Forces to Conduct Security Cooperation

Potential Advantages	Potential Disadvantages
Cost savings	Less frequent contact
Flexibility	Weaker relationships
Broader involvement of U.S. forces	Shallower expertise
	Deployment strains
	Less multinational training and other activities

tions, these forces require access to appropriate training facilities. In other regions, U.S. forces can conduct more frequent training activities for lower incremental costs with forces stationed in or near the partner country, but because it costs more to have forces permanently stationed abroad, many situations are likely to favor U.S.-based forces conducting training on a rotational basis. In some locations, rotational forces may be the only near-term option. While the costs of conducting security cooperation with rotational forces may be lower, they also may not achieve all the benefits of a permanent presence, including the depth of relationships with partners.

Currently, the United States maintains a training facility in Germany that provides a venue for sophisticated combined training: the JMTC. If interoperability with NATO partners remains a high priority, then continuing to fund the JMTC, maintaining enough forces in Europe to utilize the training facility fully, and conducting a range of combined training exercises there will also likely remain priorities. In the Pacific, expanded access to Australian training ranges will allow the United States to maintain and perhaps expand interoperability with the Australian military. Current security cooperation goals in Southeast Asia can be achieved without a greater permanent presence. In the future, a training facility that would allow complex multinational training could increase the competence of security partners in the region, particularly if some of those partners invest sufficiently in their military capabilities to make such a step feasible.

CHAPTER FIVE

Risks to Investing in Facilities Overseas

This chapter discusses various types of risk that might accompany overseas posture. It begins by discussing political risk, that is, the risk that a host nation might deny the United States use of facilities established in a given country. This might occur because the country does not support the operation that the United States is undertaking, as was the case in Operation Iraqi Freedom, when a number of countries denied the United States access for refueling or overflight. This included some longtime allies such as Turkey, which refused to allow the United States to stage forces from its soil. The second type of risk is operational, from long-range precision weapons, which might expose forward-deployed U.S. forces to heavy losses. The third type of risk to consider comes from violent extremists, who could employ a range of attack methods against foreign U.S. facilities and personnel.

Political Risks to Access

Since the outbreak of World War II in Europe, the United States has relied on access to foreign bases to project power across the globe. *Access* is a broad term that encompasses permanent basing rights, steady state transit or overflight rights, and permission to use military facilities or transit through a nation's sovereign territory or airspace for a particular operation. In the past, most states only acquired overseas bases through the process of formal empire.[1] With few exceptions, however, in the post–World War II era the United States has permanently stationed its forces only in countries that have freely agreed to host an American military presence.[2] The United States has forged various types of

[1] Robert E. Harkavy, "Thinking About Basing," in Carnes Lord, ed., *Reposturing the Force: U.S. Overseas Presence in the Twenty-First Century*, Newport, R.I.: Naval War College Press, 2006, pp. 11–12.

[2] Christopher Sandars, *America's Overseas Garrisons: The Leasehold Empire*, Oxford: Oxford University Press, 2000, pp. 126–138, 161–166. Exception occupations: Okinawa prior to its reversion to Japanese sovereignty in 1972, Panama.

arrangements—ranging from official basing agreements to informal understandings—to obtain access to military facilities in foreign countries."[3]

To varying degrees, all of the U.S. military services rely on access to foreign territory or bases to project power, but that access cannot always be assured. As the United States discovered in 2003 when it sought to move Army units from Germany to the Middle East, if other nations deny American forces the ability to move through their territory, it can create significant logistical challenges that can impede military operations.[4]

In the 1990s, the United States also had to deal with significant constraints on its ability to use Middle Eastern bases and the forces stationed at them to enforce the no-fly zone and sanctions against Iraq. Saudi Arabia and Turkey, for instance, would not permit the United States to use American fighter aircraft stationed on their soil to conduct strike operations against Iraqi forces attacking Kurds in September 1996. Similarly, during Operation Desert Fox in December 1998, Saudi Arabia and the UAE prohibited the United States from employing fighter aircraft based in their country to target Iraqi WMD infrastructure. As a result, the United States was limited to using combat aircraft stationed in Kuwait and Oman, in addition to carrier-based naval aircraft. Nevertheless, the Saudis and Emirates did permit U.S. forces to fly noncombat aircraft such as tankers during the operation.[5]

The United States also faces challenges to peacetime access. For example, in 2009, U.S. forces were expelled from Ecuador when President Rafael Correra refused to renew the lease to Manta Air base, which was a location used to interdict drug shipments from South America. As a result, the United States lost access to facilities that it spent more than $60 million improving.[6]

---

[3] Alexander Cooley and Hendrik Spruyt, *Contracting States: Sovereign Transfers in International Relations,* Princeton, N.J.: Princeton University Press, 2009, pp. 101–102, characterizes U.S. basing agreements as incomplete contracts in which the host nation retains residual sovereign rights that it uses as leverage to renegotiate basing agreements on more favorable terms.

[4] In 2003, Austria denied U.S. forces passage on its railroads or through its airspace. The Swiss also denied U.S. forces overflight rights. Consequently, U.S. Army units had to sail from northern Germany around Western Europe and through the Mediterranean to reach the Persian Gulf. According to Rumsfeld, 2005, if transit rights through southern Europe are denied, Army units in Germany take roughly the same amount of time to reach the Middle East as those in the United States. The most high-profile access problem occurred when Turkey refused to allow the U.S. 4th Infantry Division to use its territory to launch a northern offensive into Iraq. See Todd W. Fields, "Eastward Bound: The Strategy and Politics of Repositioning U.S. Military Bases in Europe," *Journal of Public and International Affairs*, Vol. 15, Spring 2004, p. 82.

[5] Christopher J. Bowie, *The Anti-Access Threat and Theater Air Bases*, Washington, D.C.: Center for Strategic and Budgetary Assessments, 2002, pp. 34–35.

[6] Clare Ribando Seelke, *Ecuador: Political and Economic Situation and U.S. Relations*, Washington, D.C.: Congressional Research Service, RS21687, May 21, 2008, pp. 5–6.

During the Bush administration, DoD emphasized that it wanted access to places "that allow[ed] our troops to be usable and flexible."[7] In other words, Secretary of Defense Rumsfeld aimed to forge access agreements that could guarantee that the United States would be given contingency access. This was an ambitious goal, as access—peacetime and contingency—is always uncertain. Because basing agreements are incomplete contracts, they are always subject to renegotiation, and host nations typically avoid making explicit promises about contingency access, preferring vague terms that provide them with flexibility and the power to veto U.S. operations from their territory.

The historical examples given above illustrate that there are no access guarantees. In part, this uncertainty is a growing concern due to the fact that there is no longer any single, overriding, and unambiguous global threat akin to the Soviet Union during the Cold War. During this period, the United States established overseas garrisons that were primarily intended to defend host nations against a common threat.[8] Today, however, host nations generally face more varied security challenges; their interests are therefore less likely to align with those of the United States. Likewise, the United States often seeks access arrangements that enable it to use bases for a range of different operations. As a result, it is more difficult to create a direct and enduring tie between U.S. bases and the security of a host nation, which complicates obtaining and preserving access.[9] Many nations are also hesitant to allow U.S. forces to be stationed on their soil to counter unspecified future threats because the host nation will be implicated in any operations that these forces conduct.[10]

---

[7] Rumsfeld, 2005.

[8] However, even during the Cold War, when the United States wanted to use its European bases or the forces stationed at these facilities for other operations it encountered problems. In 1958, Greece, Libya, and Saudi Arabia refused the U.S. overflight and basing rights for its intervention in Lebanon; in 1959, France denied the United States the right to store nuclear weapons on bases in its territory; in 1962, Portugal and France denied U.S. overflight and base access because of Washington's involvement in the Congo crisis; in 1967, Spain denied the United States use of its bases to evacuate U.S. nationals during the 1967 Arab-Israeli war; in 1973, Spain, France, Italy, and Greece refused to grant base access and overflight rights to U.S. planes lifting supplies to Israel; in 1986, Italy, Germany, France, and Spain refused to cooperate with a U.S. air strike on Libya by denying the U.S. basing rights or overflight for Operation El Dorado Canyon. See Christopher J. Bowie, Suzanne M. Holroyd, John Lund, Richard E. Stanton, James R. Hewitt, Clyde B. East, Tim Webb, and Milton Kamins, *Basing Uncertainties in the NATO Theater*, Santa Monica, Calif.: RAND Corporation, 1989, not available to the general public, pp. 4–5; Richard F. Grimmett, *U.S. Military Installations in NATO's Southern Region*, report prepared for the U.S. Congress, Washington, D.C.: U.S. Government Printing Office, 1986; Walter J. Boyne, "El Dorado Canyon," *Air Force Magazine*, Vol. 82, No. 3, March 1999; and Adam B. Siegel, *Basing and Other Constraints on Ground-Based Aviation Contributions to U.S. Contingency Operations*, Washington, D.C.: Center for Naval Analysis, March 1995.

[9] Krepinevich and Work, 2007, p. 190. Historically, the most common reason that the United States initially obtained access to foreign bases is due to a shared perception of threat. Pettyjohn, 2012, pp. 102–103.

[10] Lincoln P. Bloomfield, Jr., "Politics and Diplomacy of the Global Defense Posture Review," in Carnes Lord, ed., *Reposturing the Force: U.S. Overseas Presence in the Twenty-First Century*, Newport, R.I.: Naval War College Press, 2006, pp. 61–62.

Access comes through agreements between the United States and the host nation government, but the host nation public can often influence outcomes.[11] While public opinion most directly affects the foreign policy of democratic states, even authoritarian leaders may have to take into account their citizens' views.[12] Public opposition to an American military presence can constrain U.S. access; affect the costs of access; and, if severe enough, lead to U.S. expulsion.

In general, there are many reasons why the population of a host nation might oppose an American military presence. First, foreign bases are often seen as compromising a state's sovereignty.[13] Second, some citizens may object to foreign military bases because of the interruption of their daily lives the bases cause by taking up valuable land, polluting the environment, generating disruptive noise, and creating safety hazards.[14] Third, people may disagree with the stated mission of U.S. forces or their particular actions.[15]

Given these considerations, this section will explore when U.S. bases are at risk and when a host nation is likely to authorize the United States to use its bases for a particular operation.

Risks to Peacetime Access: Where Are U.S. Bases at Risk?

Although some host nations contribute funds to construct and maintain U.S. overseas bases (see Chapter Seven), typically the United States pays the lion's share of the costs associated with building and maintaining its overseas military facilities. Investments in military infrastructure abroad are sunk costs that the United States may not

---

[11] Alexander Cooley, *Base Politics: Democratic Change and The U.S. Military Overseas*, Ithaca, N.Y.: Cornell University Press, 2008, p. 132; Calder, Kent E., *Embattled Garrisons: Comparative Base Politics and American Globalism*, Princeton, N.J.: Princeton University Press, 2007, p. 67. For more on two-level games, see Robert D. Putnam, "Diplomacy and Domestic Politics: The Logic of Two-Level Games," *International Organization*, Vol. 42, No. 3, Summer 1998, pp. 427–460.

[12] Authoritarian leaders are not accountable to the public in the same way that democratic leaders are, yet there is considerable evidence that they do take into account and can become entrapped by domestic public opinion. See Michael N. Barnett, *Dialogues in Arab Politics*, New York: Columbia University Press, 1998, pp. 25–27; James Reilly, *Strong Society, Smart State: The Rise of Public Opinion in China's Japan Policy*, New York: Columbia University Press, 2011; and Marc Lynch, "Paint by Numbers," *The National*, May 29, 2009.

[13] Catherine Lutz, "Introduction: Bases, Empire, and Global Response," in Catherine Lutz, ed., *The Bases of Empire: The Global Struggle Against U.S. Military Posts*, New York: New York University Press, 2009, p. 30.

[14] Mark L. Gillem, *America Town: Building the Outposts of Empire*, Minneapolis, Minn.: University of Minnesota Press, 2007, pp. 34–70.

[15] James Meernik, *U.S. Foreign Policy and Regime Instability*, Carlisle, Pa.: United States Army War College Strategic Studies Institute, May 2008.

be able to recoup if it is expelled from a base.[16] For example, in 1966 French President Charles de Gaulle withdrew his nation from NATO's integrated command structure and announced that all U.S. and NATO troops had to depart France within one year. In Operation Relocation of Forces from France, the Pentagon had to quickly relocate the 70,000 U.S. military, civilian, and dependent personnel in France. As a part of this move, the United States abandoned nearly 200 installations, which represented an investment of $565 million.[17] More recently, between 2004 and 2009, the United States invested substantially in base construction in Iraq before troops were withdrawn in 2011.[18] While not precisely analogous to the France experience in 1966, it does show in a different way that the benefits of investments the U.S. military makes in facilities abroad may not endure.

A number of different studies have sought to identify the conditions under which an American military presence will generate significant opposition within a host nation that results in limitations to, or the complete loss of, peacetime access. This is not a simple question, because while there have been significant protests against U.S. bases in some countries, such as Italy, the United States has been able to maintain its presence in Italy, though sometimes operational constraints can limit the flexibility of those forces.[19] By contrast, in other locations, such as Okinawa and South Korea, the United States has faced sustained and serious public opposition that has led to adjustments, including reductions to its posture, but basing rights have not been revoked.[20] Finally, some nations completely expelled U.S. forces, including Uzbekistan in 2005 and the Philippines in 1992.[21]

These divergent outcomes are influenced by three different variables—regime type, the size of the U.S. presence, and the type of access relationship—in determining the level of access risk that the United States faces.

[16] Though in some cases the legal agreement that defines the terms of uses has provisions for compensation to the United States, should U.S. forces be asked to leave early.

[17] Simon Duke, *United States Military Forces and Installations in Europe*, Oxford: Oxford University Press, 1989, pp. 149–152.

[18] The United States spent approximately $2.1 billion on base construction in Iraq. Michael Gisick, "U.S. Base Projects Continue in Iraq Despite Plans to Leave," *Stars and Stripes*, June 1, 2010.

[19] For more on Italy, see Andrew Yeo, "Ideas and Institutions in Contentious Politics: Anti-U.S. Base Movements in Ecuador and Italy," *Comparative Politics*, Vol. 42, No. 4, July 2010b, pp. 441–446.

[20] For more on Japan and South Korea, see Andrew Yeo, *Activists, Alliances, and Anti-U.S. Base Protests*, Cambridge: Cambridge University Press, 2011, pp. 63–85, 118–148; Cooley, 2008, pp. 95–174; and Andrew Yeo, "U.S. Military Base Realignment in South Korea," *Peace Review: A Journal of Social Justice*, Vol. 22, No. 2, May 2010a, pp. 113–120.

[21] Scott G. Frickenstein, "Kicked Out of K2," *Air Force Magazine*, Vol. 93, No. 9, September 2010; Cooley, 2008, pp. 56–94, 230–232; Yeo, 2011, pp. 35–62.

Two regime types, authoritarian and transitional democracies, could be more likely to restrict or end access.[22] Authoritarian leaders who are not bound by the rule of law or constrained by a system of checks and balances can easily rescind American access or capriciously demand that basing agreements be renegotiated. In addition to being unpredictable, authoritarian regimes are also more prone to be overthrown, either in a coup or by a democratic revolution.

States that are undergoing a regime change—in particular democratization—are also unreliable host nations. While their democratization may be in the long-term interest of the United States, in the short term it could jeopardize an ongoing security relationship. In part this is due to the fact that new leaders will frequently criticize basing agreements made by their predecessors, as the United States found in the Philippines. Additionally, because democratizing states hold elections in the absence of strong institutions, aspiring leaders may politicize an American military presence in an effort to win votes. In contrast, consolidated democracies characterized by procedural legitimacy, institutional stability, and well-regulated political competition are the most reliable partners and host nations, because they cannot easily abandon their agreements.[23]

The size of a U.S. military presence and its proximity to large population centers may also contribute to access risk.[24] Local hostility to bases may be directly related to the frequency and intensity of interactions between American military personnel and local communities. According to this view, sprawling base complexes that permanently host large numbers of American forces are more visible and therefore more likely to provoke local resentment. Opposition may stem from the fact that large bases are seen as an affront to nationalism or because large, permanent facilities can create friction with local communities.[25] This rationale underpinned the George W. Bush administration's global posture review, which sought to rely on smaller, more austere bases while moving larger bases away from urban areas.[26]

Overlapping threat perceptions are another factor that can help determine the reliability of basing arrangements. If nations offer to host American forces based on a common perception of a threat, there is a fairly stable foundation for a basing agreement—so long as the U.S. military presence remains focused on countering this

---

[22] Calder, 2007, pp. 112–119; Cooley, 2008, pp. 13–18.

[23] Cooley, 2008. Also, in 1999

[24] Calder, 2007, pp.119–125.

[25] While smaller facilities may reduce contact and therefore some opposition to the base, they may also generate a different kind of criticism because of the secrecy surrounding many of these bases. In Ecuador, for example, protestors denounced Manta Air Base because of so-called NIMBY (not in my backyard) issues, but there also was rampant speculation that the United States was using the base to support operations against Columbian rebels. See Stephan Kuffner, "Ecuador Targets a U.S. Air Base," *Time*, May 14, 2008.

[26] Henry, 2006, p. 46.

other potentially vital infrastructure, such as fuel storage capacity, might take longer to arrange. In particular, this might be a problem if the United States needed to operate in areas where infrastructure development may be minimal. For instance, airfields in Africa may require substantial infrastructure upgrades to support major operations. Similarly, locations in South Asia and Southeast Asia may also have uneven levels of infrastructure development. To take full advantage of opportunities to operate from new locations, the United States needs capabilities that can bring them to a standard that will support combat operations fairly quickly.

In general, a nation's decision to grant the United States contingency access will be context-dependent. In part, this calculus will rest on the type of operation the United States wants to conduct and against whom, in addition to the expected duration of the mission. Contingency access, therefore, is not a simple yes or no question. Host nations may authorize the United States to use bases on their territory for certain types of operations but prohibit others. For instance, host nations may be more likely to permit non-lethal operations to be conducted from their territory compared with combat operations. Some nations may balk at allowing the United States to use certain types of platforms or weapons from their territory. Traditionally, bombers and nuclear weapons have been particularly sensitive issues. For example, because of New Zealand's opposition to nuclear weapons and the United States' unwillingness to confirm or deny the presence of such weapons on its naval vessels, New Zealand has long prevented USN ships from visiting its ports, a policy that led to a suspension of the security pact between the two nations in the mid-1980s. Moreover, contingency access may evolve over the course of an operation, growing more restrictive or permissive depending upon the situation. Finally, the United States may be able to influence the host nation's calculus by offering inducements or applying pressure, but it is not guaranteed that these will result in the desired outcome.

The most important factors that will influence a host nation's decision include the degree to which the host nation's interests and those of the United States overlap, whether domestic public opinion favors or opposes the particular operation, the host nation's perception about the likelihood of reprisals, and the degree to which a host nation is dependent on American security guarantees.[31]

One of the most straightforward factors is whether the host nation has similar interests to the United States with respect to the particular contingency. While a nation may generally have shared interests with the United States, these can vary depending upon the situation. Consider the example of Turkey, a longstanding NATO ally. In 2003, Turkey prevented the United States from using its territory to stage ground forces for the invasion of Iraq in part because it did not share the United States' interest in deposing the Saddam Hussein regime. Rather, the overthrow of Hussein raised

---

[31] Cooley, 2008, pp. 11–12; David A. Shlapak, John Stillion, Olga Oliker, and Tanya Charlick-Paley, *A Global Access Strategy for the U.S. Air Force*, Santa Monica, Calif.: RAND Corporation, MR-1216-AF, 2002, p. 37.

the specter of an independent Kurdistan, which could in turn exacerbate Turkey's chief security concern, the violent extremist threat posed by the Kurdistan Workers' Party (PKK). Reflecting these interests, Turkey did, however, allow the United States to station and fly Predator UAVs from Incirlik to observe—and, if needed, to launch strikes against—the PKK. In short, its willingness to grant access was conditional on its particular security interests, which were not always aligned with those of the United States.

Domestic public opinion can also influence a government's decision to allow contingency access. If the United States wants to conduct a mission that is viewed unfavorably by a host nation's public or that is perceived to lack international legitimacy, a host nation will probably be less willing to grant access for fear of alienating its constituents. This is even more likely if the host nation's legislature has to authorize access for foreign forces, as is the case in Turkey and the Philippines. Popular opposition to the ongoing sanctions against Iraq likely influenced Saudi Arabia's decision to prohibit the United States from using its bases for combat operations during the late 1990s.

Another factor that may negatively influence whether the United States is granted contingency access is the perceived likelihood of reprisals, which could take the form of kinetic attacks or economic retaliation. For example, most European nations denied the United States permission to use bases in their countries to airlift supplies to Israel during the 1973 Arab-Israel War because they feared being targeted by an oil embargo. Reprisal is a concern, but presumably the United States could take steps to mitigate the effects of retaliatory attacks with the deployment of defense capabilities in a crisis or crafting military operations to limit retaliatory options (e.g., targeting military capabilities that may not pose a direct threat to U.S. operations but may pose a threat to host nations, such as Iraqi Scud missiles in 1991).

Given this irreducible uncertainty, the more countries that provide access for the U.S. military, the more resilient the U.S. posture will be to access risks. However, the more dispersed U.S. facilities are across many different nations, the greater the costs, because each new facility has its own fixed costs. This has led the United States to seek access to facilities, without permanently stationing forces there and without a responsibility for maintaining all of the infrastructure, reflective of, for instance, the Air Force CSLs. Even when this involves some up-front investment to ensure that the infrastructure can support U.S. operations, this is significantly less costly than establishing and maintaining U.S. facilities, while still providing flexibility to the force.

Changing Operational Risks to Posture

Military forces, by their nature, accept certain threats as inherent. They plan, train, and equip to face a range of threats on the battlefield, but in recent years capabilities have been developed that require a reassessment of posture. The advent of extremely accurate ballistic and cruise missiles could cause catastrophic losses to forces exposed to them. These weapons could cripple an airbase, incapacitate an aircraft carrier, and devastate concentrated ground forces. Defenders faced with such a threat can invest in active defenses, efforts to shoot these weapons out of the sky, dispersal, hardening, and passive defenses for stationary targets and mobility for mobile targets. In the context of an assessment of posture, there can also be decisions to deploy forces out of range of these weapons, or to limit the number and kind of forces exposed.

Ballistic and cruise missiles have been operational for many years. Nazi Germany used ballistic missiles to attack London during World War II, but their accuracy was so poor that they were essentially weapons of terror. In recent years, the accuracy of ballistic missiles has become so precise as to give them the ability to achieve potent military effects far beyond what was achievable in the past with conventional long-range missiles. This qualitatively different capability necessitates consideration of such weapons in posture decisions. China appears to have invested most heavily in such weapons. It has fielded a variety of ballistic and cruise missiles with varying ranges and accuracies, and with warheads for different purposes. Iran also has fielded some of these weapons. In the future, other potential adversaries may also make such investments, so it is likely that this could become an important operational feature in future conflicts.

Many military activities benefit from centralization. Putting forces at a large air base, for example, allows for efficiencies of scale in base operations and logistics support. To the extent that a base is defended from air and ground threats, a larger base provides opportunities to concentrate defenses. Ground forces also tend to create large staging areas that are fairly far forward to keep troops engaged in combat well supplied. Airbases and large depots within range of ballistic and cruise missiles make for potentially very lucrative targets, which calls for cost-versus-risk trade-offs in pursuing basing centralization.

A country with a large number precision weapons within range of concentrated U.S. forces could inflict serious damage in a matter of minutes. The threat to maritime forces is different due to their mobility, but they still typically need to operate fairly close to land to achieve effects on shore, which exposes them to long-range threats. The mobility of maritime forces makes it more difficult for adversaries to find and target them, but there are ongoing efforts in China, Iran, and other countries to improve their sensors, command decision speed, and the in-flight guidance systems of their long-range maritime strike forces.[32]

DoD strategy documents often talk about the importance of countering anti-access challenges. The 2010 QDR highlights the desire "to improve the resiliency of U.S. forces and facilities in the region [PACOM] in order to safeguard and secure U.S., allied, and partner assets and interests in response to emerging anti-access and

---

[32] See Office of the Secretary of Defense, *Annual Report to Congress: Military Power of the People's Republic of China*, 2009, for a description of the Chinese anti-ship ballistic missile system.

area-denial capabilities."[33] The implication of the growing anti-access threats for U.S. forward presence could be substantial. Facilities and operational locations that have served the United States well in past years may now leave U.S. forces vulnerable. In areas where anti-access capabilities might be employed, the United States may have to consider much more substantial defensive protection for forward forces than has been the case. A combination of active defenses, such as missile-defense capabilities; passive defenses, such as hardening and concealment; and dispersal all can limit exposure to anti-access threats.

The presence of precision-guided weapons does not mean that the United States must concede that it cannot operate within range of these weapons. However, it does mean that if the United States wants to maintain capabilities to operate within range of these systems, it will have to change the way it operates, including its posture.

It also means that characterization of the different levels of threats to forward bases help develop tailored responses. For instance, although China, Iran, and North Korea all have missiles, the accuracy and ranges of these systems are different. To put these differences in context, RAND developed a simple categorization of threats into heavy, moderate, light, and minimal zones.[34] These threat categories represent the level of threat in a specified region based on the accuracy and density of a potential adversary's particular mix of capabilities and their ranges. The heavy threat zones are defined as areas under threat from thousands of missiles[35] with accuracy better than 50 m CEP.[36] Moderate threat zones are defined as areas exposed to hundreds of missiles with accuracies less than 1,000 m CEP, while the light threat zones lie within range of missiles with accuracies greater than 1,000 m CEP (see Table 5.1).[37]

Taking into account the number, accuracy, and range of current ballistic and cruise missiles, the United States currently operates from a number of bases within high threat zones, as shown in Figure 5.2.[38,39] All of the bases in the highest threat

---

33 DoD, 2010a, p. 66.

34 Many thanks go to RAND colleague Jacob Heim, from whose work and expertise on assessing the relative capabilities of missile forces this analysis drew heavily.

35 Also included in the high threat band are areas within reach of long-range, multiple rocket launchers where volume of fires can compensate for larger CEPs.

36 CEP is a measure of missile accuracy. It is the radius of a circle into which 50 percent of the rounds are expected to fall.

37 These criteria were derived from analysis in John Stillion and David T. Orletsky, *Airbase Vulnerability to Conventional Cruise-Missile and Ballistic-Missile Attacks: Technology, Scenarios, and U.S. Air Force Responses*, Santa Monica, Calif.: RAND Corporation, MR-1028-AF, 1999; and David A. Shlapak, David T. Orletsky, Toy I Reid, Murray Scot Taner, Barry A. Wilson, *A Question of Balance: Political Context and Military Aspects of the Cross-Strait Dispute*, Santa Monica, Calif.: RAND Corporation, MG-888-SRF, 2009.

38 For further discussion of these threat categories and supporting analysis, please see Appendix G.

39 This is based on unclassified estimates of missile performance and inventory and the locations of U.S. bases. See Jane's Strategic Weapons Systems database; Office of the Secretary of Defense, 2011; Zhang Han and

lead to decisions to either divest or repurpose facilities. Different potential strategies are explored below.

Avoidance: One option is to operate outside of these threat ranges. The implication for posture is that bases may have to be abandoned, or the forces and equipment there might have to change substantially. This change would likely have substantial force structure implications, as it would increase the value of longer-range capabilities and lower the value of shorter-range capabilities.

Harden and limit exposure: Another option is to operate from a few bases within range of threat systems but increase and harden their defenses. Some forces would remain exposed, but this exposure would be limited by the lower number of forces and the added protection afforded to them. This strategy would make bases within range of sizable ballistic- and cruise-missile arsenals more expensive than those elsewhere because more resources would be necessary to harden the facilities against attack. Such measures could involve building shelters for aircraft, bunkers for ammunition, and underground fuel-storage tanks. It could also mean thicker and longer runways and keeping equipment and personnel available to repair them when damaged by attack.

Dispersal: When geographically feasible, increasing the number of operating areas and facilities within range of these threats is another possible strategy. Given a finite number of attacking systems, increasing the number of facilities forces potential attackers to dilute attacks on all facilities or conduct larger, selective attacks that leave other facilities unmolested. Such a strategy would require dispersal wide enough to force the attacker to make tough allocation choices. In practice, it may be difficult to find enough facilities, either because of geographic constraints or political constraints. The implication of such a strategy for posture is that the United States would need numerous dispersal options in a given region. Those options may not require substantial pre-conflict preparation if there are investments in capabilities that enable the United States to bring facilities rapidly up to a standard sufficient to support the operations selected for that location. In some instances, the modifications may be minimal; in others, facilities may require construction and equipment to make them suitable for operations.

Essentially, the United States may not be able to afford to pose lucrative wartime targets to adversaries or potential adversaries. Large facilities with substantial combat capabilities or critical supply nodes, valuable as they maybe in operations short of war, should not be left vulnerable or without alternatives in times of war. The problem could be severe enough that no one solution will solve the problem. Combinations of active defenses, dispersal, hardening, passive defenses, and resilient operating concepts are all relevant contributors to solutions.

Assessing Violent Extremist Risks to Posture

U.S. military facilities have been the targets of violent extremist attacks in the past, including some tragic ones, like the 1983 attack on the USMC barracks in Lebanon. U.S. military personnel also have been the targets of attacks while off base, again sometimes resulting in terrible consequences, like the 1996 Khobar Towers attack in Saudi Arabia. Even U.S. warships have suffered attack, most notoriously the USS *Cole* in 2000. While military facilities tend to be constructed in ways that make such attacks difficult and use guards trained to disrupt such attacks, consideration of violent extremist threats to facilities is warranted when considering posture decisions.

A straightforward way of getting an initial read on violent extremism risk is to assess the number of violent extremist attacks that a host nation has suffered historically. This can give an initial indication of which facilities might be at increased danger of violent extremist attacks relative to other facilities. In general, most violent extremist groups have a fairly local agenda, so geographically based assessments of past activity provide a fairly informative picture. However, some groups, like al-Qa'ida, have agendas that extend beyond immediate geographical areas. To the extent that those groups can be thought of as able to project power—that is, accomplish attacks in new geographic areas outside of their base of support—a purely geographic assessment of historic attacks probably does not fully capture the nature of the threat from such groups.

Another way to assess vulnerability to violent extremist attacks is to focus on the methods of attack. In Iraq and Afghanistan, the U.S. military operated against terrorist groups, such as al-Qa'ida and its affiliates. In those encounters, these groups were very adaptive from an operational perspective. Confronting these groups fell into the age-old pattern of military confrontation, in which both sides continuously adapt to their opponent to try and exploit adversary vulnerabilities and to shore up one's own vulnerabilities. In this regard, an assessment of previous attack methods can be instructive. An unpublished RAND report assessed violent extremist attacks on military targets between 2000 and early 2009.[41] Of those 1,800 incidents, only about 20 percent were against military facilities (see Figure 5.4).[42] The rest were against military personnel outside of those facilities. As these data included incidents in Iraq and Afghanistan, there were numerous attacks on personnel in vehicles through IEDs (33 percent), as well as in exposed locations (43 percent), and only a small fraction occurred in non-military structures (5 percent).[43]

---

[41] Unpublished research by David R. Frelinger of RAND.

[42] The data came from two sources, the RAND Database of Worldwide Terrorism Incidents and the University of Maryland Global Terrorism Database. These databases contain attacks on military targets that meet their definitions of terrorism, and therefore represent a subset of all attacks on such targets by nonstate actors.

[43] Unpublished research by David R. Frelinger of RAND.

ity that the United States opens increases costs, so while some redundancy is necessary to guard against political access risks, some prioritization is in order. The United States could try to reduce risks by trying to increase the geographic dispersion of the peacetime access of its forces, but costs and political challenges may leave few opportunities to create new large bases, nor may such large facilities be necessary. In many cases, smaller facilities with little or no footprint may provide sufficient presence without requiring substantial U.S. investments or ongoing operational costs. While these can certainly be a less costly alternative, it is uncertain whether such peacetime efforts will pay dividends in a conflict. That is, will a country be more likely to grant operational access to a facility that the United States had previously operated from, or established a very small presence on, compared with a facility where the United States had no previous connection? While there are no guarantees about future access, that does not mean that the United States should abandon forward bases. Rather, the number of options and the alignment of U.S. and host-nation goals affect the likelihood of gaining needed access for a given conflict.

There are also factors that can mitigate the consequences of losing access—or not having it in the first place. In some circumstances, maritime presence can compensate for lost access. Another countervailing factor to mitigate some of the access risk is the historical precedent of obtaining new access during a conflict. In every major conflict of the last 40 years, the United States has been able to add new operating facilities.

The operational challenges posed by accurate and numerous precision weapons pose another risk to U.S. forward presence. The threat is currently most severe in Northeast Asia, where China has fielded numerous precision capabilities and where the geography of the region leaves few alternatives. In that region, a complex calculation regarding the benefit of assuring allies and deterring foes must be considered against the level of investment the U.S. might be willing to make to implement a mixed strategy of hardening facilities and dispersing capabilities to multiple locations. In considering different options, the United States should strive to not present very lucrative targets, which could risk an early, crippling strike against U.S. forces.

Risks against U.S. forces from violent extremists are not new, and U.S. military facilities are already designed with such strikes in mind. In the past, attacks against facilities have highlighted that attackers tend to favor indirect weapons, to fire over a fence and into a compound, or to attack the gate or perimeter. However, it is not just the facility that should concern planners, but also the safety of personnel when they are required to travel off the base. This may be the time when U.S. personnel are at greatest risk. The risk of violent extremist attacks can be assessed in part by looking at historical trends of such attacks in the area where the facility is located, as most violent extremist groups are geographically localized. This is not true for other groups that have an international agenda and have demonstrated a capability to launch attacks beyond their immediate base of support.

CHAPTER SIX

Installation Conditions

In assessing trade-offs between alternative postures, future restoration and modernization (R&M) costs at different facilities could be a factor in policy choices. Differences in estimates of current requirements for R&M among locations offer the best available insights into this question. Were these requirements to differ among locations, disparities in the physical condition of facilities would be largely responsible. Differences in facilities' conditions do not just affect current R&M needs, but could also affect future costs, including more near-term recapitalization requirements.[1] To the extent that anticipated needs differ between U.S. and overseas facilities, these differences would need to be integrated into the assessment of the relative costs—including both installation support and MILCON.[2]

In this chapter, we consider the evidence on facility conditions and, as far as possible, use that evidence to derive estimates of the requirements for R&M across locations. DoD gathers information on installation conditions in the form of a "Q-rating," depicted as a facility conditions index (FCI), scaled 0 to 100, for each structural asset that relates to the facility's R&M needs.

We find that foreign facilities appear to be no less fit and possibly slightly more fit than U.S. facilities, but shortcomings in the data hinder the analysis, including the comparison of FCIs across locations and the estimation of R&M requirements. Where we observe differences between U.S. and foreign FCIs, those differences tend to be small on average. The average FCI for all structural assets in the United States is about 84.5, and the average for all foreign locations is 86.5; if weighted by plant replacement value (PRV), the respective averages are 86.6 and 89.2. Our discussions with subject-matter experts lend support to the general conclusion that foreign facilities are at least

[1] The accumulation of deterioration, as might be reflected in a backlog of R&M, can beget further damage, accelerate depreciation, and hasten the need for recapitalization.

[2] As outlined in DoD, Office of the Under Secretary of Defense, Acquisition, Technology, and Logistics, "Facility Quality Rating Guidance," Memorandum, September 5, 2007, R&M projects are programmed as non-recurring repair and minor construction and generally budgeted and funded through the Operations and Maintenance Program Element. In addition, R&M also includes projects programmed as MILCON. Other R&M fund sources include "Military Family Housing" and "Non-Appropriated Fund."

as fit as U.S. facilities, but there are a number of reasons, pertaining largely to differences in the ways in which the services conduct assessments and construct FCIs, to treat the estimates cautiously. As a practical matter, given the small differences that we found in average conditions and our concerns about the quality of the underlying data, our posture cost modeling in Chapter Eight and estimated cost effects of the illustrative posture changes in Chapter Ten assume the same relative conditions in the United States and overseas for existing facilities.

In the following discussion, we present an introduction to the facility condition data and their limitations and a more-detailed summary of the results of our analysis.

Data on Installation Conditions

Data pertaining to installation conditions are housed in the "Real Property Database" (RPAD), which is maintained by the Office of the Under Secretary of Defense for Acquisition, Technology, and Logistics (AT&L). Ratings are assigned at the "asset level" to each building or other structure in the database.[3] The relevant data element in the RPAD is that of the "Facility Physical Quality Rate," which captures the "Q-rating."

According to the RPAD data dictionary:[4]

> That rate is a percentage used to depict the physical capability of existing facilities as measured by a condition index [*referred to as the facility condition index or "FCI"*]. The Facility Physical Quality Rate represents a facility's restoration and modernization requirement but does not represent a facility's sustainment or new footprint requirement. The Facility Physical Quality Rate estimates will not contain any annual sustainment tasks or "deferred sustainment" costs, although they may contain restoration costs caused by deferred sustainment. The Facility Physical Quality Rate will also not represent costs to build out capacity deficits. Facility Physical Quality Rate will be in terms of the estimated cost to restore and modernize facilities to full-up "90–100 percent rating" status.

On this basis, the FCI, scaled 0 to 100, represents an asset's R&M requirement compared with that asset's ability to support a particular mission. (Up until FY 2008, this data element was reported on the basis of quality rating bands: i.e., Q-1 [90–

---

3 As specified in the RPAD data dictionary (DoD, "DoD Real Property Inventory Data Element Dictionary," RPIM, Version 4.0, April 22, 2010b, p. 168), "There must be a Facility Physical Quality Rate recorded for each valid RPA [real property asset] Type Code value of "B" (Building), "S" (Structure), or "LS" (Linear Structure) entered."

4 See DoD, 2010b, p. 168.

100, good condition], Q-2 [80–89, fair condition], etc.) Moreover, the FCI should be derived from the sum of necessary R&M costs in comparison with the asset's PRV.[5]

AT&L guidance[6] provides a formula for calculating Q-ratings:

$$1 - [\text{requirements}]/\text{PRV} \times 100$$

That guidance also specifies the content of the numerator (see Table 6.1).

The RPAD does not include a data element for the underlying R&M requirement, but it does include a data element for the PRV. On that basis, it is technically possible to back out the R&M requirement from the available data, but the validity of the resulting measure depends on whether the FCI has, in fact, been calculated from an R&M estimate, in accordance with the AT&L guidance, and whether the underlying data are accurate. However, an AT&L analysis of the data, building on prior internal work and prepared at RAND's request, suggests substantial data integrity problems, relating primarily to differences in the ways in which the services assess facility conditions and construct FCIs.[7] Of particular relevance to our interest in assessing U.S. and foreign installation conditions is any methodological difference across the services that could bias comparisons of the two.

Table 6.1
AT&L Guidance on Q-Rating Calculations

Formula Numerator	Facilities Sustainment	Facilities Restoration and Modernization
Included	Sustainment requirements that at present are materially degrading the condition of a facility	Repair requirements to restore or replace facility components, services systems, or meet codes or mission needs (except conversion)
Excluded	Regularly scheduled adjustments and inspections; preventive maintenance tasks	Conversion construction; "new footprint" construction

SOURCE: DoD, Office of the Under Secretary of Defense, Acquisition, Technology, and Logistics, 2007.

NOTES: The approach to sustainment in this AT&L guidance appears to be largely consistent with that provided in the more recent RPAD Data Dictionary (2010).

---

5 As defined in the RPAD data dictionary (DoD, 2010b, p. 169), the PRV is "The cost to replace a facility using current DoD facility construction standards."

6 See DoD, Office of the Under Secretary of Defense, Acquisition, Technology, and Logistics, 2007. This memorandum offers the formula for the "Facility Physical Quality Rating" or "Q-rating," but notes that the Q-rating data field captures the index. Thus, it appears that the FCI and the Q-rating are functionally equivalent. This might not have been the case in the era of "quality bands," at which time the Q-rating (i.e., Q-1, Q-2, etc.) would have been derived from the FCI.

7 E-mail to the authors from CAPT David Berchtold, Department of Defense Office of Acquisition, Technology, and Logistics, June 28, 2012.

Among the concerns that AT&L cites the following:

- lack of consistency in how the FCIs are calculated
- lack of progress in condition assessments
- inflated condition assessments.

Our subsequent analysis, which draws additional insight from further exploration of the underlying data and from conversations with USAFE mission-support representatives and other subject-matter experts, largely confirms these concerns.

Regarding consistency, the AT&L analysis references three primary considerations: the skill sets of those undertaking the assessments; the processes employed by those undertaking the assessments; and the components' different interpretation of the FCI numerator, including the calculation of R&M. Whereas AT&L's guidance is clear on what should and should not be included in the numerator (Table 6.1), the AT&L analysis reports that the Navy uses calculates a condition index, capacity index, and configuration index and assigns an FCI that represents the worst of these three; the Air Force does similarly, except it does not include capacity requirements; and the Army is "the purest amongst the three."[8] In this context, "purity" refers to adherence to the AT&L guidance.[9]

Although we cannot attest to a lack of consistency from direct experience, other credible sources—including the terms of references for "methodology" in the AT&L guidance and discussions with practitioners—support the generalities of the claim, if not all the specifics. The AT&L guidance on rating methodology seeks "validity" and "accuracy," but clearly allows flexibility—not necessarily a bad thing—that could, ultimately, lead to inconsistency in the conduct of assessments. However, the guidance does not call for flexibility in either the application of the criteria or the definition of the FCI, both of which appear to be reflected in the service practices described above.[10]

Additionally, USAFE personnel indicated that the Air Force reports only R&M that has a place on the near-term docket. More specifically, the Air Force uses "100s" for the FCI when it has not identified any projects for a facility and validates projects

---

[8] Email to the authors from Berchtold, 2012.

[9] DoD, Office of the Under Secretary of Defense, Acquisition, Technology, and Logistics, 2007.

[10] As specified in the AT&L guidance,

> Various methodologies exist to assess facility conditions, from detailed building inspections by engineers and technicians, to occupant surveys, to computer systems that model condition. Military Services and DoD Agencies/Activities may use various methodologies, however, each is responsible for validating the accuracy of the system used and making adjustments to their methodology as needed to increase accuracy and achieve compliance with this guidance. Each Service, Agency, Activity is to conduct periodic, on-site facility inspections by qualified engineering staff for a sufficient sample of facilities to gain "benchmark" data to validate the condition assessment methodology as being consistent, repeatable and accurate within the Q-Rating bands.

for only two years beyond the current fiscal year.[11] For that reason, anything in the backlog of R&M that has not yet been formalized as a validated project in that two-year window does not affect the score. We have no evidence that any other service employs this practice.

Regarding the lack of progress in making assessments, AT&L reports that over 12,000 assets still did not have FCIs in the RPAD as of FY 2011; that many assets share just a few FCI values, which it describes as a carry-over of the mid-rating quality bands from reports pre-dating the change in reporting in FY 2008 (e.g., 95, which is the mid-point of the old Q-1 band of 90–100); and that the Army assigns many FCIs on the basis of business rules because of the high costs of assessment.

We note that although 12,000 is a large number, it represents less than 2 percent of the total pool of 648,617 assets in the AT&L analysis.[12] Moreover, the AT&L analysis, produced for RAND under considerable time pressure, included all quality-rated structural assets, regardless of their status in the inventory, even if surplus, closed, or disposed. Our own analysis of the RPAD data indicate that about 50 percent of the assets lacking FCIs were active assets, and the rest were disposed (45 percent) or closed (5 percent). Our analysis of the RPAD data on asset reviews, which involved the calculation of the average lapse of time since the most recent physical inspection or functional (adequacy) condition survey recorded in the RPAD, suggests, perhaps more importantly, that many assets might not have been reviewed in recent years. However, whether this finding speaks to a genuine lack of awareness of installation conditions on the part of the reporting components, to a lack of reporting for inclusion in the database, or to the difficulty of identifying the most relevant indicators of progress in making assessments is an open question.

Lastly, the AT&L analysis refers to "inflated condition assessments" that could impede the comparison of conditions across services and locations. In particular, it calls out the relatively large share of "100" scores among Air Force assets, describing them as placeholder values for assets without assessments. Our analysis of the RPAD data confirmed the relatively large share of 100s, which is also consistent with the practice reported to us by Air Force installation management personnel of assigning 100s to assets in the absence of any pending, validated projects.[13] Thus, the issue of inflation appears to be one of reporting methodology, such as using 100s for placeholder values, rather than one of "inflation" per se.

The analysis shown in Table 6.2 also suggests substantial differences in scoring across services—we find that the differences are just as stark when we calculate the shares of assets obtaining scores of 100 and a little less stark, but still present, when we

---

[11] Interviews with USAFE Mission Support personnel, August 30, 2012.

[12] The AT&L presentation of finding reports only on assets with FCIs, amounting to 636,617 assets; adding back in the 12,000 assets that did not have FCIs yields a total pool of 648,617 assets

[13] Interviews with USAFE Mission Support personnel, August 30, 2012.

Table 6.2
Components' Shares of Assets and PRV with FCIs Equal to 100

Component	Real Property Assets (%)	PRV (%)
Air Force	83	54
Washington headquarters	68	31
Navy	60	33
Army	18	28
Marine Corps	10	8
Total	40	35

SOURCE: Authors' analysis based on FY 2011 real property data from RPAD (Office of the Deputy Under Secretary of Defense [Installations and Environment], Business Enterprise Integration, "RPAD 2009–2011," July 31, 2012), not available to the general public.

NOTES: Washington headquarters manages the Pentagon reservation and other DoD facilities in the National Capital Region. Assets refer to active assets only.

calculate the same shares weighted by PRV (i.e., the percentage of PRV accounted for by facilities with ratings of 100), further supporting the lack of consistency in measurement and reporting across the components. The differences in how the services assess installation conditions hinder the analysis and interpretation of the RPAD data.

Findings on Installation Conditions

Having reviewed the problems associated with the RPAD data, we proceed with our analysis of installation conditions and call out the relevance of those problems at each juncture. We do so noting that we hesitate to draw firm conclusions strictly on the basis of the underlying data and so, to the extent possible, supplement those data with evidence drawn from discussions with subject-matter experts.

To compare installation conditions across the United States, U.S. territories, and foreign locations, we calculated two FCI averages from the FY 2011 data.[14] The first is PRV-weighted and second is unweighted. Although we present the results of both calculations in Tables 6.3 and 6.4 for completeness, we believe that the first calculation provides a better representation of the data; the unweighted or "simple" average overstates the importance of small assets in the mix. Regardless of the measure, be it weighted or unweighted, we find that FCIs are higher, on average, in foreign locations than in the United States. This result holds, in aggregate (Table 6.3) and for most of the "countries of interest" (Table 6.4) that we discuss in the chapter on host-nation support and U.S. payments.

[14] The AT&L analysis also employs this approach.

Table 6.3
Comparisons of Aggregate U.S., Territorial, and Foreign FCIs

	PRV-Weighted Average		Simple Average		PRV	Derived R&M Needs	Real Property Assets
	FCI	Age	FCI	Age	$ millions		Number
United States	86.6	41.7	84.5	34.6	684,093	91,952	492,861
Territories	79.2	43.0	84.4	34.8	20,662	4,305	9,178
Foreign	89.2	37.7	86.5	29.4	142,035	15,283	100,415
Total	86.8	41.1	84.8	33.7	846,790	111,540	602,454

SOURCE: Authors' analysis based on FY 2011 real property data from RPAD (Office of the Deputy Under Secretary of Defense [Installations and Environment], 2012).

NOTES: Assets refer to active assets only; age refers to the average age of active assets in each location, either PRV weighted or unweighted, as applicable; the figures for R&M have been derived from the data on FCI and PRV in accordance with the AT&L-provided formula for calculating the FCI.

The difference in U.S. and foreign FCIs could, in fact, reflect underlying differences in quality; that is, foreign assets could be "fitter" than U.S. assets. Our analysis of the data indicates that U.S. assets are, on average, older than foreign assets. If age were correlated with condition, hence FCI, this could be an explanatory factor. (We find evidence of a correlation between age and FCI values in data aggregated at the country and territory levels, but do not find evidence of such a correlation in asset-level data.) Perhaps more compellingly, we found confirmation in our discussions with U.S. military representatives in Europe and in the general logic of consolidations. Overall, discussion with EUCOM representatives suggested that closures had emphasized older facilities, leaving the remaining facilities in above-average condition, some having been recently upgraded or modernized.[15] Similarly, USAREUR representatives also noted that, in the course of consolidating facilities, USAREUR had used installation condition as a criterion, thereby shedding facilities with deferred maintenance and retaining those in relatively good shape.[16] The same could well hold true in other venues that have undergone such change.

Tables 6.3 and 6.4 provide estimates of the implied R&M needs that we derived from the RPAD data on FCI and PRV, using the AT&L-provided formula for calculating the FCI (presented above). However, we urge caution in the interpretation of those estimates because the services do not uniformly—if at all—use estimates of R&M to develop their respective FCIs. Thus, the R&M figures are mere arithmetic artifacts of the underlying FCI and PRV values in the RPAD data.

---

15 Interviews with U.S. military leaders in Europe, August 27–31, 2012.

16 Interviews with U.S. military leaders in Europe, August 27–31, 2012.

Table 6.4
Comparisons of Aggregate and Country-Level FCIs

	PRV-Weighted Average		Simple Average		PRV ($ millions)	Derived R&M Needs ($ millions)
	FCI	Age	FCI	Age		
Kuwait	51.1	33.0	54.7	33.0	7	3
U.S. Territories	**79.2**	**43.0**	**84.4**	**34.8**	**20,662**	**4,305**
Spain	84.4	43.2	86.4	41.5	3,107	485
Bahrain	85.7	17.5	88.7	14.8	1,098	157
United States	**86.6**	**41.7**	**84.5**	**34.6**	**684,093**	**91,952**
Djibouti	86.7	4.5	98.3	3.9	422	56
South Korea	88.4	28.3	79.5	26.7	15,080	1,753
Japan	88.8	34.5	86.9	28.3	46,444	5,198
Foreign (all)	**89.2**	**37.7**	**86.5**	**29.4**	**142,035**	**15,283**
Germany	89.7	47.5	84.5	33.4	40,350	4,174
Italy	92.8	25.2	90.9	25.3	8,347	601
UK	95.8	36.8	97.7	28.0	7,446	315
Turkey	95.9	34.4	94.8	27.6	2,218	90
Australia	100.0	29.8	100.0	25.1	12	—

SOURCE: Authors' analysis based on FY 2011 real property data from RPAD (Office of the Deputy Under Secretary of Defense [Installations and Environment], 2012).

NOTES: Age refers to the average age of active assets in each location, either PRV weighted or un-weighted, as applicable; the figures for R&M have been derived from the data on FCI and PRV in accordance with the AT&L-provided formula for calculating the FCI.

Having flagged the apparent differences in U.S. and foreign quality ratings, we comment on the ways in which problems in the data might accentuate those differences. Specifically, we note that the tendency to assign "100s" might be greater overseas than in the United States. In aggregate, we find that about 45 percent of all foreign assets and 38 percent of all U.S. assets have FCI values of 100 and that about 38 percent of foreign-located PRV and 35 percent of all U.S-located PRV has an FCI value of 100 (Table 6.5). It could just be that there are that many more "100s" overseas, but it could also be that methodological differences have biased the comparative results. One reason to believe that methodology has played a part is that the Air Force, which tends to assign a larger share of 100s than the other components, also has a larger share of its assets in foreign locations than the other services. The Air Force holds 23 percent of its reportable assets and 20 percent of its reportable PRV in foreign locations; comparable figures for the other services are as follows: Army, 14 and 16 percent;

Table 6.5
Locational Shares of Assets and PRV with FCIs Equal to 100

	Real Property Assets (%)		PRV (%)	
	All	Excluding Air Force	All	Excluding Air Force
United States	38	27	35	28
Territories	45	42	40	38
Foreign	45	27	38	26
Total	40	27	35	28
Ratio of foreign to United States	1.16	0.99	1.10	0.94

SOURCE: Authors' analysis based on FY 2011 real property data from RPAD (Office of the Deputy Under Secretary of Defense [Installations and Environment], 2012).

NOTE: Assets refer to active assets only.

Navy and the Marine Corps, 17 and 15 percent.[17] A comparison of the percentages of U.S. and foreign assets and PRV with scores of 100 that excludes the Air Force lends further support to the hypothesis that methodology is a contributing factor. Absent the Air Force data, the shares of U.S. and foreign assets with FCI values of 100 are almost indistinguishable—both are about 27 percent—and the share of U.S.-located PRV with an FCI value of 100 is slightly higher than that of foreign-located PRV.

In conducting this analysis, we sought to triangulate our results with an examination of relative spending on foreign, U.S., and U.S. territorial MILCON. For example, we assessed the ratio of MILCON to PRV, by location, over the past three years. Ultimately, we found this approach unsatisfying because we could neither clearly assign the totality of the MILCON estimates to particular locations nor interpret the resulting ratios. In any given year, a nonnegligible share of all MILCON funding is assigned to the category "unspecified worldwide funds"; moreover, any observable differences in MILCON spending across locations also likely reflect changes in mission and priorities and activities performed at installations (e.g., due to restationing or realignment, which drive a significant portion of MILCON funding), unrelated to relative installation conditions.[18]

---

[17] The Marine Corps, taken separately, holds slightly less than 18 percent of its reportable assets and 20 percent of its reportable PRV in foreign locations.

[18] See U.S. Air Forces in Europe, "Base Conditions and Facility Costs," Section 3, slides 20–24, August 30, 2012a.

Implications of Findings on Installation Quality for Postures

The data on installation conditions are weak, but, in combination with other qualitative evidence from U.S. military representatives overseas, they suggest that installation conditions overseas are at least as good as those in the United States and U.S. territories and possibly better. The differences do not appear to be large. If taken at face value, the FCIs on foreign soil are higher than in the United States, but on average differ by less than 3 percentage points.

What does this finding mean for locational decisions and the cost estimates that might inform them? At the very least, it suggests that it would be imprudent to assume that relocating from a foreign location to a U.S. analogue would necessarily spare the U.S. military a relatively high impending R&M bill. In some instances, relocation might imply an upgrade in the physical condition of the facilities, but in other instances it might imply a downgrade. The answer will depend on the conditions of the particular facilities in question. As a practical matter, given the small differences that we found in average U.S. and foreign conditions and our concerns about the quality of the underlying data, our cost modeling in Chapter Eight and estimated cost effects of the illustrative posture changes in Chapter Ten assume the same relative conditions in the United States and overseas for existing facilities.

CHAPTER SEVEN

Host-Nation Support and U.S. Payments to Other Countries

This chapter responds to the NDAA requirement to determine both direct payments and other contributions from host nations supporting U.S. military facilities overseas and direct payments by the United States to host nations for use of facilities, ranges, and lands. It puts these contributions and payments in the context of posture analysis by reporting on the ways in which they can affect the relative costs of maintaining an overseas presence and discussing how to factor them into analyses of alternative postures. Flows can occur bilaterally, as in the case of host-nation support (HNS) and U.S. payments to the governments of host nations, or multilaterally, as in the case of the NATO Security Investment Program (NSIP).

To assess the magnitude and fiscal implications of HNS and U.S. payments to other nations, we looked at 12 countries (Germany, South Korea, Japan, Italy, the UK, Australia, Bahrain, Turkey, Spain, Qatar, Kuwait, and Djibouti) that could be expected to provide the United States with HNS or, conversely, receive U.S. payments. Together, the 12 countries account for a large majority of the United States' overseas military presence.

We found that the data are scant and scattered for many of those countries, but sufficient to draw a few general conclusions relevant to posture decisions. For example, a close examination of historical and contemporary records suggests that HNS—in various forms—offsets a substantial share of U.S. overseas basing costs in a number of venues, even if sizable amounts of U.S.-borne costs remain. In the case of Japan, the country for which we have the most comprehensive data, we identified annual support potentially within range of $6.5 billion or more, depending on the scope of the definition of HNS and the forms of support—direct, indirect, cash, and in-kind—under consideration.[1] If we were to limit our discussion to "cost sharing," as a subset of all HNS, then that figure might be about $2.3 billion instead. South Korea and Germany, two other countries for which we have some recent quantitative information, also contribute noteworthy sums. It is also the case that some countries, such as Japan, have been more inclined to contribute directly and others, such as Germany, have been

[1] The $6.5 billion figure, which we present in Table 7.6, is an estimate of overall support, including but not limited to cost sharing. We discuss this and other estimates in detail later in this chapter.

more inclined to contribute indirectly, through forgone rent and lease payments and waivers of taxes, fees, and damage claims. We have less evidence of U.S. payments to other countries. The only concrete figure pertains to U.S. payments to Djibouti, which amount to about $30 million each year.

From an analytical perspective, the relevance of HNS and U.S. payments differs depending on the purpose of the analysis. For example, in comparing the costs of current operations in foreign and U.S. locations, we would seek to assess the difference between the bottom lines of operating in either location, i.e., the net cost of our presence with any contributions and including payments. On that basis, only some HNS and U.S. payments would affect our computations because much of the DoD accounting data that we employed for our cost analyses are "net" from the outset (i.e., costs that remain after taking these contributions and expenses into consideration). Care must be taken, however, to treat both sides of the equation—U.S. and foreign—according to the same accounting rules and to consider whether additional spending or investments would be required to change venues. If contemplating a shift from one foreign location to another, the data on HNS and U.S. payments could play a much larger part in a comparative assessment. For example, if the U.S. posture were to shift to less-developed parts of the world, it seems likely that HNS would drop and U.S. payments would increase.

We begin this chapter with a brief introduction to the terminology associated with HNS and U.S. payments to host nations and follow with an overview of historical and contemporary data sources. Next, we present the data on HNS and U.S. payments and discuss the relevance of those data to the analysis of postures. In addition to bilateral flows, i.e., HNS and U.S. payments, NATO allies—including the United States—also undertake multilateral cost sharing, through common- and joint-funded budgets, such as the NSIP. We consider the scope and relevance of NSIP funds separately. We also discuss a recent congressional proposal to increase data collection.

Terminology and Data Sources

Because the choice of a particular definition can affect the calculation of the amounts of support and U.S. payments, possibly by multiple orders of magnitude, we explain our use of the terms in this report. The *Department of Defense Dictionary of Military and Associated Terms* provides a formal definition of the overarching concept of HNS, per se, but it does not define "cost sharing," "burden sharing," or other terms that infuse discussions about bilateral flows.

Definitions of Host-Nation Support, U.S. Payments, and Related Concepts

The DoD dictionary of military terminology defines HNS as "civil and/or military assistance rendered by a nation to foreign forces within its territory during peace-

All other contributions, including but not limited to payments that foreign nations make to individuals, contractors, or other entities for goods or services provided in support of U.S. operations, thus constitute "in-kind" contributions.[7]

DoD reports[8] define "direct" and "indirect" support from the perspective of the host nation; that is, if the host nation bears the cost "on budget," the support is direct, and if the host nation bears the cost "off budget," the support is indirect.[9]

In the context of cost sharing:

> Direct cost sharing includes costs borne by host nations in support of U.S. forces for rents on privately owned land and facilities, labor, utilities, and vicinity improvements. Indirect cost sharing includes forgone rents and revenues, including rents on government-owned land and facilities occupied or used by U.S. forces at no or reduced cost to the United States, and tax concessions or customs duties waived by the host nation.[10]

Table 7.1 provides a matrix that categorizes illustrative examples of contributions (primarily cost sharing) on two dimensions: direct versus indirect support and cash versus in-kind support.

Table 7.1
Forms of Support and Illustrative Examples of Contributions

	Direct (On budget for host nation)	Indirect (Off budget for host nation)
Cash	• Compensation for local national employees, MILCON projects, and supplies and services of DoD, including refunds of utilities and payroll costs	• Not applicable
In-kind	• Direct provision of labor, structures, land, and infrastructure, as well as supplies and services, including facility administrators; construction; transportation infrastructure; utilities; and security, repairs, and maintenance • Payments for damage claims • Compensation of various kinds to local communities	• Forgone rent or lease payments • Waivers of customs duties and other taxes, fees (e.g., driver's license and airport landing and take-off), and damage claims

SOURCE: Authors' analysis.

---

[7] U.S. Forces Korea, 2010, p. 20, offers the following example of in-kind support, "a U.S. logistics contract awarded by the ROK [Republic of Korea] government in which the contractor is paid by the ROK government in won is an example of an "in-kind" contribution for which the ROK government receives cost sharing credit."

[8] DoD, *2004 Statistical Compendium on Allied Contributions to the Common Defense*, 2004b, p. A-3.

[9] In this context, the terms "on budget" and "off budget" refer to the budget of the host nation.

[10] DoD, 2004b.

We are not aware of any formal or commonly accepted definition of the term "U.S. payments" and have developed our own definition. For purposes of this analysis, we focus on direct payments to host-nation governments made explicitly for the purpose of supporting the U.S. overseas posture in accordance with the NDAA's stated focus on direct payments for use of facilities, ranges, and lands. We do not consider U.S. funding for security assistance (e.g., International Military Education and Training and Foreign Military Financing), development assistance (e.g., Overseas Development Assistance), or U.S. payments to other entities (e.g., employees and contractors) for goods or services that might affect the host-nation economy but are not paid to the host-nation government.

Sources of Data on Host-Nation Support and U.S. Payments

Up until 2004, DoD provided aggregate national-level information on direct and indirect cost sharing and other forms of "contributions toward the common defense and mutual security of the United States and its allies" in an annual report titled *Allied Contributions to the Common Defense*. This report presented DoD's assessment of the absolute and relative contributions of the United States' NATO allies, Pacific allies (Australia, Japan, and South Korea), and the Gulf Cooperation Council (GCC) nations. To gather data for the report, particularly the bilateral cost-sharing components, DoD relied heavily on contract support[11] and extensive inputs from U.S. embassies and DoD components, including the military departments and commands.[12,13] The annual reports were useful insomuch as they provided estimates of the dollar value of direct and indirect cost sharing for most NATO, Pacific, and GCC allies, but they provided little insight into the underlying composition of that support in each case and with little documentation; hence, they are virtually impossible to validate or reproduce. We found no comparable source of information on past U.S. payments.

The data for recent years are less readily available. At present, there are no centralized sources of information on contributions or U.S. payments, other than those pertaining to cash support and multilateral cost sharing. To the extent that bilateral information is available, it must be culled—oftentimes in broadly descriptive fragments—from a variety of documents, websites, and individuals, including the following:

- provisions of SOFAs that pertain to indirect support, e.g., those addressing tax and other fee waivers[14]

---

[11] RAND discussions with DoD/OSD/Comptroller, September 13, 2012, Arlington, Va.

[12] See DoD, 2004b, p. A-3.

[13] Interviews with USAFE Mission Support personnel, August 30, 2012.

[14] See, for example, NATO, "Agreement Between the Parties to the North Atlantic Treaty Regarding the Status of Their Forces," Washington, April 4, 1949. For a complete list of the international agreements and technical

- special measures agreements (SMAs) that the United States has negotiated and implemented with Japan and South Korea
- acquisition and cross-servicing agreements that are negotiated between the United States and its allies to allow U.S. forces to exchange most common types of support, including food, fuel, transportation, ammunition, and equipment
- various implementing agreements, technical arrangements, and diplomatic statements
- annual, mandatory reports that DoD provides to Congress on cash payments made from host nations to the United States under §2350j and §2350k, Title 10, U.S.C., Subtitle A, part 4, Chapter 138, Subchapter II
- official government websites of the host nations themselves
- representatives of EUCOM, USFK, USFJ, and others with on-the-ground knowledge of U.S. operations and related support.

Of these sources, only the annual reports on cash payments provide a comprehensive, regularized accounting of the flow of funds. Those reports tally cash payments to the United States from Japan, South Korea, and Kuwait.

Data on the NATO allies' NSIP contributions are more readily available, e.g., in NATO, U.S., and other budget documents, than those relating to bilateral support. We discuss those data in a separate section on multilateral flows.

Lessons from Historical Data

The final report in the *Allied Contributions to the Common Defense* series was released in 2004[15] and provides data on NATO, Pacific, and GCC allies' bilateral direct and indirect cost sharing for 2002. The data suggest that host-nation cost sharing is substantial in absolute terms and for relative basing costs, but that the roles of direct and indirect support differ by country and region. In 2002, cost sharing was reported to amount to just over 50 percent of basing costs (see Table 7. 2), on average, among the NATO, Pacific, and GCC allies; however, the bulk of all bilateral cost sharing among NATO allies—including Germany—was indirect, and the bulk of all cost sharing among Pacific allies—Japan especially—and GCC allies was direct.

As a general matter, we caution against drawing comparisons between the figures for 2002 and those available for recent years. Our reasoning is two-fold: first, lacking insight to the composition of the earlier figures—what they include and exclude in each case—we cannot assume comparability across the years; second, the world—and

arrangements consulted for this review, see reference section in this report under "United States of America and [name of other party or parties]."

[15] DoD, 2004b.

Table 7.2
U.S. Stationed Military Personnel and Bilateral Cost Sharing in 2002 (measured in nominal $ millions at 2002 exchange rates)

	Hosted U.S. Military Personnel (Dec. 31, 2002)	Derived Basing Costs	Bilateral Cost Sharing			Cost Offset Percentage		
			Direct Support	Indirect Support	Total	Direct	Indirect	Overall
NATO allies								
Belgium	1,516	74	2.2	15.6	17.8	3%	21%	24%
Canada	151	N/A	N/A	N/A	N/A	N/A	N/A	N/A
Czech Republic	15	N/A	N/A	N/A	N/A	N/A	N/A	N/A
Denmark[a]	156	20	0.0	0.1	0.1	0%	0%	1%
France	107	N/A	N/A	N/A	N/A	N/A	N/A	N/A
Germany	72,005	4,797	28.7	1,535.2	1,563.9	1%	32%	33%
Greece	598	55	2.0	15.7	17.7	4%	28%	32%
Hungary	18	*	—	3.5	3.5	*	*	*
Iceland	1,759	*	0.1	—	0.1	*	*	*
Italy	13,127	894	3.0	363.5	366.6	0%	41%	41%
Luxembourg	10	32	1.0	18.3	19.3	3%	57%	60%
Netherlands	703	N/A	N/A	N/A	N/A	N/A	N/A	N/A
Norway	120	12	10.3	—	10.3	84%	0%	84%
Poland	20	N/A	N/A	N/A	N/A	N/A	N/A	N/A
Portugal	1,041	69	1.7	0.8	2.5	2%	1%	4%
Spain	2,328	220	—	127.3	127.3	0%	58%	58%
Turkey	1,873	216	—	116.9	116.9	0%	54%	54%
UK[b]	11,351	880	27.5	211.0	238.5	3%	24%	27%
NATO allies' total[c]	106,898	7,269	76.5	2,407.8	2,484.3	1%	33%	34%

Analysis of Contemporary Data on Host-Nation Support and U.S. Payments

Contemporary data on HNS and U.S. payments are scattered across a variety of sources, including SOFAs, SMAs, and acquisition and cross-servicing agreements; annual reports on cash payments; DoD budget documents; and the official government websites of the host nations. Our review of those sources suggests that information on HNS is more readily available than information on U.S. payments.

Host Nations' Contributions to the United States

Tables 7.3, 7.4, and 7.5 collectively summarize our "state of knowledge" regarding the contributions of the 12 countries we examined. Table 7.3 indicates, in the broadest possible terms, the forms of support—cash, in-kind, direct, and indirect—that we can associate definitively with each country. That we do not associate a country with a particular form of support does not necessarily imply that the country does not, in fact, provide such support, only that we could find no corroborating evidence.[18] (Cash support is an exception. Reporting under Title 10, U.S.C. is mandatory, thus, any country that is not included in DoD's annual report to Congress, by definition, does not make cash contributions.) Table 7.4 distinguishes particular types of in-kind contributions, and Table 7.5 describes Japan, South Korea, and Kuwait's cash payments.

Much of our state of knowledge regarding bilateral contributions, especially indirect support, is only qualitative and general; for example, we can say that a country, such as Spain, adheres to the provisions of the NATO SOFA and therefore provides the United States with income tax waivers or other personnel benefits, as shown above. Only in the case of cash payments, which must be reported annually, can we confidently track the flow of funds comprehensively (see Table 7.5).

Together, the cash payments of Japan, South Korea, and Kuwait amounted to over $900 million annually from 2009 to 2011, of which about 94 percent pertained to cost sharing and 6 percent pertained to relocation.

Table 7.3
Forms of Support Associated with Countries of Interest

	Direct	Indirect
Cash	• Japan, South Korea, Kuwait	• Not applicable
In kind	• Germany, South Korea, Japan, Italy, Turkey, Spain, UK, Australia	• Germany, South Korea, Japan, Italy, Turkey, Spain, UK, Australia, Djibouti

SOURCE: Authors' analysis of international agreements, technical arrangements, etc. For a complete list of agreements and arrangements, see reference section of this report under "North Atlantic Treaty Organization" and "United States of America and [name of other party or parties]."

[18] For some Gulf allies, the information was either classified or unavailable.

Table 7.4
Types of In-Kind Contributions Associated with Countries of Interest

Type of Support	Countries
Direct	
Labor	South Korea
Structures	Japan
Land	Germany, South Korea, Japan
Infrastructure	Germany, Italy
Facility Administration	Italy
Construction	Germany
Transportation Infrastructure	Germany, Italy
Utilities	Japan
Security	Italy
Repairs/Maintenance	Italy
Payment of Damage Claims	Germany, Japan, Italy, Turkey, Spain, UK, Australia
Local Community Compensation	Germany, South Korea, Japan, Italy
Indirect	
Rent	Germany, Italy
Customs Duties	Germany, South Korea, Japan, Italy, Turkey, Spain, UK, Australia, Djibouti
Sales Tax	South Korea, Japan, Australia, Djibouti
Income Tax	Germany, South Korea, Japan, Italy, Turkey, Spain, UK, Australia, Djibouti
Avoidance of Environmental Remediation	South Korea, Japan
Tolls	South Korea, Japan, Australia
Port and Landing Fees	South Korea, Japan, Australia
Other Personnel Benefits	Germany, South Korea, Japan, Italy, Turkey, Spain, UK, Australia, Djibouti

SOURCE: Authors' analysis of international agreements, technical arrangements, etc.

NOTE: For a complete list of agreements and arrangements, see reference section of this report under "North Atlantic Treaty Organization" and "United States of America and [name of other party or parties]."

Below, we present estimates of bilateral support from Japan, South Korea, and Germany. For those countries, we have enough information to tell cogent stories about the nature of their support, but lacking any other public information on Kuwait's con-

Table 7.5
Types of Cash Payments Associated with Countries of Interest (based on U.S. fiscal year estimates and measured in nominal $ millions)

	2009	2010	2011
Japan			
Utility cost sharing	265	278	312
Training relocation contributions	9	24	20
Total cash payments	273	302	332
Percentage for relocation	3.1%	7.9%	5.9%
South Korea			
Labor cost sharing	328	244	265
South Korea-funded construction	170	144	94
South Korea relocation funds	38	46	35
Total cash payments	536	434	394
Percentage for relocation	7.0%	10.6%	9.0%
Kuwait			
Funds expended for base operations and sustainment	94	181	206
Cash payments from all countries			
Cost sharing	857	847	876
Relocation	46	70	55
Total	903	917	932
Percentage for relocation	5.4%	8.3%	6.3%

SOURCE: DoD, Under Secretary of Defense, Comptroller, "Burden Sharing Contribution Reports," 2009–2011.

NOTES: The data on cash payments are available only through FY 2011; under the U.S.-South Korea SMA, South Korea committed to a shift from cash to in-kind support for construction projects from 2009 to 2011; South Korea relocation funds are not covered under the U.S.-South Korea SMA.

tributions, we end our discussion of that country here. For readers with less interest in this amount of detail, we suggest skimming the discussion of the three countries and moving to the next section, on U.S. payments to host nations.

Japan's Contributions to the United States

Of the 12 countries under consideration, we have the most comprehensive data for Japan, partly because Japan's Ministry of Defense (MOD) and Ministry of Foreign Affairs (MOFA) post relatively detailed information about contributions on their respective websites,[19] and partly because USFJ representatives provided us with information on trends.[20] We have additional information on the cost-sharing and training relocation contributions that are covered under the U.S.-Japan SMA for the period 2011 to 2015 and on the subset of all contributions that are made in cash and thus covered under Title 10, U.S.C., mandatory reporting requirements.[21] The MOD and MOFA data are not fully consistent, but are mostly reconcilable. We have adopted the MOD estimates because they are available for four years, 2009–2012, whereas the MOFA figures are available only for 2012. We address the difference between the two agencies' estimates in the notes and text that accompany the data.

Table 7.6 reports the estimates found in the budget documents on the MOD website, using the terminology of those documents; Table 7.7 reports figures from the MOFA website, some of which might already be built into the MOD budget estimates and some of which are not because they either refer to other agencies or accrue "off budget." Because the U.S.-Japan SMA is specified in yen and the Japanese agencies report contributions in yen, we provide two segments of Table 7.6: Segment "a" shows the amounts in yen and segment "b" shows them in U.S. Dollars. Both versions provide detailed breakouts for what the MOD describes as "cost-sharing for stationing USFJ."

The amounts shown in Table 7.6, absent the addendum in Table 7.7, would all constitute direct support because they are MOD budget figures and are, by definition, "on budget." However, the aggregate figure, about $6.5 billion, represents an upper bound in that the amounts shown for base promotion include unspecified sums for purposes unrelated to U.S. facilities, and at least some of the amounts shown for the Special Action Committee on Okinawa (SACO), realignment, and relocation might fall outside the purview of HNS as we depict it in Figure 7.1. On the basis of our crosswalk of the MOD and MOFA data and information provided in the underlying documents, we do believe, however, that a significant fraction of the unspecified base-promotion amounts could involve U.S. facilities. In the case of local community expenses, the share pertaining to U.S. facilities could be almost 50 percent, and, in the case of facility rentals, compensation expenses, etc., that share could be as much as 90 percent. Applying those shares to each category, the revised aggregate figure would be about $5.6 billion.

[19] Government of Japan, Ministry of Defense, "Defense Programs and Budget of Japan," Budget documents for 2010–2012; Government of Japan, Ministry of Foreign Affairs, "U.S. Forces in Japan-Related Costs Borne by Japan (JFY 2012 Budget)," undated, Reference 6.

[20] U.S. Forces Japan, "Special Measures Agreement Overview," briefing, June 27, 2012.

[21] "Agreement Between the United States of America and Japan Concerning New Special Measures Relating to Article XXIV of the Agreement Under Article VI of the Treaty of Mutual Cooperation and Security Between the United States of America and Japan, Regarding Facilities and Areas and the Status of United States Armed Forces in Japan, Tokyo," January 21, 2011.

Most of the identified support is in-kind. Japan uses cash to pay for all—or nearly all—utility cost sharing and a substantial share of training relocation, but not the rest. The cash payments for utilities reported in Table 7.5 map reasonably well to the dollar values of those shown in Table 7.6, notwithstanding the differences in fiscal years (the Japanese fiscal year runs from April 1 to March 31). The mapping is less straightforward for the training relocation contributions; the cash payments appear to cover all of the training relocation contributions covered under the SMA and at least some of those covered elsewhere. We know less about the amounts shown in the MOFA-derived addendum (Table 7.7) except that some refer to ministries other than the MOD and some, i.e., the estimated costs of government-owned land, constitute indirect support.

Table 7.6
Japanese Funding for Cost Sharing and Other Purposes (based on Japanese fiscal year estimates)

a. In Nominal Billions of Yen

	2009	2010	2011	2012
Promotion of base measures, etc.				
(1) Expenses related to measures for local communities[a]	115.5	117.9	118.5	118.5
(2) Cost sharing for stationing USFJ				
(a) SMA				
Labor cost	116.0	114.0	113.1	113.9
Utilities	24.9	24.9	24.9	24.9
Training relocation costs[b]	0.6	0.5	0.4	0.4
Sub-subtotal for SMA	141.5	139.5	138.4	139.2
(b) Facility improvements	21.9	20.6	20.6	20.6
(c) Measures for base personnel, etc.	29.3	27.9	26.8	26.9
Subtotal for cost sharing for stationing USFJ (a)+(b)+(c)	192.8	188.1	185.8	186.7
(3) Facility rentals, compensation expenses, etc.[a]	131.6	130.5	129.3	136.6
Total promotion of base measures, etc. (1)+(2)+(3)	439.9	436.5	433.7	441.8
Total expenses related to Special Action Committee in Okinawa	11.2	16.9	10.1	8.6
Total U.S. Forces realignment-related expenses	60.2	132.0	123.0	70.7
Grand total	511.3	585.3	566.7	521.1

Table 7.6—Continued

b. In Nominal Millions of U.S. Dollars

	2009	2010	2011	2012
Promotion of base measures, etc.				
(1) Expenses related to measures for local communities[a]	1,234	1,343	1,485	1,486
(2) Cost sharing for stationing USFJ				
(a) SMA				
Labor cost	1,240	1,299	1,417	1,429
Utilities	266	284	312	312
Training relocation costs[b]	6	6	5	5
Sub-subtotal for SMA	1,512	1,589	1,734	1,746
(b) Facility improvements	234	235	258	258
(c) Measures for base personnel, etc.	313	318	336	337
Subtotal for cost sharing for stationing USFJ (a)+(b)+(c)	2,060	2,143	2,328	2,342
(3) Facility rentals, compensation expenses, etc.[a]	1,406	1,487	1,620	1,713
Total promotion of base measures, etc. (1)+(2)+(3)	4,701	4,973	5,434	5,542
Total expenses related to Special Action Committee in Okinawa	119	192	126	108
Total U.S. Forces realignment-related expenses	644	1503	1541	886
Grand total	5,464	6,668	7,101	6,536

SOURCES: Government of Japan, Ministry of Defense, budget documents for 2010–2012; the yen-denominated figures were converted to U.S. dollars using quarterly exchange rates from OECD, 2012b, and annual exchange rates from OECD, "Prices: Consumer Prices," *Main Economic Indicators,* database, 2012a, downloaded on October 21–23, 2012.

[a] The amounts shown for (1), "expenses related to measures for local communities," and (3), "facility rentals, compensation expenses, etc.," under the heading of "promotion of base measures, etc." include unspecified sums for purposes unrelated to U.S. facilities; however, on the basis of our crosswalk of the MOD and MOFA data, see the addendum, below, we believe that a significant share of these amounts could pertain to U.S. facilities. In the case of (1), the share pertaining to U.S. facilities could be almost 50 percent and, in the case of (3), that share could be as much as 90 percent.

[b] MOFA attributes another Y5.1 billion to spending on training relocation under the U.S.-Japan SMA, of which Y1.1 billion is related to the Special Action Committee in Okinawa and Y4.0 billion is related to realignment. These costs appear to be included in the MOD totals for each category but are not identified as "SMA".

Table 7.7
Addendum to Table 7.6 (in nominal billions of yen and millions of U.S. dollars, based on Japanese fiscal year estimates)

	2012	
	Yen	**U.S. Dollars**
MOFA-reported "Stationing of USFJ-Related Cost"		
(1) Costs clearly covered under MOD budget documents, labeled as "Cost Sharing for the stationing of USFJ"	186.7	2,342
(2) Costs likely covered under MOD budget documents as base promotion, including measures to improve surrounding living environments and facilities rent	182.2	2,285
(3) Costs additional to those covered under MOD budget documents		
(a) Estimated cost of government-owned land provided for use as USFJ facilities	165.8	2,080
(b) Expenditures borne by other (non-MOD) Ministries (base subsidy, etc)	38.1	478
Subtotal of additional costs (a)+(b)	203.9	2,558
Total MOFA-reported "Stationing of USFJ-Related Cost" (1)+(2)+(3)	572.8	7,185

SOURCE: Government of Japan, Ministry of Foreign Affairs, undated; the yen-denominated figures were converted to U.S. dollars using quarterly exchange rates from OECD, 2012b, and annual exchange rates from OECD, 2012a, downloaded on October 21–23, 2012.

Given the scope of coverage of the MOD and MOFA estimates (see Table 7.8 also), it seems plausible that a tally of some combination of the amounts shown in Tables 7.6 and 7.7 would capture all or nearly all of Japan's direct support. The numbers also make sense in terms of their order of magnitude relative to Japan's reported cost-sharing contributions in 2002. For 2012, the total upper bound amount shown in Table 7.6, version "b," is about $6.5 billion, of which about $2.3 billion constitutes "cost sharing for stationing USFJ." Adding the strictly additional direct component (non-MOD contributions) from the addendum in Table 7.7 brings the total to about $7.0 billion, and adding the indirect component (estimated cost of government-owned land) brings the total for all purposes, including some potentially outside the scope of our analysis, to about $9.1 billion. On the basis of the relative allocations of direct and indirect support shown in Table 7.2, it is also plausible that the MOFA estimate for the value of land provision accounts for most or all of Japan's indirect support.[22]

The explicit cost-sharing component of the base promotion measures, shown in Table 7.6, also includes most of the contributions covered under the U.S.-Japan SMA.

[22] The relative allocations of direct and indirect support shown in Table 7.2 were about 73 percent and 27 percent, respectively. In 2012, the $2.1 billion estimate of the value of land provision amounted to 23 percent of Japan's overall contributions

In our parlance, we might not adhere to the MOD's delineation of cost sharing: We might view some of the other elements of "promotion of base measures" as "cost sharing," such as those pertaining to facility rentals, but cannot readily parse them; at the same time, we might also choose to subtract some of the amounts covering training relocation costs. Another 5.1 billion yen of SMA funding—also designated for training relocation needs—appears to fall under the SACO and realignment categories.

The text of the "Security Consultative Committee Document," issued jointly by the U.S. Secretaries of State and Defense and the Japanese Ministers for Foreign Affairs and Defense in June 2011[23] (a "2+2" statement) defines HNS in terms of the funding in Table 7.6 under the heading "cost sharing for stationing USFJ":

> The Ministers confirmed that the overall level of HNS is to be maintained at the Japanese Fiscal Year (JFY) 2010 level (bearing in mind the budget of 188.1 billion yen in the JFY 2010) over the five years of the SMA period [*starting in JFY 2011*].[24] The Ministers affirmed that the two governments will implement a phased reduction of labor and utilities costs that the Government of Japan funds, while adding the amount of this reduction to the Facilities Improvement Program (FIP) funding in the current SMA period (FIP funding over the current SMA period is to be no less than 20.6 billion yen per year).

The texts of the U.S.-Japan SMA and the 2+2 statement and discussions of the content of those documents found on the MOFA website and in government-to-government exchanges provide additional detail on the allocations of Japan's contributions to different forms of cost sharing.[25] According to those sources, Japan committed to covering the costs of a progressively smaller workforce (declining from 23,055 to 22,625 workers); a progressively smaller share of stationing related utility expenditures (76 percent to 72 percent, capped at 24.9 billion yen); and a minimum level of FIP (20.6 billion yen) that could be made higher by the amounts equal to any reductions in expenditures for labor and utilities. The expenditures for labor and utility cost sharing accrue under the SMA and those related to the FIP accrue outside the SMA.

Appendix H summarizes this information and provides yen and U.S. dollar estimates for 2010 and the SMA period (2011–2015). In that appendix, we draw directly from the figures in Table 7.6 for the 2011 and 2012 estimates and extrapolate from those estimates, using data on wage growth, utility cost growth, and exchange rates,

---

[23] U.S.-Japan Security Consultative Committee, "Security Consultative Committee Document, Host Nation Support," issued by Secretary of State Clinton, Secretary of Defense Gates, Minister for Foreign Affairs Matsumoto, and Minister of Defense Kitazawa, June 21, 2011.

[24] We note, however, that the Japanese MOD reports funding less than 188.1 billion yen in 2011 and 2012.

[25] See, for examples, Government of Japan, Ministry of Foreign Affairs, "Japan-U.S. Security Arrangements," July 2012; Seiji Maehara, Minister of Foreign Affairs of Japan, letter (translated) to His Excellency, Mr. John V. Roos, Ambassador Extraordinary and Plenipotentiary of the United States of America, referring to the U.S.-Japan Special Measures Agreement, Tokyo, January 21, 2011.

for cash payments under Title 10, U.S.C., but little else.[29] The current and soon-to-expire SMA spans the period 2009–2013 and addresses labor cost sharing, logistics cost sharing, and South Korea–funded construction. The contribution of South Korea for 2009 was 760 billion Korean won, amounting to almost $600 million in then-year dollars; the contributions for 2010, 2011, 2012, and 2013 were to be determined by increasing the contributions of the previous year by the rate of inflation two years prior, not to exceed 4 percent (Table 7.9).

Some but not all of South Korea's cash payments correspond to commitments under the U.S.-South Korea SMA. Cash payments cover the labor costs associated with the SMA, amounting to no more than 71 percent of the labor costs for USFK's Korean workforce (not all of USFK's Korean-national workers are compensated under the SMA program); a declining fraction of the construction costs associated with the SMA; and relocation costs that accrue outside the SMA.[30] They do not cover any of the logistics costs associated with the SMA; those contributions are solely "in-kind." Regarding construction, South Korea committed to a shift from cash to in-kind contributions to be implemented fully by 2011, except for expenses associated with "design and construction oversight of facilities"; in 2009, the cash component of South Korea–funded construction was 70 percent, and by 2011 that share was expected to drop to 12 percent. Applying that trajectory, Appendix H provides estimates of South Korea-funded construction, cash and in-kind, for 2009–2012.

Table 7.9 reports on "estimable" direct support—from South Korea to the United States—but we cannot say whether it reports the totality of such support. A comparison of the magnitude of recent and 2002 levels suggests that it might, but the scope of coverage (see Table 7.10) looks narrow in comparison with Table 7.1.

We have only qualitative information on South Korea's indirect support—for example Article IV of the SMA references tax exemptions—but the historical record suggests that they are substantial. In 2002, for example, South Korea's indirect support amounted to more than 40 percent of the resources that it directed to cost sharing. If one were to assume a roughly similar split between direct and indirect cost sharing in the current era, admittedly a strong assumption, South Korea's total cost-sharing contribution could have amounted to about $1.27 to $1.33 billion in 2011, if Table 7.9 does, in fact, report the totality of all of South Korea's direct support, and depending

[29] "Agreement Between the United States of America and the Republic of Korea Concerning Special Measures Relating to Article V of the Agreement Under Article IV of the Mutual Defense Treaty Between the Republic of Korea and the United States of America Regarding Facilities and Areas and the Status of United States Armed Forces in the Republic of Korea," Seoul, January 15, 2009; United States of America and the Republic of Korea, *Implementing Arrangement for Special Measures Agreement*, Seoul, March 24, 2009.

[30] According to the DoD, Under Secretary of Defense, Comptroller, "Burden Sharing Contribution Report[s]" for 2009–2011, "relocation funding is not part of the SMA and is a separate negotiation between USFK and [the South Korean] Ministry of Defense for relocating troops/facilities away from the DMZ [demilitarized zone] and Seoul metropolitan area."

Table 7.9
South Korea's Cost-Sharing and Relocation Contributions (measured in nominal currency)

	Millions of Dollars			
	2009	**2010**	**2011**	**2012**
Estimable direct support				
SMA total (calendar year)	595	684	733	732
South Korea relocation funds (U.S. fiscal year)	38	46	35	N/A
Subtotal estimable direct support	633	730	768	N/A
Of which				
Cash (U.S. fiscal year)				
Labor cost sharing	328	244	265	N/A
South Korea–funded construction	170	144	94	N/A
South Korea relocation funds	38	46	35	N/A
Subtotal cash	536	434	394	N/A
In-kind (logistics and construction)	97	296	375	N/A
Total extrapolated direct and indirect support				
Including relocation	1,096	1,265	1,331	N/A
Excluding relocation	1,031	1,185	1,270	N/A
	Billions of Won			
SMA total (calendar year)	760	790	812	836

SOURCES: Author's estimates of SMA totals in billions of Won and millions of U.S. Dollars, calculated on the basis of language in Article II of the SMA, using data on inflation rates and annual and quarterly exchange rates from OECD, 2012a, and OECD, 2012b downloaded on October 17, 2012. Estimates of cash payments, including breakouts by type, are from DoD, Under Secretary of Defense, Comptroller, 2009–2011. The totals for direct and indirect support for 2009, 2010, and 2011 were extrapolated from the subtotals of estimable direct support—with and without relocation funds—in each of those years, assuming that direct support accounted for the same share of total support in each of those years that it did in 2002, i.e., approximately 58 percent. As shown in Table 7.2, in 2002, South Korea provided direct support amounting to $486.8 million and indirect support amounting to $356.5 million.

NOTES: Estimates of total direct and indirect support were extrapolated from the figures for "estimable direct support," using the relative shares of direct and indirect support that prevailed in 2002. See Table 7.2.

on the exclusion or inclusion of cash payments to cover relocation expenses. We offer this estimate not as a valid, accurate, or even plausible estimate of the current level of South Korea's total contributions, but as indicative of the level of our uncertainty around that total: It could be as little as $733 million, if considering only estimable direct support, excluding relocation (Table 7.9); as much as $1.3 billion, if looking at the extrapolated figure for direct and indirect support, including relocation (Table 7.9); or perhaps some other number entirely.

Table 7.10 summarizes our understanding of South Korea's contributions to the United States in terms of the categories presented in Table 7.1.

Germany's Contributions to the United States

For the Federal Republic of Germany, we have incomplete data on HNS for a single year—2009—drawn largely from correspondence between U.S. and German government officials. By inference and supported by discussions with EUCOM personnel, we know that all of that HNS, be it direct or indirect, is "in-kind"—if Germany were to provide the United States with any cash support, that support would be listed in DoD's annual report to Congress. A survey of documents related to the stationing of U.S. forces in Germany, including the NATO SOFA, as well as the officials' correspondence provided by EUCOM and USAREUR, indicate that Germany's indirect support includes forgone customs duties, forgone income taxes, personnel benefits (such as accepting U.S. Drivers Licenses without fees), and forgone rent on facilities provided by Germany for use by the United States. Additionally, direct support takes the form of land, infrastructure, construction, damage claim offsets, and compensation to local communities.

In Appendix H, we present a compilation of the available quantitative data, most if not all of which refers to direct support, with breakouts for each line item. In Table 7.11, we present the compilation in three, aggregated categories.

The data compilation provides evidence of contributions amounting to about 600 million euros or $830 million in 2009. The total is subject to at least two noteworthy caveats. First, much if not all of Germany's indirect support, such as the value of for-

Table 7.10
Summary of Estimable Contributions of South Korea to the United States

	Direct (On budget for host nation)	Indirect (Off budget for host nation)
Cash	• Labor cost sharing[a] • South Korea-funded construction[a] • South Korea relocation funds	• Not applicable
In-kind	• Logistics cost sharing[a] • South Korea-funded construction[a]	• By extrapolation, using historical shares of direct and indirect contributions

SOURCE: Authors' analysis.

[a] Covered in part or whole under the U.S.–South Korea SMA.

Table 7.11
Summary Compilation of Data on Germany's Contributions to the United States in 2009

Type of Contribution	Millions of Euros	Millions of U.S. Dollars
Construction		
Estimated value of construction work	450	625.1
Construction-related "reimbursements"[a]	70	97.2
Subtotal construction	520	722.3
Payments to third parties for accommodations leased for the U.S. Forces	51.1	71.0
Various other costs, including benefits for former U.S. forces employees	26.8	37.2
Total	597.9	830.6

SOURCES: Ralf Poss, "Subject: German Financial Contribution to U.S. Forces Construction Work in Germany," letter (as translated) to Glendon Pitts, November 8, 2010; Michael Schlaufmann, "U.S. Forces Stationed in Germany; Direct Support," letter (as translated) to Glendon Pitts, November 22, 2010; exchange rate for 2009 from OECD, 2012a, downloaded on October 22, 2012.

NOTE: Most if not all of these contributions appear to be direct contributions.

[a] As worded in the source document, but not constituting a cash payment.

gone sales tax associated with commissaries located on U.S. bases, is not captured in the table because we lack the data to do so. Thus, we can add the numbers to yield a total for 2009, but it is not the actual total. Looking back to the historical data presented in Table 7.2, the vast majority of Germany's contributions a decade ago were indirect, suggesting that today's actual total could be much higher. In addition, the officials' correspondence makes reference to an estimated value of the land provided by the Federal Republic of Germany to U.S. forces of about 5 billion euros. However, the correspondence goes on to indicate that these lands "have special value for the military user" and no market value calculations have been made for the areas.[31] Because we have no information about the method used to calculate the 5 billion euro estimate, we cannot offer a firm interpretation and do not include the figure in Table 7.11. Second, 87 percent of the total—some 520 million euros or $722 million—relates to construction, an activity that is inherently "lumpy" as one-time charges, such as for consolidation and relocations, heavily impact this category. For that reason, we cannot say what the numbers might look like in another year.

Remaining Countries' Contributions to the United States

For the remaining countries, we have little or no quantitative information, and in only the rarest of circumstances do we have much qualitative information. In the case of

[31] Schlaufmann, 2010.

Italy, for example, we have an overview of HNS provided by a EUCOM expert and some detail—albeit almost entirely qualitative—on contributions associated with a specific facility, i.e., Sigonella, that are spelled out in a technical arrangement.

Our information about Italy's contributions—also entirely "in-kind"—derives from a survey of documents (e.g., the NATO SOFA, the memoranda of understanding concerning the use of installations and infrastructure by U.S. forces in Italy, and the technical arrangement regarding installations and infrastructure in use by the U.S. forces in Sigonella) and from an issue paper prepared by a EUCOM subject-matter expert on "Italian Contributions to Forward Deployed U.S. Forces."[32] According to the documents, Italy provides a variety of indirect support (forgone customs duties, forgone income tax receipts, and other minor benefits to U.S. personnel) and direct support (facility administration, repairs and maintenance, and security personnel). In addition, the EUCOM issue paper indicates that the government of Italy forgoes rent on facilities and provides utility infrastructure, transportation infrastructure, payments of a portion of damage claims, and compensation to local communities to help offset U.S. forces–related costs. We have little information about the magnitude of Italy's contributions, but we have enough to affirm that the aggregate figure would be non-negligible. For example, among its non-recurring contributions, the EUCOM issue paper[33] indicates that the government of Italy has spent over 200 million euros to bolster the roads around Aviano Air Base and 10 million euros planning road upgrades around facilities in the Dal Molin and Vicenza areas, and that another 200 million euros will be needed to complete the upgrades.

Another example of the fragmentary nature of the information on HNS can be found in a USAFE-provided issue paper.[34] In that paper, USAFE indicates that they benefit from force protection savings that are attributable to co-location at host-nation facilities. We can infer that such benefits might accrue, for example, in Spain and the UK, where U.S. forces co-locate at facilities, such as Rota, RAF Mildenhall, and RAF Lakenheath, but we have no estimates of the dollar value of that co-location. Developing a dollar-value estimate for force protection, for example, would require specific information about the service requirement for each location, the share of that requirement attributable to the U.S. presence, and the costs of provision.

U.S. Payments to Other Countries

The data on U.S. payments to other countries are even scarcer than those on HNS. They might be scarcer because such payments are less common than HNS or because

[32] Paul Quintal, "Italian Contributions to Forward Deployed U.S. Forces," issue paper provided to the authors by OSD point of contact, June 5, 2012.

[33] Quintal, 2012.

[34] U.S. Air Forces in Europe, "Host Nation Support, Financial Arrangements, and Effects on Costs," Chapter 5 of briefing book, provided to RAND staff during site visit, August 2012b.

they are more scattered and sensitive and, therefore, more difficult to obtain.[35] Two types of data sources provide a small amount of insight. The first consists of the body of agreements that the United States maintains with countries in which it has security interests; the second consists of EUCOM, USFJ, USFK, and other representatives with on-the-ground knowledge of operations and financial practices.

Drawing on those sources, we find that the United States

- makes payments of about $30 million annually to Djibouti[36]
- makes no payments to any NATO allies
- likely makes few or no payments to any European non-NATO allies
- may or may not make any payments to any other host nations.[37]

The $30 million payment to Djibouti is made in consideration for the access to and use of the areas and facilities described in the *Arrangement in Implementation of the 'Agreement Between the Government of the United States of America and the Government of the Republic of Djibouti on Access to and Use of Facilities in the Republic of Djibouti' of February 19, 2003, Concerning the Use of Camp Lemonier and Other Facilities and Areas in the Republic of Djibouti* and its annexes. Under the terms of that implementing arrangement, any payments made by the United States to Djibouti, consistent with the land lease of 2003, would be credited against the $30 million. By comparison, a recent press report notes that Djibouti, which hosts France's largest military base in Africa, receives another $36.75 million per year in rent payments from that country.[38]

---

[35] Data on U.S. payments to nongovernmental entities, which we do not consider here, are likely subsumed in data on standard operating and support and MILCON costs insomuch as the United States undertakes ordinary business transactions in the countries in which it operates its facilities.

[36] For the $30 million figure, see United States of America and the Republic of Djibouti, *Arrangement in Implementation of the 'Agreement Between the Government of the United States of America and the Government of the Republic of Djibouti on Access to and Use of Facilities in the Republic of Djibouti' of February 19, 2003, Concerning the Use of Camp Lemonier and Other Facilities and Areas in the Republic of Djibouti*, May 11, 2006; and William Maclean, "Djibouti: Western Bases Pose Manageable Risk," *Chicago Tribune*, July 11, 2012.. A more recent document, the *Report to Congress on Camp Lemonier, Djibouti, Master Plan*, prepared by the Department of the Navy, the Assistant Secretary of the Navy (Energy, Installations, and Environment), August 12, 2012, indicates (p. 8) that the United States currently pays $38 million to the government of Djibouti for Camp Lemonier, including the use of Ouaramous Island, with overseas contingency operations funding.

[37] For example, although outside the scope of our 12-country assessment, we are aware that the U.S. government makes payments to the government of the Kyrgyz Republic, including reimbursement for access to and use of the Transit Center at Manas International Airport, related facilities, and mutually agreed logistic support, amounting to $60 million annually. See Jim Nichol, *Central Asia: Regional Developments and Implications for U.S. Interest*, Washington, D.C.: Congressional Research Service, RL33458, September 19, 2012; and United States of America and the Kyrgyz Republic, *Agreement Between the Government of the United States of America and the Government of the Kyrgyz Republic Regarding the Transit Center at Manas International Airport and any Related Facilities/Real Estate*, Bishkek, May 13, 2009.

[38] See Maclean, 2012.

Treatment of Bilateral Flows in Posture Analysis

In this section, we discuss the interpretation of the data on host-nation contributions and U.S. payments. We find that the role of the various bilateral flows in analysis depends on the context; for example, whether the purpose of the analysis is to

- estimate host nation costs
- evaluate U.S. fiscal exposure
- compare costs of alternative basing locations.

Differences among analytical contexts necessitate case-by-case consideration. To that end, we offer a simple framework for assessing the relevance of different forms of support to particular analytical efforts. The framework decomposes costs by bearer. The total cost of maintaining U.S. forces in a particular location can be described in terms of costs borne by the United States and costs borne by others, specifically the host nation:[39]

$$\text{Total cost} = \text{U.S.-borne costs} + \text{U.S. payments} + \text{HNS}.$$

"U.S.-borne costs" consist of all direct expenses of maintaining the U.S. presence, except those reflected in "U.S. payments"; "U.S. payments" consist of payments made by the United States to the host nation for use of facilities, ranges, and lands; and "HNS" consists of direct (cash and in-kind) support and indirect (in-kind) support.

Estimating Host-Nation Costs

The cost to the host nation is the resource value of the HNS that it contributes, less some fraction of the amount of U.S. payments that it receives. Regardless of the value that the United States places on a particular good or service, the provision of that good or service, as embodied in HNS, represents a real cost to the host nation if the host nation could have put the underlying resources to use elsewhere and if the value attributed to HNS reflects the value of those resources in the alternative uses, i.e., the market value or opportunity cost associated with the resources.[40] Similarly, the value of underlying resources comes into play in assessing the relationship between U.S. payments and host-nation costs. If the amount of U.S. payments exceeds the market value or opportunity cost of the facilities, ranges, or land under consideration, then that additional amount would offset the host nation's cost.

---

[39] For the purposes of this discussion, we are setting aside multilateral cost sharing, which, for completeness, would require the insertion of an additional term in the equation that follows. We address the implications of those flows in a separate discussion of NSIP funds.

[40] The accuracy of the HNS value attribution likely differs by type of support; for example, it might be more difficult to estimate the value of indirect support than direct support.

The host nation's cost could be taken as an indicator of that country's commitment to the presence of U.S. forces; however, a country lacking financial ability may have a high level of commitment to the U.S. presence, even if it cannot formalize that commitment through HNS. Indeed, that country might require remuneration to counter the strain that hosting would otherwise place on its limited resources. Recognizing the role of financial ability in establishing the terms of support and payments, we caution against over-interpreting either a lack of support or, conversely, a request for payments.

Evaluating Fiscal Exposure

Fiscal exposure, as we use the term, refers to the amount that the United States could be called upon to pay, were HNS to decline in a particular venue. If a host nation were to determine that it would or could not continue to provide support at the current level, e.g., for lack of budgetary and other resources, the United States could find itself "on the hook" for the amount of the shortfall, if it were to remain in that country. Alternatively, the United States might choose to shed some activities, hence costs, but if all the activities are necessary to maintain the U.S. presence, it could lack the flexibility to do so. Thus, HNS, or some fraction of HNS, can represent a financial risk.

Comparing Costs of Illustrative Postures

To compare the costs of illustrative postures—as in Chapter Ten, based on the models developed in Chapter Eight—we would look at the bottom lines for U.S. costs associated with maintaining a U.S. presence at each location under each configuration of U.S. forces. This bottom line can be represented as either the total cost, less HNS, or the sum of U.S.-borne costs and U.S. payments; however, whether we would need to factor any HNS and U.S. payments into our calculations depends on the composition of the data that we start with, the methods that we use to make our calculations, and how we define "maintaining a U.S. presence."

Because we work mostly with DoD cost data—which are primarily but not entirely accounting data—we are especially interested in the ways in which host-nation contributions and U.S. payments filter through the DoD cost accounts. In many or most instances, the DoD cost data are already net of HNS. Netting occurs automatically when the host nation's contribution is in-kind and takes the form of a good or service, so that the contribution is not recorded directly in the DoD cost accounts, or when the host nation reimburses U.S. expenditures.[41,42] In short, DoD cost data and

[41] RAND discussions with DoD/OSD/Comptroller, Arlington, Va., September 13, 2012. For more information about methods of accounting for receipts and expenditures, see also DoD, 2012b.

[42] If the reimbursement were timed or credited in such way that it did not immediately offset the corresponding U.S. expenditure, the initial spending might still be visible in the "bottom line," at least for a short while, and we might need to remove it. While timely offsets are reported as the norm, we cannot rule out the possibility of a lag. However, as a practical matter, the three-year averaging method that we use to calculate installation support and estimated facility recapitalization costs in Chapter Eight almost certainly serves to smooth the data and elimi-

the cost modeling in Chapter Eight largely reflect net DoD spending, with accounting for HNS "built in" either through the absence of DoD spending for a service or good or host-nation reimbursement for the same. Albeit less often, the data might also include elements of U.S. payments that flow through ordinary program accounts.

In those instances in which the data are already net of host-nation contributions or include U.S. payments, we do not need to draw the bilateral flows into our calculations; if we did, we would be double-counting them. For example, were we to subtract Japan's $2.3 billion cost-sharing contributions from a data-derived estimate of the costs of U.S. operations in Japan, our "bottom line" could be as much as $2.3 billion too low because much if not all of that $2.3 billion flows directly from the government of Japan to goods and services providers, as is the case of in-kind support for labor cost sharing and facility improvements, or reimburses a U.S. expenditure, as occurs in the context of utility cost sharing. Similarly, if some fraction of the U.S. payment to Djibouti were to enter into the DoD cost accounts as an ordinary expense, e.g., an amount used to cover a lease, we would not want to count it twice.

It is also possible that some host-nation contributions serve a purpose that is outside the scope of our analysis; that is, they do not line up with our definition of "maintaining a U.S. presence." The contributions might, for example, serve to expand capacity, enhance capabilities, or enable a change of mission. Some of the Japan's contributions toward facility improvements or South Korea's contributions toward construction might do more than restore, modernize, or recapitalize existing facilities; some relocation contributions might also fall outside the bounds of an estimate of the ongoing costs of sustaining overseas posture.

By and large, we conclude that a tally of the DoD cost data can come close to approximating the sum of U.S.-borne costs and U.S. payments in many environments, absent any further adjustments. Nevertheless, it is still possible that some of the data are not fully net of HNS, as should be typical, or inclusive of U.S. payments, thus necessitating adjustments in our calculations. In such cases, we would need to subtract the un-netted HNS from the cost tally to credit the contribution and add the excluded U.S. payments to recognize them.[43]

If comparing the costs of maintaining a presence in two different foreign locations, one might need to look more closely at the types and amounts of HNS offered—

nate the effects of any lags that could occur in a particular accounting period. DoD, Under Secretary of Defense, Comptroller, "Burdensharing Contribution Report[s]" FYs 2005–2011, note occasional lags across accounting periods, but the 2011 report indicates that changes in accounting practices have been implemented that would decrease the likelihood of lags in the future: "Due to changes in the accounting process that were implemented in FY2011, all transactions that are reported to Treasury are now captured and processed in the month they occur."

[43] The methods we use in Chapter Eight to estimate MILCON costs for recapitalization yield a best estimate net of HNS based on actual spending (these estimates are used in the overall illustrative posture cost estimates) and an approximate upper bound on recapitalization costs based on estimated requirements and reported levels of HNS for facility recapitalization. However, the MILCON spending data are quite variable from year to year, so significant uncertainty remains with regard to our estimates of U.S. MILCON recapitalization costs net of HNS.

or expected to be offered—and U.S. payments made—or expected to be made—at the two locations. Just because one nation offers certain types of support or requires particular payments does not necessitate that another will follow suit. (The indirect benefits associated with many SOFAs might present an exception.) When the United States already maintains a presence in one of those locations, we might have the benefit of tapping into existing data to calculate a bottom line; if the United States is considering a new location, we would need to construct a bottom line, using our best estimates of U.S.-borne costs, U.S. payments, and HNS. Should we observe lower-than-expected costs in some foreign locations, the provision of HNS (cost sharing in particular) might then explain the gaps, but it would not negate them.

Moreover, the bottom lines that we calculate for each location, whether foreign or U.S., must address not just the costs of ongoing operations, but those of additional spending or investments that might be required in the course of a change of posture. If a move would require investments in new capacity, e.g., for training or other purposes, then those costs—and the feasibility of building out that capacity—must be factored into the analysis.[44]

NATO Security Investment Program Contributions

The NSIP is financed by the ministries of defense of each NATO member country, according to a cost-sharing formula that is based on relative income and other factors,[45] and is supervised by the NATO Investment Committee.

NATO describes the NSIP as follows:[46]

> This programme covers major construction and command and control system investments, which are beyond the national defence requirements of individual member countries. It supports the roles of the NATO strategic commands by providing installations and facilities such as air defence communication and information systems, military headquarters for the integrated structure and for deployed operations, and critical airfield, fuel systems and harbour facilities needed in support of deployed forces.

Whereas the data on bilateral contributions and U.S. payments are highly fragmented, we are able to derive rough estimates of the NATO allies' contributions to

[44] This would also be true in the case of a move to a different foreign venue, but, in that case, one would also need to net out the contributions of the new host nation to the provision of that capacity.

[45] See NATO, "Paying for NATO," undated; and the related discussion in Carl Ek, *NATO Common Funds Burdensharing: Background and Current Issues*, Congressional Research Service, RL30150, February 15, 2012.

[46] See NATO, undated.

the NSIP—stated as "requirements"—from a combination of U.S. and NATO budget documents and congressional reports (see Table 7.12).

Under the current cost-sharing formula, the United States covers just over 22 percent of the total NSIP requirement. The U.S. gross requirement averaged $257 million annually from 2008 to 2012, with outlays averaging $219 million; other NATO allies' NSIP requirements averaged about $895 million in aggregate.

Because of the NSIP, the United States bears a smaller share of the infrastructural costs associated with facilities in which it has a substantial interest than it might otherwise bear. The DoD budget justification[47] for FY 2013 cites three specific examples:

Table 7.12
U.S. and Other Nations' Contributions to the NATO Security Infrastructure Program (Measured in nominal $US millions unless stated otherwise)

	2008	2009	2010	2011	2012
U.S. share of total	23.2%	23.2%	21.7%	21.7%	22.2%
U.S. gross requirement[a]	207	244	293	259	283
Derived gross requirements of					
All NATO allies	894	1,051	1,352	1,193	1,274
Non-US NATO allies	687	807	1,058	934	991
U.S. net requirement[b]	201	241	276	259	273
U.S. budget authority[c]	201	331	197	258	248
U.S. outlays[b]	249	231	192	194	229

SOURCE: Authors' estimates based on DoD, "Military Construction Program Budget: North Atlantic Treaty Organization Security Investment Program, Justification Data Submitted to Congress," various years; DoD, Under Secretary of Defense, Comptroller, *National Defense Budget Estimates*, Washington, D.C., various years; and Ek, 2012.

NOTE: All years are U.S. fiscal years.

[a] The U.S. gross requirement is the dollar value of the U.S. share of the total NISP requirement, as reported by DoD, based on the formula in the existing cost-sharing agreement and budgeted exchange rates. The gross requirements of all NATO allies and of non-U.S. NATO allies were derived from the U.S. gross requirement, using the inverse of the U.S. share in each year.

[b] The U.S. net requirement is the request for new appropriations, as reported by DoD, after accounting for amounts expected to be available from recoupment of prior year work funded by the United States.

[c] The amounts of U.S. budget authority and outlays for 2012 are estimates. U.S. budget authority, which is provided by law to enter into obligations that will result in immediate or future outlays of federal funds, and U.S. outlays may differ in any year because of differences in the timing of the provision of budget authority, primarily through an appropriation, and actual spending.

---

[47] DoD, 2012b, pp. 5–7.

- At Aviano Air Base, Italy, NATO funded over $465 million for the beddown of two fighter squadrons. The projects include both operational and community support facilities.
- At Ramstein Air Base, Germany, NATO has invested over $210 million to provide strategic air transport infrastructure to include parking aprons, freight and passenger terminal facilities, and a C-5–capable hangar.
- At Naval Station Rota, Spain, NATO has invested $151 million in port infrastructure upgrades to provide logistics support and resupply facilities for NATO maritime forces. Additional NSIP investment at Rota will manifest itself through the approved NATO capability package for strategic air transport. This NATO capability package includes nearly $83 million for infrastructure upgrades and recoupment eligibility to support NATO's Southern European Strategic Air Transport requirements.

Applying the U.S. share of the overall NSIP requirement, approximately 22 percent, to the NSIP activities in each location, the U.S. share of the NSIP funds in Aviano, Ramstein, and Rota would have amounted to about $102 million, $46 million, and $33 million, respectively.

The implications of the NSIP funding for our posture cost analysis are not clear-cut. Depending on the specific nature of the NSIP program activity, the effects of NSIP funding on the bottom lines at facilities that benefit from NSIP could differ markedly.[48] A key issue is whether the NSIP project activity constitutes recapitalization or something more. If recapitalization, we would need to make downward adjustments in our recapitalization estimates in Chapter Eight, amounting to 78 percent of the NSIP funding amount; if capacity expansion or capability enhancement, we would not need to make such adjustments. In the latter case, we would treat the funding as outside the scope of our cost analysis. On the basis of NATO's and DoD's characterizations of the program, it is plausible that the program activity could constitute either, but that expansions and enhancements might dominate. Thus, even though NSIP cost sharing represents a cost savings for the United States—assuming that the United States would have carried either the full burden or a larger share of the burden of the NSIP-funded infrastructure cost in the absence of NSIP—the flows of funds through the program might not be analytically "relevant" for estimating the costs of maintaining the U.S. overseas posture.

[48] As a practical matter, the United States makes its contribution to the NSIP through the MILCON account under the rubric of "Unspecified Worldwide" activity.

Host-Nation Support Data Collection in the Future

At present, the task of reporting comprehensively on either HNS or U.S. payments is nearly impossible, owing to the scarcity and fragmentation of the data. By analogy, the task resembles that of an archeologist attempting to reassemble a shattered vase: We have unearthed a few pieces that are large enough to tell us that we do, indeed, have a vase, but not enough to know its true dimensions or contours.

In looking to the future, we might first look to the past for insight. The historical record, specifically DoD's *Allied Contributions to the Common Defense* annual reports, provides insight in terms of both the value it adds and the value it does not add. On the one hand, the figures contained in those reports suggest the potential importance of HNS in funding U.S. requirements overseas. By all appearances, U.S. allies have contributed substantially over the years, both directly and indirectly. On the other hand, the difficulty of interpreting and replicating those figures speaks to the importance of thorough documentation. With hindsight, it is apparent that those reports have less value than they could have had because of the lack of documentation of methods and sources.

The historical record also provides insight insomuch as it speaks to the possible costs of any new effort to collect and compile data in the future. As noted previously, the resources required to generate the annual reports on allied contributions were non-negligible.

The NDAA for Fiscal Year 2013 calls for a return to annual reporting on host nations' contributions from 2013 through 2015. Specifically, language in that legislation requires "the Secretary of Defense, in consultation with the Secretary of State, [to] submit to the appropriate congressional committees a report on the direct, indirect, and burden sharing contributions[49] made by host nations to support overseas United States military installations and United States Armed Forces deployed in country." The report should include "at least" a description of all costs associated with stationing U.S. Armed Forces in the host nation, a description of the host nation's contributions,[50] and the methodology and accounting procedures used to measure and track host-nation contributions.

In requiring DoD to document thoroughly its methodology and accounting procedures, the legislation stands to increase the value of the end product; however, that product—and its documentation—will come at a cost. Whether the value of the endeavor outweighs that cost will depend partly on the reasons for collecting the data.

---

[49] The provision defines "contribution" as "cash and in-kind contributions made by a host nation that replace expenditures that would otherwise be made by the Secretary of Defense using funds appropriated or otherwise made available in defense appropriations Acts."

[50] Such contributions might include those made to cover or defray the costs of compensation for local national employees of DoD, military construction projects of DoD, and other costs, such as loan guarantees on public-private venture housing and payment-in-kind for facilities returned to the host nation.

The data could contribute positively to an evaluation of the costs borne by host nations or of the fiscal exposure associated with the U.S. military's overseas presence, but might not contribute as positively to a comparative assessment of the costs of operating at different locations. For a comparative analysis of the costs to the United States of basing in foreign and U.S. locations, the estimates that emerge from our and other analyses employing DoD cost data may be sufficient, in that they provide reasonable approximations of the bottom lines of operating at the different locations. The additional data could play a more important part in a comparison of the costs of basing in different foreign locations, particularly if those locations differ in either their willingness or ability to provide support. If the U.S. overseas posture shifts toward less-developed parts of the world where resources are scarcer, it seems likely that HNS will decline and U.S. payments will increase. Regardless of the intention, it will be worth gauging the necessary degree of granularity in the data collection and the merits of further refinements.

Implications of Findings on Host-Nation Support and U.S. Payments for Postures

This chapter responded to the legislative requirement to determine both direct payments and other contributions from host nations supporting U.S. military facilities overseas and direct payments by the United States to host nations for use of facilities, ranges, and lands. It put these contributions and payments in the context of posture analysis by reporting on the ways in which they can affect the relative costs of maintaining an overseas presence and discussing how to factor them into analyses of illustrative postures. Looking at 12 countries that together account for a large majority of the United States' overseas military presence, we found that the data are scant and scattered, but still sufficient to draw a few general conclusions of relevance to posture decisions.

A close examination of the available data suggests that HNS offsets a meaningful share of U.S. overseas basing costs in a number of venues, but that sizable amounts of U.S.-borne costs remain and that contributions and, to a lesser extent, payments, vary markedly across locations. The data also suggest that Japan, South Korea, and Germany are among the biggest contributors of HNS to the United States, but the limitations of the data, including the lack of availability for other countries, make it difficult to draw cross-country comparisons. We note that some countries, such as Japan, have been more inclined to make direct contributions and others, such as Germany, have been more inclined to make indirect contributions. While direct contributions can be difficult to quantify and value, it can be even more difficult to quantify and value indirect contributions, which include forgone rent and lease payments and waivers of taxes, fees, and damage claims.

In contemplating a move from an existing foreign facility to a U.S. location, the data that we use to develop our cost estimates in Chapters Eight and Ten should yield reasonable approximations of the bottom lines of operating at either location, with little need for adjustment. The DoD spending data that we employ in our analyses are by definition net of most if not all host-nation contributions. In most instances, the data are either inherently net (i.e., the host nation pays for the good or service directly) or already netted for us. Nevertheless, we might still need to factor in some U.S. payments, as would be the case were we to develop a cost estimate for Djibouti.

For a posited move from one foreign location to another foreign location, or from the United States to a new foreign location, key considerations would include the willingness and ability of alternative hosts to provide HNS and their need for payments. Countries experiencing greater fiscal hardship might be less willing or able to provide support and might be more needful of payments, resulting in higher costs to the United States, assuming all else to be equal. Although not a focus of this chapter, the last point merits comment. In reality, all else may not be equal. The calculation of the bottom line for the United States—that is, the sum of U.S.-borne costs and U.S. payments—will depend not just on the amount of HNS and U.S. payments that prevail in a particular location, but also on the costs of obtaining goods and services in that location. If a country is relatively expensive to operate in because labor, energy, and other inputs are especially costly, the bottom line for the United States could still be relatively high, even with substantial HNS and no U.S. payments.

Thus, for the purposes of examining the cost effects of posture changes with respect to those locations in which we currently operate, the existing cost data can be used to effectively support the analyses. In contrast, where posture options would contemplate facilities in nations in which the United States does not have existing facilities, cost estimating would be more uncertain and have to rely on either analogous situations or discussions with the prospective host nation. Moreover, it would be important to consider whether additional spending or investments would be required to change venues, either from a foreign to U.S. location or one foreign to another foreign location. If one location presents a full complement of facilities in good condition and another requires ground-up construction or major R&M or recapitalization, the costs of shifting from one location to another could be greater than an analysis of current operating costs might suggest.

CHAPTER EIGHT

Relative Costs of Overseas Basing and Rotational Presence

This chapter lays out the analyses we conducted to determine the incremental costs of maintaining the broad range of U.S. installations and forces overseas—by region and type—versus in the United States, and it explains how we embedded the results of these analyses in models that provide a policy-relevant way of comparing how total DoD costs are likely to change as the result of posited posture changes. In doing so, this chapter describes three cost-assessment methodologies and the derived cost models developed to comprehensively estimate the cost effects of changes in posture. The first methodology determines the relative recurring costs of stationing forces and operating military facilities and training ranges overseas compared with the United States, accounting for any costs that would simply transfer from one installation or region to another versus those that would actually change from a DoD standpoint.[1] The second assesses the recurring costs of rotational presence, a possible alternative to permanent presence, for different types of operational units to different regions. The third methodology determines one-time investment costs to transition from one posture to another. Together, the three resulting models enable estimates of how changes to posture would affect total DoD costs. The models and their parameters are also informative by themselves for understanding how overseas posture, and changes to it, affects costs.

To conduct the analyses and develop the models, we collected data from each service across all relevant budget accounts—not just those directly associated with maintaining installations, such as operations and maintenance funding for installation support—and from relevant DoD agencies to get a comprehensive picture that is broader than that typically generated when examining the costs of maintaining facilities and forces overseas. Additionally, we developed the cost models to get insights into the incremental costs of overseas presence, as opposed to only determining the actual costs in given locations. This was necessary to estimate the cost effects of different

[1] In this chapter, we generally compare costs in the United States with those on foreign soil. For the purposes of cost estimation, we usually exclude Hawaii from both groups and include Guam with other PACOM bases, with a few exceptions that are noted. We use the terms *U.S.* and *overseas* to distinguish the two sets of bases, and we differentiate specific regions, countries, and bases as necessary. When we use the terms *CONUS* and *OCONUS*, it is because a particular data set makes that distinction.

posture policy options from a total DoD standpoint. It is only through these models that we could separate out the costs that would likely change with a modification to posture versus those costs that would shift to another location, which necessitated the decomposition of several costs components into fixed and variable cost elements, as described in the chapter. Budget data do not neatly categorize into these two categories, which is why could not just use raw budget data to assess the costs associated with overseas presence. And the data in some budget accounts that are affected by location and assignment policies, such as allowances in the military personnel accounts, are not binned and aggregated by location, further necessitating analysis of budget data and model development.

In this chapter's introduction, we first define several terms, summarize our analytic approach, discuss the limitations of our approach, and preview the findings. We then spend the bulk of the chapter describing our three methodologies and the associated cost models. Finally, we end the chapter with a discussion of the implications of this cost analysis for the U.S. military posture. Note that all costs discussed in this chapter are net of host-nation support.

Introduction to Cost Analysis

Definitions of Terms

The following key terms are used throughout the report to characterize costs.

- *Incremental cost*, or *incremental cost difference*, refers to the relative difference in recurring costs (for similar units, populations, or installations) between the United States and an overseas region or between two different overseas regions.
- *Investment cost*, or *one-time investment cost*, refers to the monies needed to shift the U.S. posture by relocating forces or closing (or vacating) installations.
- *Recurring costs* refers to the annual costs of operating and sustaining forces and installations, whether in the United States or overseas. For recurring costs, we draw a further distinction between fixed and variable recurring costs:
  - *Fixed costs* refers to the recurring costs that are not sensitive to base population and is basically invariant to the number of people assigned to the base.
  - *Variable costs* refers to those costs that vary with base population.[2]

To further explain fixed and variable recurring costs, if forces shift from overseas to the United States, the variable costs at the "sending" base decrease, and those at the "receiving" base increase, by the respective variable costs per person for the associated

[2] Economists also refer to this variable component as *marginal costs*. We use the more colloquial term *variable cost* in this report.

base types and regions. The fixed costs, because they are insensitive to population, change (i.e., are eliminated or newly added) *only if a base is closed or opened.*

Approach to Overall Cost Analysis: Constructing Cost Models

This cost analysis is designed to provide: (1) cost assessments of the illustrative posture options developed in this report, (2) insights about the key drivers of the costs of overseas posture to facilitate decisionmaking, and (3) a toolkit—or at least a methodology—that could potentially be used by DoD to make a rough estimate the cost effects of other change options. To achieve these goals, we had to develop cost models. There are three reasons why we take this approach and develop cost models.

First, we needed to evaluate a range of posture options that modify the numbers and locations of overseas forces and bases in different ways to assess the cost impacts. To do so, our approach had to be flexible enough to estimate the effects of a range of changes to posture for which not all of the proposed base locations and their hosted forces would precisely match the attributes of existing bases and the forces assigned to them—and thus would be expected to have the same costs as one of them. It also had to be able to produce results for different posture options quickly as we explored a range of options. This led us to produce generalized models that we could employ to estimate the likely cost effects of changing base populations, opening or closing bases, and adding or eliminating rotations—the three inputs to our models—in ways that do not match existing cases. In essence, we had to be able to "extrapolate" from existing cases. In creating a general model, we lose some of the fidelity possible from more granular analysis specific to each unit and location. But in exchange for some loss of fidelity, we gain flexibility and explanatory power, while preserving sufficient accuracy to compare the relative costs of different options to determine whether they offer sufficient merit from a cost-benefit standpoint to analyze in further depth. In short, the cost models are designed to provide a decision-support tool for the first phase of a posture options analysis—what options to develop in more depth with sufficient detail to more precisely determine cost and operational impacts—not to provide the final specific costs to inform a specific decision or develop budget estimates for posture change implementation plans.

Second, we sought to provide decisionmakers with tools to quickly develop and assess a range of options. To readily develop a range of possibilities, it is valuable for decisionmakers to understand the major factors that drive costs. By understanding the relative impact of different factors on cost, decisionmakers can better determine what options are likely to reduce costs, if that is an aim in the development of posture options, and what options are likely to increase costs the most if they are seeking to develop posture options to improve or add new capabilities in a region. To do this, our cost models must parse the important features of overseas posture, but still allow rapid computations and be easily understood.

Third, we developed cost models to serve analytical purposes. We believe our approach accurately enables assessment of the relative cost differences for DoD between various posture options based on how total costs actually change, which cannot be done simply using raw budget data for bases or forces. The simplest case of reducing overseas posture is removing the forces from a given base and closing the base altogether. In that case, one would not actually save the full scope of annual operating costs at the base. A portion of the costs are fixed costs and would be saved, but another portion includes variable costs that would still accrue somewhere in DoD but possibly at a different level. Whether DoD's annual recurring variable costs associated with the personnel at the closing base would change depends on whether there is a difference between the variable costs at the closing base and those at the base(s) receiving the forces. Thus, one must also understand the cost structure of the receiving base and how the fixed and variable costs at a base are split. In short, how the costs in a region change when a base is closed is not what is of interest. Rather, what is of interest is how total DoD costs change, involving additions and subtractions for every move. Currently, the DoD data do not readily decompose into these fixed versus variable costs; thus a modeling approach to distinguish the two is crucial.

To scale total costs to personnel, one must develop some kind of per-person or per-unit costs to apply to the force types relocating to the United States. Simply dividing the total support costs at an installation by the total personnel without taking into account the fixed and variable cost dynamic would overestimate per-person variable costs and understate fixed costs dramatically. Just assessing the gross costs of maintaining the current U.S. overseas posture base by base would have been insufficient, because this would not enable the evaluation of changes at a base except for full closure and would not enable evaluation of changes to net DoD costs instead of changes at a regional or base level. This necessitated the development of generalized cost models.

Approach to Individual Cost Models

Recurring Costs of Permanent Posture

To conduct this analysis, we analyzed the costs incurred by forces and installations to produce cost models based on the ways those activities actually incur costs. To do so, we grouped costs into four categories: training, personnel-related, installation-related, and regional logistics:

- Training activity costs are simply the operations and sustainment costs that enable operational units' day-to-day operations, excluding the military pay for unit personnel; they vary in our modeling based on where units are and by unit type.

- Personnel-related costs are allowances and permanent-change-of-station (PCS) moving costs incurred by DoD.[3] In our modeling, we estimate how these costs vary based on location and service.
- Installation-related costs include facility operations, maintenance, R&M; family services; morale, welfare, and recreation; medical services; and dependents' education. We also include the construction costs to periodically recapitalize infrastructure, but, because of the cost dynamics, we take a slightly different approach than for other installation-related costs. In our modeling, installation-related costs vary based on whether installations open or close and how many people are at an installation.
- Regional logistics costs are additional logistics costs borne by DoD to support permanent overseas forces and bases. These costs include additional transportation costs for overseas delivery of supplies and overseas regional distribution centers, which form an additional layer of logistics. In our modeling, these costs vary with the number of people in a region.

Training costs are driven by and assignable to operational units (e.g., a fighter wing or an infantry brigade) themselves. Data provided by the services indicate that training costs sometimes vary for like units located in different regions. Some differences may have to do with differences in the types and frequencies of activities they conduct (such as partner-nation exercises and training), restrictions on those activities, or actual cost differentials for the same activities. The annual training cost difference between the United States and region r for a single unit type j can be expressed as $c_{jr} - c_{jUS}$, where c_{jr} is the annual training cost in region r for unit type j, and r = *{Pacific, Europe}*. We derive these different costs from service data and models. We apply the difference in the annual per-unit cost to each instance of a unit relocating from an overseas region to the United States in our cost estimates for the illustrative postures.

Personnel-related costs are specific to each person based on where they are located, assignment characteristics, and family status. Allowances are monies provided for specific needs adjusted for costs specific to a location and the servicemember's situation (i.e., living in barracks, on-base housing, or off-base housing, as well as whether accompanied by dependents) and include housing, the overall cost of living, and compensation for living apart from their families.[4] Additional PCS costs for stationing service members overseas are incurred because it is more expensive to transport people and their household goods to and from overseas locations than within the United States.

[3] We exclude basic pay from our analysis, because, basic pay is, by definition, the same across all regions.

[4] One category we exclude is Special and Incentive pays. These pays provide the services with flexible additional pays that can be used to address specific manning needs and other force management issues that cannot be efficiently addressed through basic pay increases. These include, but are not limited to reenlistment bonuses, danger and hardship pay, and a variety of incentives for medical practitioners. We exclude these from our analysis of permanently stationed forces but apply some as necessary to our cost assessment of rotational deployments.

The rate of PCS moves per person per year may also be higher overseas, depending on assignment policies. The cost impact from shifting personnel to or from the United States is $\sum(c_{jr} - c_{jUS})x_{jr}$, where $c_{jr} - c_{jUS}$ is the differential cost arising from the difference between overseas and U.S. cost per person for a person in service j in region r, and x_{jr} is the number of personnel of service j shifted from or to region r.

For installation-related annual recurring costs, there are fixed-cost components and variable-cost components. Within installation-related costs, we include installation support, construction to recapitalize existing infrastructure, and DoD Education Agency (DODEA) support. For each of these categories of cost in a given region, the basic model for cost is estimated using a regression model and has the form $c_{isr} = a_{sr} + b_{sr}p_{isr}$. Here, c_{isr} is the total installation-support cost of installation i of service s in region r, a_{sr} is the fixed-cost component of an installation of service s in region r, and $b_{sr}p_{isr}$ is the product of the variable cost per person for service s in region r and the number of personnel at installation i of service s in region r. The personnel we include as an independent variable (p_{isr}) are "cost drivers" (the people on an installation who, by virtue of being assigned there, create demand for installation support), and the personnel we exclude are part of the installation-support cost dependent variable, c_{isr}. In applying this model, we explored different variants that employ additional independent variables. For instance, in the category of installation-support cost, we find that the fixed cost at an installation sometimes depends on whether operational forces are present at the installation.

Using this approach for installation-related costs, the annual cost savings (if any) from relocating forces to the United States depends on whether the sending installation remains open or closes and whether there is reason to believe the annual fixed cost at the receiving installation remains the same. In the most common case in our analysis, the sending installation closes and the fixed cost at the receiving installation is estimated to not change. In this case, the annual fixed cost at the sending installation would be saved by DoD. In addition, there might be a savings or increase in variable cost. However, for installation-support costs, we often find little or no statistically significant difference between overseas and U.S. estimates of the variable-cost coefficient, b_{sr}, so the potential savings in the installation-support portion of installation-related variable costs, $(b_{sr} - b_{US})p_{isr}$, is often zero or relatively low. In other possible cases, the sending installation remains open and its fixed costs continue to be incurred—hence no savings in fixed cost.

Finally, regional logistics costs are those borne by DoD to support permanent overseas forces and bases within a specific region. These costs include the higher cost of transportation for the delivery of supplies overseas relative to within the United States and regional distribution centers. Most supplies are brought into the DoD supply system through U.S. distribution centers or delivered directly from U.S. suppliers. It simply costs more to ship items overseas than around the United States, and sometimes it is necessary or costs less to stock them overseas, necessitating regional distribution centers.[5] We determine the incremental logistics costs for the delivery of supplies overseas on a regional basis and then convert this to cost per person.

We aggregate these various cost assessments into a single, recurring-cost model that estimates the cost impacts of changes in postures. When personnel are relocated from overseas to the United States as the result of a unit restationing, we apply the relative training cost differences, if any, for their unit type. We also apply the per-person costs aggregated from personnel-related costs, the variable-cost component of installation-related costs support, and per-person regional logistics costs. When a base is closed, we also apply the fixed-cost component of the installation-support cost to determine the change in costs.

Before we step into the details of our recurring cost analysis, we briefly discuss some of the statistical concepts relevant to our analysis. First, sample size was often a constraining factor for the portions of our analysis where we used regression analysis, thereby limiting the ability to assess too many different factors and in some cases limiting the ability to produce significant modeling results. Generally, the sample sizes for U.S. bases (several dozen data points for each service) were adequate to develop parameter estimates with reasonable standard errors for statistical significance and that fit the data well to provide strong explanatory power. The case was rarely the same for overseas bases. For each service, there were usually 5–10 data points per region, or even fewer—too few from which to perform separate regression analyses, and usually too few from which to derive consistent (and reasonable) parameter estimates for each region separately. Sometimes we overcame this by combining regions to develop an aggregate overseas estimate, sometimes we excluded one or more outliers in a set typically associated with unique installation characteristics, and sometimes the data points were simply too few, and we sought to estimate costs for individual data points (i.e., bases) instead of using generalized modeling results.

Finally, we used a linear specification for the relationship of personnel to support costs, rather than an alternative, such as logarithmic or quadratic. There were several reasons for this. The primary reason is that military planning factors for personnel and facility planning explicitly include linear formulations for requirements determination. Personnel planning factors that define the number of personnel authorized for given activities, such as food service, medical support, and facility maintenance, often use linear relationships. These are formulated in the same way as the regression equation we discussed above: $c_{isr} = a_{sr} + b_{sr}p_{isr}$, where the fixed component, a_{sr}, is the management

[5] Regional distribution centers enable the delivery of supplies to a region using inexpensive sealift. Then the supplies can be delivered rapidly and on demand within the region. Without these distribution centers, such orders would have to be filled using airlift from the United States to the overseas region, which can be quite costly. Thus, even though the regional distribution centers add a distinct cost to overseas support, without them, the overseas transportation bill would be even higher.

overhead element, and the variable component is driven by the base population.[6] Likewise, DoD facilities planning factors include linear requirements, such as an amount of space (e.g., square feet) required per person. Second, we found that most data plots suggest linear relationships. These data plots can be seen in Appendix A. Finally, we tested alternative specifications for several cases. In cases where the plots did not appear as convincingly to be linear, we tested alternative specifications. We found these alternatives to provide no better, and in almost all cases worse, fit to the data and to produce counterintuitive (and in some cases absurd) cost equations.

Costs of Rotational Presence

To develop a cost model for estimating the costs of rotational presence, we examined the costs associated with the deployment of personnel and equipment, prepositioning equipment, supporting personnel, and special compensation for deployment. We built the model to estimate how changes in presence levels and policy options would affect costs. Presence level is a function of rotation frequency and length, which both affect costs in the model. Policy options include whether or not equipment is prepositioned, and, if not, whether airlift or sealift is used to transport unit equipment. The final cost factor is whether or not the United States has or maintains a base in the region to which the units rotate, whether the host nation provides base support in other cases, and whether base operating support would have to be established and provided for the rotating unit.

Investment Costs for Restationing Forces

To determine one-time investment costs for moving forces to a new location or closing bases, we developed an investment cost model incorporating personnel-related movement and separation; base-related closure costs, including support contract termination, mothball/shutdown, severance to foreign civilians, labor contract settlement costs, and a range of one-time unique costs; and construction of new facilities to host the units. Because of the uncertainties inherent in potential environmental cleanup costs and in the potential gains from the residual value of facilities returned to host nations, we excluded these costs from our assessment of investment costs. We used standard DoD rates for the personnel-related costs, and we mined past reports to develop estimates of base closure costs. We used regression modeling similar to that used for the recurring-cost analysis to determine new facility requirements. In this modeling, we determined relationships between the number of people at a base and the value of the facilities there, again decomposing the facility value into fixed and variable costs, with fixed costs, for example, representing facilities for which only one on a base might be needed (e.g., commissary).

---

[6] Directorate of Manpower, Air Force Personnel Center, *USAF Manpower Standards*, not available to the general public.

Limitations of This Analysis

This analysis is intended to produce estimates of the impact of posture changes. It is sufficiently accurate for cost-benefit analysis for decisionmakers to determine what courses of action should be considered in more depth. If changes were to then be considered as possibilities to pursue, more detailed, budget-quality cost estimates that account for all location-, unit-, and option-specific factors would need to be done.

Preview of Cost Analysis Findings

In the course of our analysis, several costs arose as the primary drivers of incremental differences between U.S. and overseas forces and bases. We found the largest single driver of the incremental cost difference in variable per-person costs between the U.S. and overseas locations to be personnel-related costs, specifically overseas allowances and PCS moving costs. These costs comprise 60–80 percent of the incremental cost difference in per-person costs (which ranged from around $10,000 to $40,000 per person, depending on the service and region). Dependents' education was next in influence, followed by additional overseas logistics costs. These categories comprised most of the per-person cost difference.

Unit training costs (which are captured as per-unit, not per-person costs) sometimes drive incremental cost differences, but they do not feature as largely as per-person costs in the impact of posture changes on costs.

Often, but not always, we found differences in the fixed-cost component of recurring costs between U.S. and overseas locations. USAF bases appear to have systematically higher fixed costs overseas when compared with similar U.S. bases. Two drivers of these higher costs appear to be that these overseas bases have systematically more facilities than bases in the United States relative to their populations and that they have higher numbers of military personnel providing installation support. We discuss these differences in detail in Appendix A. For all other services, the message is mixed. Many overseas bases have fixed costs that appear to be in line or just a little higher than their U.S. counterparts.

The more important point with regard to fixed costs is that our modeling does suggest fixed and variable cost components to installation-support costs. Thus, if a base is closed and the personnel are moved to an existing base, either using new facilities built to accommodate them (with an attendant investment cost) or using any existing underutilized facilities, then one set of fixed costs is eliminated, producing savings. This is a benefit of consolidation, whether it was to occur overseas or in the United States, although there is more flexibility to expand facilities and forces on U.S. bases than overseas, since agreements may need to be worked out with the host nation. Additionally, consolidation from overseas back to the United States produces more DoD savings because of the higher overseas variable costs, and, in some cases, higher overseas fixed costs. On the other hand, when overseas bases are closed, ownership reverts to the host nation. In the United States, closed bases can be repurposed for other uses, providing economic benefit within the United States that accrues outside of DoD. These additional potential benefits are not accounted for in this study.

In addition to recurring costs for permanently stationed forces, we also assessed the costs associated with rotational deployments. A key finding is that rotational costs are very dependent on the frequency and duration of deployments. For deployments to the same region for the same unit type, higher deployment frequency combined with lower deployment duration greatly increases costs relative to lower-frequency, longer deployments to achieve the same degree of presence. For longer, less frequent deployments (e.g., one unit replacing another every 12 months), personnel-related costs, such as food, housing, and special allowances, tend to dominate the costs, driving roughly two-thirds to three-fourths of the costs and potentially enabling some cost savings over permanent presence. For shorter, more frequent deployments, the deployment costs themselves—transporting personnel, equipment in some cases, and aircraft for the USAF—tend to dominate. In these cases, equipment prepositioning, when sufficient equipment is already available in the force, can mitigate some of those costs. For ground forces, when the equipment is available, prepositioning offers some savings over sealift, which also has a transit time disadvantage, with airlift extremely expensive for ground-unit rotations with equipment due to the size and weight of units.[7] For the USAF, prepositioning can save over airlift, but because air forces have relatively little equipment to transport, the difference is not too significant.

As a result, rotational presence is not necessarily less expensive than permanent presence. Generally, if forces are realigned to U.S. bases but their original overseas base stays open, the costs of even small, infrequent rotational deployments can exceed the savings. If the sending base is closed (greatly increasing the savings), substituting full rotational presence (12 months out of the year) for permanent presence sometimes saves money, sometimes costs money, and sometimes roughly breaks even, depending on the service, unit type, region, frequency and length of rotations, and equipment policy options (transporting versus prepositioning unit equipment). In particular, achieving extensive presence through high-frequency, short rotations would greatly increase costs, leaving longer rotations as the only option that enables some savings or avoidance of increased costs while maintaining high presence. If only partial, rotational presence is substituted for permanent presence, then, depending upon the rotational design, savings can be more substantial. In other words, we found no single, definitive comparison for permanent versus rotational presence. Each case must be examined individually.

Furthermore, if a base were to be closed and its forces realigned, another permanent base in that country or region must be maintained to support the rotating forces

[7] Army prepositioned equipment in CENTCOM is currently stored outdoors, significantly increasing the periodic maintenance requirements, and thus costs. This makes prepositioning more expensive than all other equipment options, but prepositioning may be favored for reasons other than cost.

(including considerations of facilities, space, and other accommodations) or a host nation must agree to provide access to one of its bases. This is a potential constraint on some posture options.

Finally, we estimate the potential investment costs necessary to shift from the current posture to an alternative posture. The largest single driver of these costs is by far the new construction necessary to accommodate forces relocating to the United States. Some available capacity may exist at U.S. bases (whether today or after planned force reductions), thus obviating the need for some portion of that new construction. In particular, this is most likely to apply to the Army, due to significant planned force reductions.[8] So, for any illustrative options that realign Army forces from overseas to the United States, we estimate an upper and lower bound for these costs. In our upper bound estimate, we assume all relocating forces will require such construction. This is the standard assumption for bases of other services. In this standard/upper-bound estimate, new construction comprises about 85 percent of the total cost estimate for posture changes. Thus, the lower-bound cases that would reflect the existence of available capacity in the United States would be less, given that the need for new construction would drop. There is more uncertainty around our estimate of investment costs than for other elements of our cost analysis. So our investment estimates should be considered as rough guides for the costs DoD might incur to shift its overseas posture.

Recurring Permanent Presence Cost Analysis

This section describes our methodology for determining the annual operating and support cost difference between bases and permanently stationed forces in the United States and those overseas, as well as for decomposing the fixed and variable cost portions of operating and support costs to estimate the potential costs or savings that might accrue from changes in the overseas permanent posture.[9]

Training Costs

The services sometimes operate their forces differently overseas than when they are stationed in the United States. This may be as a result of partner-nation training and exercises, constraints on U.S. activities, or a desire to generate more presence or combat

---

[8] While the Marine Corps is downsizing somewhat as well, information provided by the Marine Corps suggests that this will not produce available capacity that could be used by forces were they to realign from overseas, necessitating facilities construction in all cases. U.S. Marine Corps, Budget and Execution Division, Programs and Resources Department, correspondence to the author, December 14, 2012.

[9] This analysis built on previous RAND research for the U.S. Air Force documented in Patrick Mills, Adam Grissom, Jennifer Kavanagh, Leila Mahnad, and Steven Worman, *The Costs of Commitment: Cost Analysis of Overseas Basing*, Santa Monica, Calif.: RAND Corporation, RR-150-AF, forthcoming.

capability. To capture these differences, we directly use service-provided data and models to compute training costs for this portion of the cost analysis.

For the USAF, we used data from the Air Force Total Ownership Cost (AFTOC) database to determine unit training costs by aircraft type and command.[10] We estimated the total training cost per aircraft per year using data from FY 2009 to FY 2011. The results are shown in Table 8.1, which shows the average training cost per aircraft per year for the primary aircraft currently stationed at overseas installations.[11] For each aircraft type, the left two columns show the unit operating costs at Air Combat Command and Air Mobility Command (AMC), both located in the United States. The right two columns show the costs for the same aircraft located in either the Pacific Air Forces (PACAF) or USAFE. We use these costs as inputs when calculating the costs and savings of changes to the overseas posture based on the number of aircraft moving from one command or region to another.

Some costs are comparable across regions (within 5 or 10 percent), and some are significantly higher or lower. Recent RAND research has analyzed these costs in detail for USAFE and sheds some light on these differences. In particular, costs for depot-level reparables and munitions for USAFE units were generally found to be lower for

Table 8.1
Air Force Aircraft Annual Training Costs, by Type and Command

	Total Flying Costs per Aircraft ($ millions)			
Aircraft	**ACC**	**AMC**	**PACAF**	**USAFE**
A-10	8.2	—	6.8	6.5
C-130	—	9.2	13.7	9.7
F-15	11.0	—	8.5	11.4
F-16	6.0	—	6.2	5.7
KC-135	—	11.7	8.7	11.4

SOURCE: Author's analysis of AFTOC data.

NOTE: ACC = Air Combat Command

10 U.S. Air Force, AFTOC database. Provided to RAND on June 11, 2012, by the Deputy Assistant Secretary of the Air Force for Cost and Economics.

11 These calculations assume that overseas aircraft repositioned to the U.S. would eventually be operated and based as their Air Combat Command and AMC counterparts are today. That assumption seems reasonable for basing, as smaller squadrons of aircraft would likely be absorbed by larger wings and would gain their natural economies of scale. It also is reasonable to think that the aircraft would be operated the same. Further, any restrictions present overseas would no longer apply to the repositioned forces, so their training regimens, if currently different, would presumably fall in line with those more typical of Air Combat Command and AMC.

fighter/attack aircraft than for their U.S. counterparts. These are believed to be lower due to flying restrictions in Europe that reduce consumption rates for those assets.[12]

For the Army, we used the Force and Organization Cost Estimating System (FORCES) Cost Model (FCM) to determine the training costs for Army units in different regions.[13] FCM uses past obligations data to produce its estimates. Table 8.2 shows select unit types and their average annual unit training costs as derived from FCM. While FCM contains estimates for a broad range of cost elements, we included only three in our computation of training costs: Direct Equipment Parts and Fuel Cost, Post Production Software Support, and Indirect Support Cost. Appendix A contains a complete list of the cost elements in FCM and how we utilize them in this report.

Most Army units have noticeably higher training costs in EUCOM and slightly lower costs in PACOM. About half of the higher costs in EUCOM stem from contractual services, and the other half of the difference is from other Indirect Support costs.

For the Navy, rotational operations are built into its structure and day-to-day operations. The Navy has a sea/shore policy that governs the frequency and length of deployments, and the ships rotate on regular schedules. Its force structure is designed to provide a certain level of forces deployed on rotations at all times. While few Navy units are actually home-ported overseas, most spend significant time overseas on

Table 8.2
Army Unit Training Costs, by Unit Type and Region

	Annual Training Cost ($ millions)		
Unit type	**United States**	**EUCOM**	**PACOM**
Armored BCT	37.2	58.5	34.3
Brigade HQ	3.8	4.8	3.7
Combat aviation brigade	97.0	115.2	94.8
Infantry BCT	40.2	63.1	37.0
Stryker BCT	54.5	80.8	51.7
Theater sustainment command HQ	2.3	5.1	2.0

SOURCE: FORCES Cost Model (FCM).

[12] Moroney et al., 2012.

[13] The FORCES model is owned and managed by the Office of the Assistant Secretary for the Army for Financial Management and Comptroller. According to the Army Financial Management website, "The FCM provides realistic, current and supportable force cost estimates of Active and Reserve Component Table of Organization and Equipment units for Acquisition, Activation, Operations, Movement, Modification and Inactivation. The model is sensitive to Operating Tempo (OPTEMPO), authorized level of organization, geographic location, year and component." See Assistant Secretary of the Army for Financial Management and Comptroller, "Financial Management," 2010. FORCES model provided to RAND in June 2012.

deployments from U.S.-based homeports. Thus, this cost category is not as relevant to most Navy operations for the purpose of determining the relative costs of U.S. and overseas basing. Shore-based Navy aviation could be subject to similar regional cost differences as their USAF counterparts, but our illustrative postures did not consider the realignment of naval air assets, so we did not include such cost factors.

Similar unit training cost factors were not available to the RAND study team for the Marine Corps. We therefore assume no difference in training costs of Marine Corps units between the U.S. and overseas locations.

Allowances and PCS Costs

Several categories of allowances for military personnel are higher for overseas stations, including the following. We used the Work Experience and monthly Active Duty Pay files to estimate the net cost differences in allowances between CONUS and OCONUS locations.[14] This database includes the pay and allowances for all individual service members by month. It includes their unit identifier, which we could use to map their locations. Our use of unit identifier data is described in more depth in the installation-related costs section. For this analysis, we used FY 2011 data to reflect relatively recent rates for these allowances. These files include the following categories of allowances:

- basic allowance for housing (BAH)/Overseas Housing Allowance (OHA)
- cost of living allowance (COLA), U.S. and overseas
- family separation allowance.

We first determined which personnel were in each region, separating Japan and South Korea because of substantial differences in allowances between the two countries, especially depending on the service. We also separately determined rates for Marine Corps personnel stationed at Camps Pendleton and Lejuene, given the concentration of Marine Corps stationing at these two locations and the significant allowance differences between these two locations. Finally, we include Hawaii and Guam for reference because of the substantial U.S. forces in the former and current plans to increase forces at Guam and their relatively high rates. We aggregated allowances for all personnel stationed in the selected areas to develop region-, country-, state-, territory-, and base-specific estimates on a per-person basis. To determine per-person rates for a location and service combination, we included only personnel that were in the location for the full year. We excluded records where personnel were in more than one region during the year, as they received allowances for more than one region. The results are shown in Table 8.3. The top portion of this table shows the annual per-person cost for

[14] Active Duty Pay files are also referred to as "JUMPS". Both files are provided by Defense Manpower Data Center (DMDC).

Table 8.3
U.S. and Overseas Personnel Allowances

	Annual Cost of Allowances ($)			
Country/Region	**Army**	**Air Force**	**Marine Corps**	**Navy**
CONUS	17,900	17,800	15,400	21,000
Pendleton			19,000	
Lejeune			12,400	
Europe	20,600	32,500		35,100
Japan	29,400	29,500	20,400	29,200
South Korea	26,000	22,500		27,100
Hawaii	34,500	39,600	24,200	34,700
Guam		27,800		27,700

	Annual Cost of Allowances Relative to CONUS ($)			
Country/Region	**Army**	**Air Force**	**Marine Corps[a]**	**Navy**
Europe	2,700	14,700		14,100
Japan	11,500	11,700	5,000	8,200
vs. Pendleton			1,400	
vs. Lejeune			8,000	
South Korea	8,100	4,700		6,100
Hawaii	16,600	21,800	8,800	13,700
Guam		10,000		6,700

SOURCE: Author's analysis of FY 2011 Work Experience and Active Duty Pay files.

[a] Japan comparison for the Marine Corps is for CONUS; subsequent rows show comparison to specific Marine Corps locations.

the above allowances. The bottom portion of the table shows the difference between each region and CONUS.

Generally, most of the cost differences are driven by higher overseas COLA, with OHA sometimes making a significant difference as well. One can notice the especially small difference for the Army in Europe. The BAH paid to Army personnel is actually much smaller in Europe than the CONUS average, which could be due to more people living on base. For the marines in Japan, most personnel receive barracks COLA, which is much less than standard COLA, and very few receive OHA, limiting the difference for marines in Japan versus CONUS. That difference is even smaller when comparing Japan with Camp Pendleton, which is an area with relatively high BAH.[15]

Additionally, overseas PCS costs are higher than for PCS moves within CONUS. PCS move costs are incurred any time a service member must move, whether for normal duties; schooling; overseas assignments; or, in some cases, separation. Moves between CONUS and OCONUS (referred to as transoceanic travel) cost more than those within CONUS, so OCONUS positions incur that additional cost.[16] These occur as a single movement of one service member (and his or her family and household belongings) from one station to another. Additionally, OCONUS assignments tend to be shorter, increasing the number of moves per assigned person, magnifying the effect of the higher costs per PCS move.

To derive PCS move costs, we looked to each service's budget estimate, which list the total expenditures for PCS move, called PCS travel, which is divided into several categories.[17] Operational travel includes moves to and from permanent-duty stations located within the United States or within another country or region involving no transoceanic travel. Training travel includes moves to and from permanent-duty stations located within CONUS to service or civilian schools. Rotational travel includes moves from permanent-duty stations in CONUS to permanent-duty stations overseas, and vice versa, as well movements among overseas regions involving transoceanic travel. For accessions and separation travel, which include travel to and from first and last duty station, respectively, regardless of whether in CONUS or OCONUS, we allocated portions of each to CONUS and OCONUS categories based on the relative proportions of assigned personnel to each for the respective service. We used FY 2011 costs, because they reflect actual spending, not estimates of spending, and were the most recent such actuals available at the time of this study.

---

15 The Marine Corps also separately provided a detailed analysis of allowances for personnel stationed in Japan and Camp Pendleton. It shows a difference between Japan overall and Camp Pendleton of $656 per marine and $1,292 per marine when comparing just those stationed on Okinawa versus Camp Pendleton (the cost per marine at Iwakuni is quite low). Given that these numbers are relatively close to our results, we elected to use the Work Experience/Active Duty pay files–derived rate in our modeling for consistency across services, The effect on the overall results is very small, particularly as we are considering only Okinawa versus Camp Pendleton in our illustrative posture changes, using $1,433 per marine versus the Marine Corps–provided $1,292. The Marine Corps analysis was based on data from the Marine Corps Total Force System data and data from the Marine Corps' Operational Data Store Enterprise. U.S. Marine Corps Budget and Execution Division, 2012.

16 These costs are paid from each command HQ's operating budget, not the operational unit where the personnel actually work, so these costs are not reflected in the accounting of unit costs.

17 Department of the Air Force, *Fiscal Year (FY) 2013 Budget Estimate: Military Personnel Appropriation,* February 2012; Department of the Army, *Fiscal Year (FY) 2013 Budget Estimate: Military Personnel, Army, Justification Book,* February 2012; Department of the Navy, *Fiscal Year (FY) 2013 Budget Estimates: Justification of Estimates, Military Personnel, Navy,* February 2012; Department of the Navy, *Fiscal Year (FY) 2013 Budget Estimates: Justification of Estimates, Military Personnel, Marine Corps,* February 2012.

fixed costs. The fixed cost also represents the opportunity for savings from closure of a base and consolidation of forces at other bases.

Next, even though our regression analysis has estimated the fixed and variable parameters with a standard, accepted level of statistical confidence, it does not perfectly explain the costs of every data point. Our statistical outputs tell us that this regression model (the combination of fixed and variable parameter estimates) explains about 79 percent of the variation in installation-support costs based on the number of military personnel on a base. The dots in the oval to the right show that even two bases with virtually identical numbers of military personnel can have significantly different installation-support costs. There is some natural variation inherent in systems like this, but there could also be knowable specific local attributes that could explain some of that variation. However, in this study we have not attempted to determine all of the specific local factors that would cause all of the variation, and it may not even be possible to do so. Regardless, the model is a generalization to quickly estimate the effects of changes and will rarely predict any single data point perfectly, but should predict costs well enough to determine whether an option is worth further consideration. If an option appears promising to pursue given all considerations—estimated costs, risks, and benefits—then one would need to determine the exact costs at the specific bases under consideration to finalize the decision. But the model should produce reasonable estimates for comparing potential courses of action to consider.

Next, we highlight the EUCOM bases, the red triangles in Figure 8.2. It appears that four of the five EUCOM bases fall more or less in the range of the U.S. bases. At the same time, there is an extreme outlier, which is circled. This is Grafenwoehr, Germany. Again, there are good reasons, similar to those for Ft. Benning, why this is the case, given the significant training facilities located at Grafenwoehr, which incur costs to train transient populations. There are too few data points to perform a separate regression analysis on EUCOM bases alone, and any generalization should be approached with caution. It does appear that the four EUCOM bases in the range of U.S. bases in fact have similar cost relationships, and that Grafenwoehr is an outlier with installation-support costs not explained by military personnel alone. We can say that from these data, there is not enough evidence to show that, *in general*, EUCOM bases have higher or lower fixed or variable costs for installation support than U.S. bases.

Finally, we have plotted the regression lines that would result from analyzing these U.S. and EUCOM bases separately and forcing the y-intercept through the origin. These results are shown by the thin lines. The purpose of this is to demonstrate the result of taking the simple average costs per person of these two data sets, instead of trying to estimate fixed versus variable costs. To simply divide the total support costs by the total military personnel in these two data sets would result in per person costs of about $9,000 per person for U.S. bases and $42,000 per person for EUCOM bases. (Here, Grafenwoehr does not significantly change the regression line.)

Aside from problems with statistical fit or significance, this approach creates two problems. First, it misrepresents the cost dynamics at Army bases. We know that there are fixed costs (i.e., overhead) built in to some military planning factors—and virtually all facilities and organizations—and the data support this. To use a simple average is to overstate per-person costs, potentially overstating savings from realigning personnel, and would not identify those costs that would likely only be saved if the base were closed.[18]

The second shortcoming is that this approach (that of taking a simple average cost per person) actually makes overseas locations appear relatively more expensive than U.S. bases when they may not actually be so. This is because for most of the services, overseas bases have smaller base populations than those in the United States. (This can be seen in the data plots in Appendix A.) If we include the four clustered EUCOM data points with the U.S. data points in Figure 8.2, it does not significantly change the regression results compared with the U.S. base–only regression. However, because EUCOM bases are smaller and therefore closer to the origin than U.S. bases, the slope of the regression line if using only these four bases and if forced through the origin is much steeper, implying higher per-person costs. It may be that EUCOM bases do have higher per-person installation-support costs, and that the difference between EUCOM and U.S. bases would be saved were units relocated to the United States from EUCOM. But from the data shown in Figure 8.2, the case is not convincing. In short, the EUCOM bases are smaller, have a higher proportion of their installation-support costs in fixed costs than U.S. bases with likely similar variable costs per-person, and have relatively similar fixed costs per base. This can be thought of as akin to spreading the overhead across a larger population.

This does not mean that there are no cost differences between U.S. and overseas bases. Installation-support costs are only one of the components we analyze, and we do find significant incremental overseas costs in other areas.

For all the reasons above, when possible, we develop regression-based models to explain the relative costs of overseas posture. Given the discussion above, one can also see that the case is not always clear-cut. As we proceed with our analysis, we do so cautiously, explaining our assumptions clearly, so the reader can understand when we utilized our own analytic judgment in ways that influence the results. In addition, we

---

[18] It is important to note here that the Base Realignment and Closure (BRAC) process does attribute cost savings to closing bases. However, after analyzing the model used for BRAC (Cost of Base Realignment Actions [COBRA]), and the data provided in the BRAC commission reports and in data outputs, it appears that most, if not all, of the costs savings accrued from closed bases was due to the cutting of military and civilians from the force due to *consolidations*. Often multiple units or organizations with similar functions were consolidated, gaining economies of scale and shedding personnel. Our analysis seeks to be neutral to force structure and assess the savings from closing bases due to their fixed-cost components. However, savings associated with reductions in fixed costs would be associated with reductions in base support personnel, whether civilian or military, and contract services.

bers are average values for FYs 2009–2011. We used multiple years of data to smooth out the impacts of events unique to a given year for a base. For these data, we excluded most U.S. territories, and considered only Guam and bases on foreign soil for inclusion in PACOM. The purpose of this was to create data sets from which we could estimate regional cost differences, and traditional definitions of some COCOM AORs include a mix of U.S. soil, U.S. territories, and foreign soil. We differentiated among these to keep cost categories clear.[20]

Compile Personnel Data

Our second step was to analyze personnel data. We obtained personnel data from DMDC.[21] The 2012 posture tables in Chapter One summarize these personnel data, with totals for each major installation (permanently assigned personnel only). In addition to raw personnel numbers, an important element in these data is the unit, specified by the unit identification code (UIC). A UIC is a unique identifier assigned to a unit or organization in DoD. Table 8.7 summarizes the personnel data we obtained. It shows the total number of locations and UICs worldwide in the DMDC data, by service, for locations with greater than 100 active duty personnel. These UICs are important to our analysis because of our approach to analyzing the personnel and cost data. Forces stationed at an installation demand base services, and this drives support costs. However, not all personnel on base demand the same level of support, and some personnel are providing those base services. We therefore introduce a classification system to categorize military activities on an installation to inform our analysis.

For the purposes of our analysis, we classified all military units (by UIC) in one of three broad categories based on whether (and how) they drive installation support or provide it. These categories are mutually exclusive and completely exhaustive.

Table 8.7
UIC Data Summary

	Army	Air Force	Marine Corps	Navy
Number of locations with more than 100 active duty personnel[a]	124	123	68	135
Number of service-unique UICs	29,042	7,613	1,642	10,224[b]

SOURCE: Author's analysis of DMDC data.

[a] Including locations operated by another service.

[b] Total includes some Marine Corps units that have Navy UICs.

---

[20] Table 8.6 includes bases with more than 100 military personnel and more than $1 million average support costs. This table excludes bases outside of the three regions listed. There were few of these bases, and we did not utilize them for our cost modeling.

[21] DMDC provided a report using data as of May 31, 2012. DMDC data are updated every month.

Referring back to Figure 8.1, we classified two types of activities as drivers of installation-support costs, i.e., as independent variables: operational and institutional forces. Within operational forces, we include both deployable forces—e.g., fighter wings, BCTs, battalion landing teams, and carrier groups—and nondeployable operational forces such as space, missile, and cyber forces that have ongoing operational missions but employ them from home-station. One way to encapsulate operational forces is to ask whether these units or activities are directly employed by COCOMs for operations, whether from the U.S. or foreign soil.

The second subclass of installation-support drivers we call institutional. Institutional activities are those that sustain the organization as a whole (including the needs of the operational forces—other than installation support) but do not themselves directly execute operational missions assigned to COCOMs. Examples of institutional forces are support activities such as recruitment, training, and supply chain management. These activities are generally found in commands managing materiel (e.g., Army and USAF material commands), executing training, and conducting R&D, along with entities such as service headquarters.

These operational and institutional activities must be performed somewhere, and they incur costs wherever they are located. Thus, these two activities drive installation-support costs (often called base operating support), our dependent variable. Installation support includes all the facilities, equipment, and personnel needed to run the installation where those operational and institutional units and activities are located: to feed, protect, provide medical support to, and otherwise service the population, whatever task that population performs on a day-to-day basis. Their purpose is essentially directed at supporting other units and activities *on the installation*. In general, the location of institutional forces is not determined by the size and shape of operational forces *on that base*. So, at base level, operational and institutional forces are independently located and do not, on the whole, affect one another on a given base. We treat operational and institutional activities separately, even though they are both drivers, or demanders, of base support, because they place different types of demands on installation support with different cost implications.[22] We applied this framework to the personnel data to derive cost relationships for base support. We assigned all UICs to one of those three categories.

We now walk through a couple of examples of how we classified UICs at different bases (see Table 8.8). The first set is from Ramstein Air Base, in Germany. The UICs are all for Air Force units. The first UIC is for the 37th Airlift Squadron, which operates C-130Js to conduct airlift, airdrop, and aeromedical evacuation operations. The second unit shown is an element of the USAFE headquarters, which we classify as institutional. The third unit there, the 86th Mission Support Group, provides installa-

22 We could further divide these two activities into sub-types (e.g., different brigade types), but in many cases the sample sizes are not big enough for developing models.

Table 8.8
Example UIC Classification

Base	Service	UIC Code	UIC Name	Category
Ramstein Air Base				
	Air Force	FFLZB	0037 Airlift Squadron	Operational
	Air Force	FF5BH	USAFE Headquarters	Institutional
	Air Force	FFBRT	0086 Mission Support Group	Support
Fort Bragg				
	Army	WABJAA	1st Battalion, 319th Field Artillery Regiment	Operational
	Army	W3YBAA	Headquarters, U.S. Army Forces Command	Institutional
	Army	WOU3AA	U.S. Army Garrison, Fort Bragg	Support

tion support to Ramstein, including mission support, communications, security, and base services.

The second base is an example from the Army: Fort Bragg. The first UIC is for the 1st Battalion, 319th Field Artillery Regiment, which we classify as operational. The second unit shown is Headquarters of the Army Forces Command, which we classify as institutional. The third unit is the U.S. Army Garrison, Fort Bragg. This unit, composed mostly of civilians, provides installation support at Fort Bragg.

We used this approach to classify the over 38,000 UICs from the DMDC data. To do so, we drew on service documentation about the structure of UICs and information contained in the UIC data (such as the title and composition). We produced a single data set with all personnel and UIC classifications for all four services, for U.S. and overseas locations. We note here that we exclude personnel from one service living at the base of another service, as well as their support costs.[23] This data set fed the next step of our analysis.

---

[23] Some services capture in their spending the cost for their personnel at other bases, e.g., USAF personnel at Ft. Rucker, an Army installation. Arguably, the personnel accrue installation costs at the rate of other service personnel, not the rates of their own service. Further, in some cases the host service captures reimbursable expenses paid by tenant units. We are not confident that even if we included full reimbursed costs and all other service personnel that (1) we would have captured all the money flows from one service to another being accrued by host service for tenants, (2) we would have counted all the tenant units and personnel (we have other service and other DoD organization tenants, but we do not have other government agencies and cannot parse reimbursable expenses to pinpoint each service or tenant organization), and (3) the resulting calculations would add sufficient fidelity to be worth the difficulty of accounting for all the comings and goings. Thus, when a service is the host/manager of a base, we count only the host's forces and non-reimbursable support costs. We thus exclude other service personnel and costs, whether accounted for by the tenant or host cost data.

Integrate Personnel and Cost Data

The third step in our analysis was to integrate the cost and personnel data. This essentially amounted to matching the cost and personnel data for the same installations, across services and regions. We were left with very few mismatches between the data sources.

Analyze Personnel and Cost Data

The fourth step in our approach was to analyze the personnel and cost data and create the installation-support cost model. To do so, we performed several least-squares regression analyses to isolate the fixed and variable components of installation support. In addition to the general relationships between military forces and support costs, we sought to differentiate among the services and regions to the degree possible. This entailed modeling each service separately (i.e., conducting separate regression analyses on each) and creating variables to differentiate among regions and types of bases.[24]

For our mathematical models, we sought to capture the basic dynamics of each subset (e.g., region/service). We thus excluded bases that were far outside the range of typical values.[25] An example of this is that some Air Force training bases in the United States have significant training facilities that bring with them additional costs out of proportion with the aircraft and personnel located there, relative to bases without training operations.[26] In general, when separating classes of installations, we filtered to develop sets that had viable analogs from set to set (e.g., we excluded large depot maintenance facilities, because they do not have analogs in overseas locations).

We focused on two primary dependent variables: total installation-support costs and numbers of support personnel. We iteratively tested a range of models that specified different independent and control variables to identify the primary drivers and best predictors of support costs. We used statistical fit methods to determine the best regression models.[27] Table 8.9 summarizes the results of the regression analysis, with

---

[24] These are referred to as dummy variables. For base type, we coded bases with permanently stationed operational forces to test whether the fixed cost of running and maintaining unit training operations (e.g., runway, training grounds, additional security) changes the cost profile of a base compared with bases without such forces. We coded both PACOM and EUCOM bases to test whether there is a significant difference in fixed cost between CONUS and overseas bases, and between the two regions. We treated U.S. territories separately from the U.S. mainland and bases on foreign soil.

[25] These bases often had an outsized impact on the model results and fit.

[26] We could also capture this dynamic with a dummy variables, but having too many dummy variables, given the small data sets we had, can dilute the effects of each variable, impeding the power of the cost model. Another reason not to use too many dummy variables is that the classes often have too few data points to contribute anything valuable to the model.

[27] We used the overall fit of the model as measured by the F-statistic and total R-squared (which quantifies the percent of variation in the dependent variable captured in the independent variable) to determine which combination of independent variables best explain variation in support costs across the bases in the dataset. We used a p-value of < 0.1 as a threshold for the statistical significance of each effect or variable to include in a model.

Table 8.9
Installation-Support Costs Regression Analysis Results

Region	Army	Air Force[a]	Marine Corps	Navy
	Fixed-Cost Component per Base ($ millions)			
United States	65.0	73.4	49.2	55.9
EUCOM–No forces	—	78.2	—	26.8
EUCOM–Forces	65.0	146.1	—	73.8
PACOM–No forces	—	—	—	26.8
PACOM–Forces	65.0	123.0	49.2	73.8
	Variable-Cost Component per Person ($ thousands)			
United States	5.7	25.7	8.5	12.7
EUCOM	5.7	25.7	8.5	12.7
PACOM	5.7	25.7	8.5	12.7

SOURCE: Authors' analysis.

[a] Air Force estimates include additional personnel-related costs for military personnel providing installation support. This is described in depth in Appendix A.

the top section of the table showing the fixed costs in millions of dollars per base by service and region and the bottom section showing the additional variable costs in thousands of dollars per person. We use the values shown in Table 8.9 in our cost models to assess comprehensive posture options. In cases where bases are outliers in terms of cost versus personnel, we treat them separately rather than using these generalized values to develop posture cost estimates. We discuss these exceptions in more detail in Appendix A.

The top row of the upper table in Table 8.9 shows the estimated fixed cost for installations on U.S. soil.[28] These figures represent the net annual installation-support costs that would be estimated to be saved were that service to close an installation (notwithstanding any investments made to close the base). The fixed-cost components for U.S. bases are similar across the services.

The next two rows show the estimated fixed costs for EUCOM bases, both without and with operational forces. This differentiation, between bases with or without operational forces, only mattered for Air Force and Navy bases in our regression analyses. The Air Force, for example, occupies some small installations in the UK with units (e.g., communications) but no aircraft. The regression analysis showed these smaller installations to have significantly lower fixed costs than those with fully functioning airfields and flying units. The Navy installations showed similar differentiation

[28] These figures are for bases with operational forces, for services where that variable made a difference.

between those with and without operational forces, while that designation did not make a significant difference for costs at Army and Marine Corps locations (cells with "–" for not applicable).[29]

In general, we found little difference in the fixed-cost component between U.S. and overseas bases (comparing the first, third, and fifth rows). The Marine Corps and Army are statistically indistinguishable between the U.S. and overseas, and the Navy's fixed installation-support costs are estimated to be slightly higher overseas. In effect, this means that overseas bases with base populations similar to those in the United States had installation-support costs that were also similar. For the Navy, our regression model was not able to differentiate between EUCOM and PACOM with statistical significance, but was able to differentiate them from the United States when grouped together, which is why the table shows the same values for EUCOM and PACOM.

The Air Force installation-support fixed cost estimate is significantly higher for overseas bases for several reasons. First, overseas Air Force installations appear to have significantly more facilities (higher number and more square footage) than their U.S. counterparts with similar base populations. These facilities drive operations, sustainment, and other costs. In addition, overseas Air Force bases have a significantly higher number of military personnel providing installation support than do U.S. bases of similar size. These additional support personnel in turn require more on-base services, and they incur additional overseas allowances and PCS costs. The difference between EUCOM and PACOM Air Force installations is driven by slightly different proportions of military personnel providing installation support.

The fact that our analysis did isolate a fixed-cost component for installation support is quite important. Having a fixed cost for an installation presents an opportunity for efficiency through consolidation or concentration, regardless of whether installations are in the United States or overseas. If a base is closed and its forces moved to existing bases, the recurring fixed costs of the closed bases become recurring savings (even if one-time investment costs for new facilities are required). If, instead, moving a force from one region to another requires opening a new base, then the savings or cost increase would be the fixed cost differential for similar base types in the two regions. When we apply these numbers to develop estimates of the cost effects of changes in posture, the fixed component would be saved were a base in that region closed and altogether vacated. Again, note that is not the total installation-support cost for the base, just the fixed cost portion.

The bottom table in Table 8.9 shows the variable cost per person, in thousands of dollars, for installation support. The values shown in this table are for the variable costs per person for operational personnel only, following the categories we described

---

[29] Appendix A contains the technical details of this element of our cost study.

earlier—operational, institutional, and support. Operational personnel are those that populate the operational units that our posture options consider relocating.[30]

Notice that the installation-support variable costs do not differ by region. This means that for the installation-support cost data provided to us, the variable costs associated with installation support were not measurably different when comparing overseas to U.S. installations. The variable costs per person are roughly comparable for the Army, Navy, and Marine Corps, but are significantly higher for the USAF. We show only the parameter estimates, but each one has an associated error (shown in Appendix A), so some of these estimates could be closer to one another. These results include the pay for USAF military personnel providing installation support, which greatly increases the overall support costs for the Air Force. If one excludes the PE category with that military pay, the Air Force installation-support variable cost decreases to about $8,300 per person, which is comparable to the other services.[31]

The fact that the USAF chooses to provide a significant portion of its installation support with military personnel is a policy choice that has costs and benefits. These military personnel receive experience directly relevant to many wartime tasks on an everyday basis. In addition, using military personnel for installation support who will then deploy in wartime arguably saves money over a policy that provides installation support solely with civilians and contractors and requires a completely separate military population for warfighting. There are disadvantages of this policy choice from a force management perspective, but the net results relevant to this discussion are two. First, this use of military personnel makes USAF installation-support costs, when viewed in isolation, *appear* to be more expensive than the other services'. Second, and more important, because these military personnel offset some number of civilians who would be doing those jobs and the military personnel would still be needed for wartime missions, arguably the *total* costs of the USAF providing its total inventory of capabilities is lower.[32]

In contrast to the fixed-cost component for bases, our calculations of cost differences for personnel use only the difference between U.S. and overseas installations (i.e., the difference between the variable cost of personnel in the United States and the vari-

[30] Generally, variable costs are higher for institutional personnel than for operational personnel. In the United States, institutional personnel are concentrated at bases with large training and/or material or logistics operations. There are few institutional personnel at overseas installations, usually just at major command headquarters.

[31] In fact, given the error bands surrounding the parameter estimates, the variable cost estimates for all four services nearly overlap. In a statistical sense, then, they all have similar per-person variable costs for installation support.

[32] While the Air Force's strategy of using military personnel for installation support may decrease the total cost of providing its current inventory of capabilities, it is less clear whether that is the right set of capabilities. By shaping part of its military force around home-station, rather than expeditionary, requirements, the Air Force may create support capability imbalances relative to different operational requirements. It is beyond the scope of this report to make such assessments. Unpublished research by Patrick Mills of RAND.

able cost of personnel overseas). In a posture option, if a unit, say a fighter squadron, were moved from PACOM to the United States, installation support would have to be provided there, too. Because the rate (the variable cost) is the same, there would be no net cost savings on just the installation support. That is, unlike for fixed costs where absolute levels matter for estimating the cost effects of posture changes, for variable costs only the relative costs among regions matter.

It is generally believed that overseas bases do cost more to operate than those in the United States. Indeed, the two previous sections showed that for training and personnel-related variable costs, along with installation-support fixed costs in some cases, there is a premium to be paid for stationing forces overseas. Recall that this section analyzed only operating and sustainment installation-support costs as accounted for in service accounting data. One explanation for the lack of differentiation in variable installation-support costs among regions could be that host-nation contributions to installation-support costs, as discussed in Chapter Seven, may suppress the "real" cost of basing overseas. HNS may contribute enough to "level the playing field," thus making overseas installation-support variable costs indistinguishable from those in the United States. Recall that the costs we show in this section are net of HNS, showing the actual costs to DoD after HNS is accounted for. We discuss these regression analyses and results in greater detail in Appendix A.

Medical Service Costs

While medical services are provided at installations, we separate the costs from installation support for our analysis. We obtained detailed medical-support cost data from the Air Force and Navy, and more aggregated data from the Army. In addition, the Army FCM contains cost factors for Defense Health Program operating costs. The net result of our analysis of medical cost data is that medical costs do not appear to be significantly different between U.S. and overseas locations, so we exclude them from our analysis and cost models.[33]

Construction Costs to Recapitalize Infrastructure

The majority of DoD installation-support costs go to supporting the many facilities on bases. DoD defines three categories of facility support. *Operations* refers to day-to-day operational expenses, e.g., utilities, service contracts. *Sustainment* refers to the maintenance and repair activities on real property that are necessary to keep facilities in good working order. *Recapitalization* refers to major renovation or reconstruction activities (including replacement facility construction) needed to modernize facilities and prevent obsolescence.[34] For the purposes of this discussion, when we use the term *facility*

---

[33] The Air Force, Navy, and Army each manage medical facilities in the United States and overseas. The Marine Corps depends on the Navy for medical support.

[34] DoD, *Facilities Recapitalization Front-End Assessment*, August 2002.

recapitalization, we mean the complete replacement of a facility—at a comparable level of capability—that has reached the end of its serviceable life.

We seek to capture the full range of these recurring costs to DoD, but these definitions of facility support activities do not align precisely with accounting data. The spending data we obtained from each of the services includes operations and maintenance spending, and covers facilities operations, sustainment, and non-MILCON restoration and modernization. To calculate the sum of DoD spending on the range of facility support activities, one could add to that the spending from the MILCON appropriation.

However, MILCON expenditure data would be problematic to use to estimate recurring construction costs for recapitalization, for several reasons. Only one application of MILCON spending is simply to replace aging facilities—recapitalization—that have reached the end of their serviceable lives. Given no changes in mission (e.g., introduction of a new weapon system) or size of the force, these expenses would be relatively constant over a long enough time period. It is important for us to include these in the recurring costs of maintaining military presence. However, MILCON is performed on a project basis, and a range of factors drive which projects are planned and performed from year to year; it is not steady based on a long-term annual requirement. These factors include, but are not limited to, availability of funding, criticality of need (e.g., mission needs, facility conditions), the step or periodic function of conduction recapitalization of a building rather than the same amount every year, and political considerations.

Beyond the replacement of facilities, changes in mission or activities at a base drive MILCON. New weapon systems drive the need for different kinds of facilities or simply additional ones. These are introduced irregularly, and thus create spikes in construction activity. Also, DoD has reduced its forces and consolidated installations since the end of the Cold War. Both of these trends, sometimes operating independently of each other, also affect the total demand for and funding for MILCON from year to year, as BRAC and other restationing often require investments at installations at which consolidation is occurring.

Finally, while MILCON spending is proposed and funded on the basis of specific projects at named locations, actual MILCON data do not make completely clear exactly which projects occurred at which locations. As discussed in Chapter Six, we could not assign the totality of the MILCON estimates to particular locations. Without doing so, we cannot confidently use installation-level MILCON spending data to estimate the relative cost differences among regions. Thus, it is not possible to develop cost models of the sort—based on actual spending at the installation level—in which we are interested.

Considering all these factors, actual MILCON expenditures are not a reliable source from which to develop cost models to predict, based on a (potentially small) shift in the overseas posture, what the net effect on MILCON would be. Analysis of

MILCON data shows large increases and decreases in spending from year to year, many of which cannot be definitively explained by the factors listed above, or associated cleanly with recurring facility recapitalization costs.

Nonetheless, for completeness, we need to estimate the recurring costs of maintaining DoD's inventory of buildings in an operational status. Rather than using MILCON, we instead chose to model these costs by estimating the total dollar value of DoD facilities and incorporating a periodic recapitalization rate. This leads to what is termed a requirements model. The main advantage of this approach is that it avoids the variations in spending described above. There are potential disadvantages. First, as described in Chapter Seven, host nations often share a substantial cost burden for DoD's overseas installations. Some of these contributions go to construction, whether directly or indirectly. The United States does not currently bear the full cost burden of construction costs at some of its foreign installations. In that sense, a pure PRV-based cost estimate may overestimate the effective cost burden for facility recapitalization. We discuss below how we seek to control for this factor. Second, the requirement is not always executed based on the factors listed previously. Nevertheless, given the challenges associated with the alternative, we believe the benefits of our approach outweigh the costs.

To perform the analysis, we used the RPAD described in Chapter Six. First, we tallied the PRV of facilities at DoD installations, categorized by service and region. We excluded from our calculations all closed or disposed facilities, all land-only infrastructure, and all Guard- or Reserve-only installations. Table 8.10 summarizes these data.

Table 8.10 shows the total number of unique installations and the total associated PRV for each service and region. We matched these installations and their PRV values to the personnel data described above. Similar to our installation-support cost analysis, we performed a least-squares regression analysis to determine the fixed PRV per installation and the variable PRV per person by service and region. We used total military and civilian personnel as the independent variable and the PRV for each installation as the dependent variable. To then estimate the amount of annual MILCON necessary to recapitalize these facility levels, we used a recapitalization rate. Given a life span for

Table 8.10
RPAD PRV Data Summary

	Army		Air Force		Marine Corps		Navy	
Region	Bases	PRV ($ billions)	Bases	PRV ($ billions)	Bases	PRV ($ billions)	Bases	PRV ($ billions)
United States	31	142.8	50	137.7	11	43.8	28	105.3
EUCOM	12	33.4	11	21.6			3	5.7
PACOM	4	7.5	4	13.5	2	11.4	5	27.5

SOURCE: Author's analysis of RPAD data.

facilities, we can estimate what fraction of the inventory of facilities would, on average, need to be recapitalized each year. We used a 67-year factor recapitalization rate drawn from a 2002 DoD analysis on the topic.[35] This means that 1/67th of the total value of facilities on each installation would need to be replaced per year. One can think of this as an upper bound estimate of potential recapitalization spending, i.e., a requirement.

There are several factors that would affect actual levels of spending. First, HNS offsets a substantial portion of the DoD's cost burden, including for facility construction. HNS would likely reduce the actual recapitalization cost burden on the DoD, but because of the difficulty inherent in fully parsing all HNS contributions, we are unable to estimate exactly how much of this total requirement would be offset by HNS. Second, the services can simply choose to defer recapitalization beyond the 67-year policy, accepting the risks and impacts of deteriorating facilities.[36] Third, it is generally believed that restoration and modernization impacts recapitalization such that maintaining facilities at too low a level can shorten their serviceable lives. The services could choose to take that risk, and if this belief is true, this would cause facilities to need to be replaced earlier than policy would suggest, which would make our estimates too low.

Considering these caveats, this method is a replicable way to produce a rough estimate of recapitalization requirements. Later, we discuss a method to adjust our estimates to better estimate likely spending, but first we show the results of our regression analysis. Table 8.11 summarizes the results of the PRV regression analysis.

Table 8.11 shares the same format as the one in the previous section for installation-support costs. The top section shows values for the fixed PRV component per base; the bottom section shows values for the variable PRV component (again, this is PRV per person, not recapitalization cost per person). We divide these PRVs by 67 to approximate the annual recapitalization requirements for each category.

In some cases, overseas bases show a significant difference in fixed cost per base when compared to U.S. bases. Air Force installations overseas have a fixed-cost component about twice as high for overseas installations than for the United States. To achieve statistical significance with this small sample size, we combined EUCOM and PACOM into one overseas estimate. This does not mean that their PRVs or costs would be exactly the same, but we use this as a general factor.

---

[35] DoD, 2002.

[36] We note that this method ignores facility condition in the calculations. In the short term, facility condition matters greatly: Recapitalization can generally be deferred for longer the better the condition. Our data analysis in Chapter Six found slightly higher condition values for EUCOM than for the United States. One could conclude that in the near term, EUCOM ought to be able to spend less on recapitalization than comparable U.S. installations. This may indeed be true, and could be one of the benefits of the consolidation that has occurred in EUCOM. However, we are endeavoring to estimate long-term costs, and in the long term, every facility must be recapitalized. Thus, we assume an infinite time horizon for our recapitalization calculations (as we do for all of our cost models).

Table 8.11
Regression Results for PRV Data Analysis

	Fixed Component of PRV, per Base ($ millions)			
Region	**Army**	**Air Force**	**Marine Corps**	**Navy**
United States[b]	2,400	1,200	2,440	1,960
EUCOM	2,400	2,370	—	1,960
PACOM	2,400	2,370	4,510[a]	1,960
	Variable Component of PRV, per Person ($)			
Region	**Army**	**Air Force**	**Marine Corps**	**Navy**
All	159,000	180,000	191,000	134,000

SOURCE: Author's analysis of RPAD and DMDC data.

[a] PACOM estimate shown for the Marine Corps is for Okinawa only.

[b] Includes locations in Alaska and Hawaii.

For the Army, most overseas bases' PRVs are in the range of comparably sized U.S. bases. The exceptions to this are Grafenwoehr and Kaiserslautern. These are outliers relative to U.S. bases. This is consistent with the findings from our installation-support cost section, where those two bases were also outliers.[37] It is likely that the activities they support, training and central logistics for USAREUR, respectively, drive both high installation-support costs and PRV relative to their assigned populations, with the higher PRV in turn also magnifying the support costs.

For the Marine Corps, the only significant permanent overseas presence is in Japan, and most of that is on Okinawa. The results for PRV are driven by the Camp Butler Complex. PRV relative to assigned personnel for Iwakuni is in line with comparable U.S. bases. For Camp Butler, the PRV is an outlier relative to U.S. bases given its population (this includes, as does the estimate for PRV in Table 8.11, the 5,600 personnel on Unit Deployment Program [UDP][38] rotations to Okinawa in the total base population), but roughly half the additional PRV for Camp Butler can be attributed to its relatively high area cost factor relative to the United States. We estimated the difference between Camp Butler and the U.S. predicted value from our regression model as described in Appendix A. The illustrative posture options described in Chapter Nine

[37] In our installation-support cost analysis, Kaiserslautern is accounted for under the parent location of Baden-Wurttemburg.

[38] The Marine Corps Unit Deployment Program (UDP), which commenced in 1977, provides for ongoing presence in the Western Pacific via unit deployments of approximately six months instead of using short 12-month PCS assignments, thereby reducing unaccompanied tours and improving unit stability. Department of the Navy, Headquarters United States Marine Corps, Marine Corps Order P3000.15B, *Manpower Unit Deployment Program Standing Operating Procedures*, Marine Corps Order P3000.15B, October 11, 2001.

The same trend is consistent in the bottom section, which shows the one-year maximum MILCON spending versus the estimated requirement. The United States, UK, and Germany are all roughly comparable, Italy is noticeably lower, and Japan and South Korea are still lower. We deduce several things from this. First, the similarity among the United States, UK, and Germany is consistent with our understanding that the latter two countries provide little, if any, MILCON support purely for recapitalization.[42] Second, DoD MILCON spending for U.S. installations in these three countries is, on average, no more than about 80 percent of a simple 67-year PRV-based recapitalization calculation.

Further, South Korea and Japan appear to offset a significant amount of MILCON spending the DoD would normally provide for facility recapitalization. We do not want to make too much of the exact percentages, because of the small sample size and variation within only three years of data (e.g., MILCON cost in Japan varied by an order of magnitude from year to year for these three years, reflecting the factors discussed above). However, the trend seems clear and is consistent with what is reported on HNS: In relative terms, DoD spends significantly less on MILCON for installations in Japan and South Korea as compared with those in Europe and the United States. We assume this is due to HNS provided by these countries, as discussed in Chapter Seven.

Because of what these findings suggest, we adjust our PRV-derived recapitalization requirements to better capture what *actual DoD spending* would be for installations in these regions. Therefore, for our cost models, we make the following adjustments to the estimates:

- baseline estimate
  - U.S. and Europe: 80 percent of calculated value
  - Japan and South Korea: 40 percent of calculated value
- upper bound estimate
  - U.S. and Europe: 80 percent of calculated value
  - Japan and South Korea: 60 percent of calculated value.

For the U.S. and Europe estimates, we simply adjust them to be more in line with what actual MILCON spending looks like on average. For Japan and South Korea, our logic is as follows. As a baseline for our cost calculations, we use a 40 percent adjustment to mirror the data analysis shown above and to parallel the adjustment made to U.S. and European bases. The MILCON portion for Japan is lower on average, but the few values we have varied widely in just a few years. Thus, we use a more conservative estimate and use 40 percent for both South Korea and Japan. For the upper bound, we use a different adjustment for Japan and South Korea. We were told by Marine Corps

---

[42] It is our understanding that these two countries do sometimes provide labor or funding to mitigate the costs of construction for new capability or capacity.

personnel that Japan offsets 25 percent of its MILCON recapitalization requirement (but a little more than 25 percent of the actual construction costs since Japan budgets to only about 90 percent of the requirement).[43] If we use 80 percent as a starting point for all PRV-generated MILCON estimates, and then multiply that estimate by 75 percent (per the Marine Corps), the resulting factor is 0.8 x 0.75 = 60 percent. We use the lower, baseline estimate for Japan and South Korea in our posture cost estimates in Chapter Ten but point out the potential difference if the higher factor is used.

Dependent Education

DoD, by means of the DoD Education Activity (DoDEA), administers and delivers programs that provide education for the dependents of military personnel both in the United States and overseas. Two of these programs are relevant to our cost analysis. DoD Domestic Dependent Elementary and Secondary Schools (DDESS) support those in the United States, and DoD Dependents Schools (DoDDS) provide education to those overseas.[44] We excluded impact aid from our analysis.[45]

We obtained data from DoDEA operating costs for FYs 2009–2011 at the installation level. Table 8.12 summarizes the data.[46]

Table 8.12 shows the total number of installations with DDESS and DoDDS support and the total annual operating costs (average FYs 2009–2011) for those facilities for the United States, Europe, Japan, and South Korea. Most U.S. installations do not have DDESS support. Half of those that do are Army installations. DDESS has a presence at seven Army installations in the United States, and those installations have about a third of Army personnel stationed in the United States. Thus, it is possible that Army forces realigning to the United States will relocate to a base with DDESS support. Most overseas installations do have DoDDS support (28 out of 34 locations with

---

43 Facilities Directorate (GF-2), Marine Corps Installations Command Correspondence to the author, December 19, 2012.

44 The other programs are the Management Headquarters, the Consolidated School Support, and the Educational Partnership Program.

45 Federal impact aid provided through the Department of Education "provides assistance to local school districts with concentrations of children residing on Indian lands, military bases, low-rent housing properties, or other Federal properties and, to a lesser extent, concentrations of children who have parents in the uniformed services or employed on eligible Federal properties who do not live on Federal property." (Department of Education, Office of Elementary and Secondary Education, "About Impact Aid: Impact Aid Programs," August 27, 2008). The purpose of impact aid is to compensate local education agencies where the schools bear a burden of educating such children but the property tax pool is reduced as a result of where their parents live. The total budget in FY 2011 and FY 2012 was just under $1.3 billion per year (U.S. Department of Education, Fiscal Year 2013 Budget). About 40 percent of this aid goes to school districts to serve children of service members (Military Impacted Schools Association, "DoD Impact Aid Funding for Military Children", undated; Richard J. Buddin, Brian P. Gill, and Ron W. Zimmer, *Impact Aid and the Education of Military Children*, Santa Monica, Calif.: RAND Corporation, MR-1272-OSD, 2001). Thus, the level of support provided is approximately $450 per CONUS-based service member.

46 We excluded costs for activities in Bahrain and Guantanamo Bay, Cuba.

Table 8.12
Summary of DoDEA Cost Data, by Region

	Number of Locations	Average Costs FYs 2009–2011 ($ millions)
Europe	26	739
Japan	12	314
South Korea	6	79
United States	13	340

SOURCE: Author's analysis of DoDEA data.

more than 1,000 active duty personnel, 40 out of 79 installations with more than 100 active duty personnel).

To derive cost factors for dependents' education, we performed a least-squares regression analysis, using the DMDC personnel data applied to each installation in the data set summarized in Table 8.12. Table 8.13 shows the results of our regression analysis.

In this table, the fixed cost represents the cost that would be saved by DoD for DoDEA services if a single installation where DoDEA is active were closed. Note that this is not the full cost to the U.S. taxpayer, because most dependents in the United States attend non-DoD schools, with service members contributing local taxes when appropriate and federal impact aid compensating when the service members and their families live on base on federal property and thus do not pay property tax.

In the course of our regression analysis, we found that Europe and Japan had similar dependents' education costs, so we grouped them into one category for the purpose of estimating cost parameters. It is apparent from looking at the data points for bases in South Korea that their costs differ from other countries and regions, but there were too few data points to develop a meaningful regression model. But this is also consistent with the fact that accompanied tours in South Korea remain limited. We therefore do not include a cost estimate for dependents' education activities at Korean bases. Our

Table 8.13
Regression Analysis Results for DoDEA Annual Costs, by Region

Region	Fixed Cost (per base)	Variable Cost (per person)
United States	$1.9 million	$1,100
Europe or Japan	$22 million	$4,900
South Korea	—	—

posture options in Chapter Nine do not explore closing any bases in South Korea, so this does not affect our net cost results.

The variable cost represents the cost per person borne for each service member. Because most overseas installations do have DoDDS support, for our cost calculations, we assume all forces incur overseas DoDDS costs. Because few U.S. installations have DDESS support, when we do our cost calculations, we assume the following as a baseline for forces realigning to the United States:

- One-third of Army forces move to installations with DDESS support. We thus subtract one-third of the U.S. DDESS variable cost from the appropriate region's variable cost differential to estimate the savings per person.
- All other forces move to installations with no DDESS support. We thus count the full DoDDS variable cost for the appropriate region toward the savings for realigning forces from overseas to the United States.

Regional Logistics Costs

This section discusses regional logistics costs not assigned to personnel or installations, but which must be allocated to reflect the real cost burden on DoD to support overseas forces. The most relevant activities in this category are Defense Logistics Agency (DLA) distribution centers and transoceanic transportation costs. DoD activities that need supplies to conduct operations or other activities purchase supplies from the service and DLA working capital funds, transferring operations and maintenance funding to do so. The prices they pay are based on the acquisition cost of the material, which covers the cost of transportation for delivery to DoD or first destination transportation, and a cost recovery rate or surcharge to cover the working capital fund organization's overhead and operational costs; the working capital fund organizations have to recover their costs on a reimbursable basis. The surcharges do include second destination transportation (SDT) for the delivery of supplies within CONUS; however, they do not include over-the-ocean SDT, called "over the ocean transportation" (OOT), which is ultimately paid through central service accounts. Therefore, overseas customers actually slightly subsidize U.S. customers' SDT whenever they order an item since overall CONUS SDT is included in item prices. Thus, the costs reflected in overseas operations and maintenance accounts are slightly inflated with U.S. SDT costs. Conversely, since those overseas customers are not directly charged for their own OOT, with these costs paid centrally, their spending does not reflect the full cost of operations. Therefore, it is necessary to estimate the costs of transportation for the delivery of supplies overseas as well as to estimate the amount by which these overseas customers subsidize CONUS sustainment. The difference between these two costs is the amount of additional overseas transportation we need to account for in determining the variable costs of overseas logistics with respect to transportation.

Additionally, OCONUS support is based on a combination of direct delivery from CONUS defense distribution depots and regional forward distribution depots (FDDs), adding a layer to the support of OCONUS operational forces. These OCONUS FDDs play a significant role in minimizing the OCONUS transportation cost difference but add facility fixed and variable operating costs and require additional total inventory in the system to be effective. In sum, delivering material to overseas forces costs more than delivering to forces in the United States, consisting of the transportation and FDDs—and the requisite inventory—used in tandem to do this in the lowest cost manner possible.

Incremental Transportation Costs for the Delivery of Supplies Overseas

To estimate annual transportation costs within CONUS and OCONUS for the delivery of supplies, we extracted data on CONUS and OOT costs from the Strategic Distribution Database at RAND and used the data to directly compute the cost of transportation by region.[47] We calculated the total cost of OOT for supplies by service and region and subtracted the amount of U.S. SDT they would have subsidized (by using regional transaction data), and then we scaled the difference to a per-person cost. We show the results of this calculation in Table 8.14, which shows the incremental OOT cost per person.

DLA Forward Distribution Depots Costs

We then estimated costs for the second component of DLA overseas support, the FDDs themselves. This includes the transactional costs of the overseas facilities. We obtained data from DLA on the annual operating costs of its distribution centers. Table 8.15 summarizes the cost data.

Table 8.14
Annual Over the Ocean Transportation Costs per Person, by Service and Region

	EUCOM	PACOM
Air Force	$460	$816
Army	$775	$2,265
Navy	$2,431	$1,537
Marine Corps	—	$397

SOURCE: Author's analysis of RAND Strategic Distribution Database.

[47] RAND Strategic Distribution Database, 2012.

Table 8.15
DLA Forward Distribution Depot Data

Location	Annual Cost ($ millions)	Military Personnel	Cost per Person ($)
Europe	18.6	79,291	235
South Korea	8.2	27,709	297
Guam	8.7	5,346	1,636
Sigonella	10.7	3,612	2,965
Japan	30.3	45,996	659

SOURCE: Authors' analysis of DLA depot cost data, RAND Strategic Distribution Database.

NOTES: Europe costs reduced to adjust for contingency support.

In Table 8.15, each FDD is shown on a separate row, and U.S. locations are totaled for comparison. The first column shows the average operating costs (millions of dollars), from the DLA data; the next column shows the total military personnel from the DMDC data, and the last column simply divides the former by the latter. Here, the Guam FDD supports only forces on Guam. The FDD at Sigonella mainly supports the Navy and transiting ships.

The main FDD in Europe, in Germersheim, Germany, also provides support to contingency operations in the Middle East, so it is important to separate the costs incurred to support forces in the region where the FDD is located from the costs incurred to support forces deployed to contingency operations in another theater.[48] In our cost model, we add the cost per person rate for Europe, Japan, and South Korea. We assign Guam costs only to Guam, and Sigonella costs only to naval forces in that region.

Summary of Recurring Fixed Costs

To provide an integrated view of some of these costs, we now show all recurring fixed costs that inform our cost model. Figure 8.5 shows each component of the fixed cost differences by region for bases with operational forces.

In Figure 8.5, the height of each column shows the total fixed-cost component of overseas military installations by service and region, along the x-axis, i.e., the approximate annual cost that could be saved by closing an installation of a given service

[48] To adjust for Europe's support to contingency operations, we used DLA cost data from FY 2008 to FY 2009 only. The surge in Afghanistan (begun in 2009) and the switching of Defense Distribution Depot Europe to be the primary overseas distribution center in support of operations in Afghanistan around September 2011 significantly changed the workloads in the European depot in support of Afghanistan operations. We then reduced the total cost of Defense Distribution Depot Europe by the proportion of its transactions that supported CENTCOM, which was about 25 percent in 2009, before the surge and transition of support.

Personnel support costs during deployment include food, lodging, and special compensation. Policy option costs are those associated with whether or not to preposition equipment sets and the level of installation support present at rotationally deployed locations. While a unit does incur training costs while deployed, we exclude training costs during the operation. We assume that the training each unit receives during the deployment is used to satisfy a portion of its annual training requirement. Additionally, we assume no change in the force structure would be needed to support rotations.[51]

The box on the right articulates the level of detail of the outputs. The model calculates the total cost to sustain rotations by the specific unit type to the specified location under the conditions (frequency, duration, and policy options) specified. For example, it would cost approximately $100 million per year to sustain two three-month deployments per year of an IBCT to South Korea.

The following sections explain more about the components of rotational operations (including our data sources and methods), policy options important to the cost calculations, and some example rotational deployment packages to illustrate our approach. We note here that this cost model applies to the Army, Air Force, and Marine Corps but not the Navy, with the partial exception of aircraft units.

Components of Rotational Presence

We break down a rotational deployment by its component parts and calculate the costs of each element. We then sum these to compute a total cost for a rotational deployment under each set of policies. We seek to capture the total costs that DoD would incur to support each cost element and deployment. For a given rotational package, a unit could incur a range of costs. We divide these costs into five categories: personnel deployment; aircraft deployment; unit equipment deployment, prepositioning, and maintenance; personnel support during a deployment; and special compensation. In the following sections, we discuss these cost elements, data sources, and our cost estimation methods.

Personnel Deployment and Redeployment

The cost to deploy and redeploy unit personnel to and from a deployment simply translates to their airfare. For each unit type specified in our posture options, we derive force packages from standard service-specific data sets or models. For the USAF, we use the Manpower and Equipment Force Packaging System (MEFPAK),[52] for the Army, we used the Military Surface Deployment and Distribution Command's *Deployment*

---

[51] Sustaining rotational presence in a location requires a "rotation" base in the force structure to enable personnel tempo goals, such as time at home between deployments, to be met. If additional units had to be added to the force structure to support rotations, this would add substantially to the rotational presence costs presented in this report.

[52] MEFPAK v2.4 file, September, 2010.

Planning Guide,[53] and for the Marine Corps, we drew unit personnel and equipment requirements from the *Organization of Marine Corps Forces*.[54]

Aircraft Deployment and Redeployment

Calculations for the USAF require the additional component of flying aircraft to and from home-station. To do this, we derived cost per flying hour for each aircraft from the AFTOC database.[55] For each aircraft type, we took the cost elements that vary by flying hour and divided them by the total flying hours for that mission design series (MDS).[56] We averaged the costs from 2009 through 2011.

We calculated the average distance from the United States to each overseas region or country of interest to approximate the deployment distance for a typical aircraft deployment. We then determined the typical block speed of each MDS to determine the average flight time for each MDS.[57] With the flight time and cost per flying hour, we could then calculate the cost to deploy each MDS. The model multiplies this cost times the number of aircraft in each deployment for each location. For fighters, we also add a tanker air bridge to help deploy (and redeploy) the aircraft.[58]

Unit Equipment Deployment, Prepositioning, and Maintenance

We include three options in the model to provide equipment for rotationally deployed units: sealift deployment, airlift deployment, and prepositioning. Each method has costs and benefits (e.g., financial cost, speed, flexibility), and all may not be possible in all circumstances. We tally the costs of each for each rotational deployment, and use the range of estimates to define upper and lower bounds for our aggregate cost calculations.

For sealift and airlift deployment, we use the same data sources to derive force packages as for unit personnel but to derive equipment lists instead. When necessary, we cross-referenced these data with the DoD's Federal Logistics (FEDLOG) database to obtain weight and cube data for calculating shipping costs.[59] To estimate the cost

---

53 Surface Deployment and Distribution Command Transportation Engineering Agency, 2012.

54 U.S. Marine Corps, *Organization of Marine Corps Forces*, MCRP 512-D, October 1998.

55 AFTOC database, 2009–2011.

56 We adapted our cost analysis approach from Moroney et al., 2012. See report for more detail on which costs we include.

57 Aircraft block speed is an aircraft's true airspeed adjusted in relation to length of sortie to compensate for take-off, climbout, letdown, instrument approach, and landing. AFPAM 10-1403 lists the block speeds for mobility aircraft. We took these as stated for mobility aircraft, and adjusted the combat aircraft flight speeds based on the rough proportions the document laid out. See Air Force Pamphlet 10-1403, 2003.

58 The planning factors and assumptions for the air bridge are adapted from a RAND analysis of alternatives for KC-135 recapitalization. The analysis is not releasable to the public.

59 For the USAF, we assume that mobility aircraft can self-deploy their personnel and equipment, so only fighters incur these additional costs.

to DoD of deploying these personnel and equipment to and from home station, we used standard TRANSCOM rates and applied them to the cargo being transported.[60] When estimating the costs for rotating an entire MEU, we assumed an Amphibious Squadron (PHIBRON) would transport the MEU, thus obviating the need to pay separately for personnel and equipment transportation.[61]

Prepositioning entails each service keeping sets of equipment in storage locations and periodically maintaining them to ensure mission readiness. A given service may have enough equipment sets (especially in light of planned drawdowns) to place an "extra" set in storage for this purpose, thus saving the purchase price of such equipment. However, equipment must be replaced at some point, so prepositioned stocks would eventually have to be recapitalized. For the calculations in this report, we assume that each service owns adequate equipment sets to support prepositioning, and we do not incorporate equipment recapitalization. The services provided information we were able to use to estimate the cost to have contractors store and maintain the equipment sets.[62] If equipment purchase or recapitalization were needed, that would add an additional cost to the prepositioning options.

Personnel Support During Deployment

DoD personnel receive essentially one of two types of support or compensation when forward deployed in peacetime. Food, lodging, transportation, etc., can be provided by the destination military base, if those services are available, local contractor support, or by per diem with which personnel can procure those services on the local market. For short deployments (e.g., a 1–2 week exercise), units often make do with more austere conditions, such as sleeping in the field and using field kitchens. The rotational deployments our posture options explore are typically too long to assume austere conditions, so we make several assumptions about providing personnel support, based on the region and country.

For all regions, destination bases are assumed to provide support by means of permanent facilities and personnel. In regions with a current permanent DoD presence (e.g., Western Europe, Japan, South Korea), this assumes that the service overseeing the installation will keep facilities and other support in-place in the event of any unit realignments, or more will be constructed to handle rotational forces if needed. In short, at least one base able to host rotational forces will be maintained in these regions. In other regions, in practice, it may be DoD or the host nation that provides or pays for these services. In Romania and Bulgaria, for example, these nations provide

---

[60] FY 2012 TRANSCOM tariff rates, derived from data provided to RAND.

[61] We assume that the Navy PHIBRONs would be fully utilized, and thus adding the transport of a MEU for rotational deployment would simply offset other missions, but would not add to the total operating cost of the PHIBRON.

[62] Data on prepositioned equipment storage and maintenance costs were provided via email by HQ PACAF/A4, HQ USAFE/A4, and U.S. Army Materiel Command.

some support for U.S. rotational forces. Thus, as an upper-bound cost, we assume that DoD will pay the host nation for this facility support. For a lower-bound estimate, we exclude these base operating costs from our calculations, assuming that DoD maintains ongoing capabilities in the region able to also support rotational forces. To pay for the day-to-day installation support and upkeep of those facilities, we apply the cost factors for installation support that we developed earlier in this chapter. We include in this the per-person variable costs for installation support, MILCON for recapitalization, and overseas logistics support. We exclude dependents' education, because service members' dependents are living at home-station. Further, we decrement these installation-support costs by 25 percent to represent reduced needs since dependents will not be living on base (e.g., family housing, family and youth programs).

Beyond day-to-day facility operations and sustainment, we also assess the costs to provide food and other incidentals. For forces in regions without current permanent presence, we assume that food service is provided by local contractors, as has been done for contingency operations or in countries in which the United States performs rotational deployments on an ongoing basis. To estimate these costs, we use the local food per diem rate as a proxy for this contractor support. The DoD Defense Travel Management Office provides detailed per diem rates for travel to a range of countries and cities in the world.[63] We utilized these per diem factors for the overseas locations identified in our posture options. When costs for specific locations were not available, we used average regional costs.[64]

For forces deploying to regions with current permanent presence, we apply a COLA as compensation. However, COLA factors available from sources like the FCM or the aggregate figures we developed from Work Experience/Active Duty Pay files are averages across the entire population for a given service and region. Since these averages include some personnel living with dependents and off base, and the rotational deployments for which we calculate costs assume personnel deploy unaccompanied and live on base, the average values would overestimate COLAs needed to compensate deploying personnel. These deployments would also be shorter and more austere than multiyear PCS tours, where personnel would have a certain anticipated standard of living. For these reasons, we apply only 50 percent of the COLA costs for each person for the region of their deployment.[65] Finally, because we already make provision for food and

---

63 Defense Travel Management Office, "Per Diem Rates Query," online, August 26, 2011.

64 Joint Federal Travel Regulations, Volume 1, *Uniformed Service Members*, Alexandria, Va.: The Per Diem, Travel and Transportation Allowance Committee, October 1, 2012. Includes units traveling in support of combat missions, peacekeeping, and disaster relief. It also includes field or maneuver training and sea duty when troops involved are not permanently assigned to a ship. The Government provides all transportation, lodging, and eating facilities when personnel traveling together are under orders directing no/limited reimbursement. (Joint Federal Travel Regulation)

65 An exploratory assessment of Defense Travel Management Office, "Overseas COLA Calculator," online, undated, showed that for a given location, grade, and number of years of service, the COLA rate for a single

lodging, we apply the local incidental per diem rate to all personnel in all regions for the duration of their deployment.

Special Compensation

DoD provides special compensation to those deploying overseas away from their home-station location. This includes family separation and hardship pay. We apply family separation for all deployments and hardship pay for all deployments outside of Western Europe, Japan, and South Korea.

Example Rotational Deployment Cost Calculations

In this section, we review the details of a few example rotational deployment packages to demonstrate how the cost analysis works. First, we show a detailed example of an Army deployment to explain how we apply our methodology to each of the components of rotational deployment. Table 8.17 shows three example deployment cost calculations of an ABCT deploying from the United States to Germany, varying the frequency and duration of rotations such that the examples all produce 12 months of presence per year.

In Table 8.17, the top section shows the basic information about the unit and its deployment, including the frequency and duration of deployment. The middle section shows the component cost calculations. The bottom section totals the costs for each of the three equipment support options.

In the middle section, one can see how the deployment costs vary for air and sealift, but not prepositioning. The rest of the costs, for personnel and installation support, are static, because the total duration of presence is always 12 months. The bottom section, with totals, shows that airlift is significantly more expensive than sealift or prepositioning. The ABCT is an extreme case because of its equipment weight, but in general, airlift to transport equipment is not a realistic option for recurring peacetime rotations for brigade-sized ground forces. Further, the left-most column shows that for a single 12-month deployment, prepositioning is more expensive than sealift. On the right side, however, as deployment costs increase with increased frequency, the prepositioning option is slightly less expensive than using airlift. Sealift requires much more time in transit, which has its own non-monetary costs.

For each service, unit type, and region, the trade-offs look slightly different, depending on unit size, transportation rates, and prepositioning costs. For our cost analysis, we use this same approach of calculating the cost of each rotational component and then aggregating into total costs. We now show two more examples to show the broader trade-offs involved with rotational operations, and how they compare with savings from reducing permanent presence. We show another ground force example using an Army Stryker brigade, and an Air Force example using an F-16 squadron.

---

person living in barracks is roughly half the rate for someone with one or two dependents living off base.

Table 8.17
Costs of Rotational Components for ABCT Deployment from United States to Germany

	Rotation length		
Category	**12 Months**	**Six Months**	**Three Months**
Rotation frequency (per year)	1	2	4
Unit personnel	3,266	3,266	3,266
Equipment cube (MTONs)	70,373	70,373	70,373
Equipment weight (short tons)	21,690	21,690	21,690
Personnel airfare ($ millions)	9.5	19.1	28.6
Equipment sealift ($ millions)	13.4	26.9	40.3
Equiment airlift ($ millions)	252.5	505.0	757.5
Equipment prepo ($ millions)	22.3	22.3	22.3
COLA ($ millions)	10.6	10.6	10.6
Family separation ($ millions)	9.8	9.8	9.8
Per diem ($ millions)	24.5	24.5	24.5
Installation support ($ millions)	22.9	22.9	22.9
Total cost sealift ($ millions)	90.7	113.7	136.7
Total cost airlift ($ millions)	329.8	591.8	853.9
Total cost prepositioned ($ millions)	99.5	109.1	118.6

SOURCE: Authors' analysis.
NOTES: MTON = measurement ton; prepo = prepositioning.

Figure 8.8 shows the cost breakdown for an SBCT rotation from the United States to Germany. In this chart, the total time spent deployed is shown on the x-axis, and the total cost per year is on the y-axis. The blue areas show the amount saved by realigning an SBCT from Europe (the lower shaded area only) and closing a base there as well (the top of the shaded areas). The lines and dots depict the cost of each of several options.

Figure 8.8 shows two equipment options—sealift and prepositioning—and three deployment frequency/duration pairs. The solid purple line shows the cost relationship of sending the SBCT on three-month deployments using airlift for personnel and sealift for equipment. At the bottom left, the cost of one three-month deployment per year is about $63 million per year, including all the costs described above. To increase the total time deployed forward (y-axis), there must be more deployments, with the gaps between them determining the average amount of annual presence, as indicated

Table 8.19
Example Rotational Deployment Cost Calculations for Air Force Units ($ millions)

				24 x F-16		24 x F-15		18 x C-130		15 x KC-135	
Region	Frequency (per year)	Duration (months)	Total Duration (months)	Airlift	Prepo	Airlift	Prepo	Airlift	Prepo	Airlift	Prepo
EUCOM	1	3	3	22	19	29	25	14	13	9	9
	1	6	6	24	21	31	27	15	14	10	10
	2	6	12	37	31	50	41	23	20	16	15
PACOM[a]	1	12	12	29	26	37	34	18	17	13	13
	1	6	6	25	23	33	30	16	15	11	11
	2	3	6	37	30	51	42	24	20	16	14
CENTCOM	1	12	12	30	26	39	33	19	17	13	13
	1	6	6	27	22	35	29	17	15	11	11
	2	3	6	40	30	55	42	25	20	16	14

SOURCE: Author's analysis.

[a] Results for PACOM are for deployment to Japan only.

Torrejon, Spain,[68] the European drawdown (specifically highlighting base closures in Germany from the late 1960s through the early 1990s),[69] and U.S. bases in Thailand.[70]

The second source we looked to was BRAC documents and data. While the specific cost factors used for the various rounds of BRAC do not always mirror the costs of similar actions for overseas locations, these sources provided a framework for the general cost categories relevant to relocating forces and closing installations, and provide a starting point for estimating some cost factors. The BRAC cost-estimation model itself was not readily available for use, and even the detailed commission reports did not provide enough granularity to back-out rough cost factors for realignment and closure costs. We thus looked to detailed data documentation to establish the specific inputs to each BRAC action and their associated costs.[71]

After consulting these sources and synthesizing their findings, we created cost factors for three broad categories: personnel-related movement and separation, base-related closure costs, and new construction for realigning units. We now describe the cost elements in each category and our methods for estimating the costs.

Personnel-Related Movement and Separation

This category of personnel-related costs includes the one-time costs associated with moving personnel and their dependents for a unit realignment, paying for persons separated as part of a base or unit consolidation, and transportation of unit equipment.

For the movement of military personnel and their families, we use the standard overseas PCS rates for overseas to U.S. movements (i.e., transoceanic) for a single move (i.e., not annual rates). We referenced these earlier in this chapter. We used the same cost factors for civilian moves as well.[72] We thus apply these additional cost factors for civilian personnel.

For realignments within a region, we use U.S. PCS rates for military and civilian personnel, assuming movement will take place over land. For realignments from one overseas region to another, we use the overseas PCS move rates for both military and civilian personnel.

---

68 GAO, *Overseas Basing: Costs of Relocating the 401st Tactical Fighter Wing*, GAO/NSIAD-89-225, September 1989.

69 Bonn International Center for Conversion, *Restructuring the US Military Presence in Germany: Scope, Impacts, and Opportunities*, June 1995.

70 GAO, *Overseas Basing: Withdrawal of U.S. Forces from Thailand: Ways to Improve Future Withdrawal Operations*, LCD-77-446, November 1977.

71 U.S. Code, 2013.

72 Civilian move rates were more expensive per person than military moves in the BRAC data we analyzed, and included a category called the Priority Placement Program, which matches civilians in need of posting after a realignment/closure/RIF with openings elsewhere. For simplicity's sake, we replicated the military personnel PCS rate.

For civilians who are separated in a realignment (e.g., headquarters units no longer needed, base or unit consolidations), we include severance pay, early retirement, and unemployment compensation based on information from the most recent BRAC round.[73]

In the past, some of the PCS moves accompanying a base closure have been phased in as a part of normal periodic personnel rotations, so no additional PCS movement is needed. So we consider our cost estimates to be an upper bound on the one-time move costs, assuming all personnel require a move outside the timing of their normal restationing.

One additional per-person cost found in the BRAC documentation is called program management. This represents the costs of administrative support for movements of personnel and equipment and is estimated as a per-person percentage overhead fee on top of the individual personnel expenses described above. For this category, we looked to major U.S. installation closures and realignments from the 2005 BRAC round from which to derive costs. We developed per-person cost estimates from each example, and then averaged these costs across the examples (dismissing the highest and lowest values in each set). The result was an average per-person cost for program management costs. Table 8.20 shows these cost factors.

We used these per-person costs as inputs to our total cost calculations. For each person realigned in one of the posture options, whether to/from the United States or across regions, we apply these cost factors to estimate the total one-time cost to realign these personnel and their equipment. For unit moves, we use TRANSCOM sealift shipping rates to and from the appropriate region. As an upper bound, we assume military and civilian moves pay both PCS and program management costs. As a lower bound, we exclude the PCS move costs, assuming that those moves can be made over time as a part of normal personnel rotations.

Table 8.20
Personnel-Related Closure and Realignment Cost Factors

Factor	Between Overseas and United States	Within Overseas Region
Military move	$7,424	$5,575
Civilian move	$7,424	$5,575
Civilian separation	$20,370	$20,370
Program management	$2,380	$2,380

[73] U.S. Code, 2013.

Base-Related Closure Costs

Base closure costs include, but are not limited to, support contract termination, mothball/shutdown,[74] severance to foreign civilians, labor contract settlement costs, and a range of one-time unique costs.[75] These are the key costs we include in our analysis.

The closure of a military installation also inevitably includes environmental mitigation costs. The overall scope of these costs vary widely from installation to installation, and cleanup requirements have changed significantly over the past few decades during which many closures have occurred. In addition, the cost-sharing burden for these kinds of costs varies depending on the region and country, and the United States' relations with the host nation government.

A 2004 GAO report had this to say about environmental remediation costs:

> Historically, overseas regional commands have incurred limited costs for environmental remediation as a result of these property returns. For example, the component commands in South Korea and Japan have incurred limited costs to date, while EUCOM currently estimates its potential costs for environmental remediation at about $90 million, regardless of whether the property is returned in the future. However, in the future, there is less certainty regarding potential costs for environmental remediation because these issues are becoming an increasing concern in South Korea and Japan. For example, according to PACOM officials, South Korea has established procedures for addressing environmental remediation, and the Government of Japan is enacting more stringent environmental laws.[76]

For these reasons, we exclude environmental mitigation costs from our total base closure cost estimates.

For data sources, we looked to the reports on the closure process for the overseas closures listed above, and the BRAC data for major installation closures.[77] In particular, the overseas closure reports provided cost estimates due to local national employee litigations, as well as residual value negotiated settlements. Some of the U.S. costs we use may be different from those that would actually occur during an overseas closure. The data available to inform these estimates were sparse, so we looked to the best available sources. We intend these calculations as illustrative of the kinds of costs that could accompany an overseas base closure.

---

[74] This is the cost to mothball facilities at a closing base where the facilities will not have a future re-use. This represents the minimum cost to close-up a facility in preparation for disposal/demolition.

[75] These are the unique non-recurring expenditures during each year that cannot be portrayed properly elsewhere. Includes such costs as land purchase costs and lease terminations.

[76] GAO, *Defense Infrastructure: Factors Affecting U.S. Infrastructure Costs Overseas and the Development of Comprehensive Master Plans*, GAO-04-609, July 2004.

[77] U.S. bases used to inform these estimates are Cannon Air Force Base, Onizuka Air Force Station, Kulis Air Guard Station, and Brooks City Base.

We developed per-base cost estimates from each example and then averaged these costs across the examples (dismissing the highest and lowest values in each set). The result was an average per-base cost for each category. The list below shows the cost factors we developed for base closures:

- support contract termination: $0–10.9 million
- mothball/shutdown: $312,000 (average)
- one-time unique costs: $0–10.5 million
- litigation and local employee settlements: $19.2 million
- total: $312,000–40.9 million.

We used these per-base costs as inputs to our total cost calculations.

New Construction

For each unit realigned to the United States, or within a region, there is a potential need for new construction to provide barracks; training operations morale, welfare, and recreation; and more. However, recent and planned force reductions will free up some facilities for use by units being realigned to the United States. For example, the Army is planning to draw down the force by 80,000 troops, and our posture options contemplate only a fraction of that number of soldiers being realigned to the United States. Thus, one could argue that in many cases excess capacity will be freed up to accommodate these moves or the units could inactivate in place. We estimate the potential construction costs necessary to accommodate all realigning forces and consider these an *upper bound* to the expense DoD might incur to perform these realignments. As a *lower bound*, we assume the Army can absorb all realigning forces at existing U.S. bases with existing infrastructure, thus eliminating one-time MILCON expenditures for force realignments. While the Marine Corps is downsizing as well, it reports that it would not have slack capacity to absorb realigned units because it is already short of capacity to support the 2012 level of 202,000 personnel. During the buildup, temporary facilities were employed, and MILCON was not programmed for expansion of facilities.[78] We assume that the Air Force and Navy would also have to expand their U.S. facilities if forces were realigned to the United States. If these two services, especially the Air Force, did have some slack capacity at appropriate U.S. installations, it would reduce the amount needed for new MILCON below the level of the overall upper bound estimates shown for the illustrative postures in Chapter Ten.

To derive construction cost estimates, we again looked to the RPAD for data on total PRV of military installations. Instead of seeking specific MILCON data on unit moves and construction projects to find analogs to each of our unit moves (which number up to several dozen), we used a similar approach to the one we developed for

[78] U.S. Marine Corps, Budget and Execution Division, 2012.

estimating annual MILCON recapitalization requirements described earlier in this chapter.

We took the same inputs we used for the MILCON recapitalization variable cost estimates developed earlier (Table 8.3), and used the PRVs instead of translating them to an annual recapitalization expense. In other words, we calculated the total PRV, by service, but did not divide by the 67-year recapitalization figure. Then we used the regression results to determine the fixed level of PRV per U.S. base and the variable level of PRV person for U.S. bases by service. Thus, we arrived at an estimate of the total variable one-time cost per person to construct new facilities to accommodate units and personnel arriving at new locations. These estimates ignore constraints or costs on things like training ranges and airspace:

- Air Force: $180,000
- Army: $159,000
- Navy: $134,000
- Marine Corps: $191,000.

As one can see from these figures, the one-time MILCON expense to accommodate a single person realigning to a U.S. location ranges from $134,000 to $191,000 per person across the services. These figures comport well with a 2004 Congressional Budget Office study of overseas Army posture changes.[79] That report included MILCON estimates for unit training facilities (e.g., headquarters, operations); morale, welfare, and recreation; barracks; and DoD schools. If one aggregates the Congressional Budget Office study's costs and inflates them to today's dollars, the cost is roughly $135,000 per person.

The Marine Corps also provided us with data to estimate MILCON costs for realignment of personnel, derived from its recent analysis of construction costs to accommodate the relocation of forces from Japan to Guam. For reasons we explain in Appendix A, we believe the estimates derived from this assessment, while based on detailed facilities analysis and accurate for estimating the MILCON needs in Guam and somewhat similar circumstances, would overstate the investment costs for a relocation to a large U.S. base rather than Guam. We therefore use our PRV-calculated estimate as the baseline value for illustrative posture estimates. Nevertheless, in Chapter Ten, we show the cost implications of using the Marine Corps' per-person cost factor instead of ours, as well as explore some of the cost differences involved with relocating Marine Corps forces to Guam and Hawaii as opposed to CONUS.

---

[79] Congressional Budget Office, *Options for Changing the Army's Overseas Basing*, May 2004.

Example Investment Cost Calculations

Now we apply these cost factors to two illustrative example base closures and realignments. Table 8.21 shows the upper bound of the cost estimate for the investment necessary to implement the closures of Misawa Air Base in Japan and Ansbach/Illesheim, Germany, along with the full realignment of all their personnel and unit equipment to the United States.

In Table 8.21, for each example overseas base, we show the inputs at the top, including personnel and equipment measurement tons. For the equipment shipping calculations, we use standard TRANSCOM sealift rates from Japan and Germany to the United States.[80] In the second section, we show the movement cost elements and totals. In the third section, we show the base closure cost. Finally, we show the MILCON cost for new facilities in the United States, followed by the total implementation cost.

One can see that MILCON costs, which depend upon whether there would be sufficient available capacity in CONUS, dominate the total, driving about 90 percent of the total investment cost. This percentage, for these two examples, is consistent with the range of calculations we perform for our aggregate cost assessments.

Table 8.21
Example Investment Cost Calculations

	Misawa, Japan	Ansbach/Illesheim, Germany
Major units	35th Fighter Wing	12th Combat Aviation Brigade
Military personnel	3,519	2,834
Civilian personnel	154	230
Unit equipment (MTONs)	2,524	70,674
Move military cost ($ millions)	34.5	27.8
Move civilian cost ($ millions)	1.5	2.3
Move equipment cost ($ millions)	0.5	6.1
Move total cost ($ millions)	36.5	36.1
Bases closed	1	1
Base close cost ($ millions)	40.9	40.9
MILCON cost ($ millions)	633.4	450.6
Total investment cost ($ millions)	710.8	527.6

[80] We exclude the ground legs needed at to transport equipment from origin to sea point of embarkation (SPOE) and sea point of debarkation (SPOD) to destination, as these are an order of magnitude less than the SPOE-to-SPOD cost.

These investment costs should be considered rough estimates, for several reasons. First, there is quite a bit of uncertainty surrounding the per-person MILCON cost estimates (details can be seen in Appendix A). Since MILCON drives the overall investment cost estimates, then, the estimates provided in Chapter Ten should be treated as rough guides for the costs DoD might incur. Second, because the needs are very site-specific and can vary greatly, we excluded environmental cleanup from our total estimates. For similar reasons, we did not consider potential gains from the residual value of installations returned to the host nation. GAO reported that, between 1989 and 2007, the U.S. government received approximately $592 million in residual value and payment-in-kind compensation from property returns in EUCOM's AOR.[81] Given these considerations, more detailed analysis would be needed to develop budget-quality investment cost estimates for overseas posture changes.

Implications for Posture

This chapter assessed the relative costs of maintaining and modifying overseas posture by utilizing methodologies that estimate the recurring costs of overseas bases and stationed forces, the costs of rotational presence, and one-time investment requirements to shift from one posture to another. Beyond creating a useful tool for estimating the financial effects of posture changes, though, the modeling effort also produced some general conclusions that can be useful for thinking about how posture affects costs.

Our modeling results support our hypothesis that there are fixed and variable costs to operating installations and stationing forces. Given fixed costs, to the extent there are unit inactivations or realignments that can be accomplished by moving forces to existing bases (even if some MILCON is necessary), consolidation produces savings. This would apply whether in the United States or overseas, although it may be much less feasible to consolidate further overseas, given already smaller bases without excess land capacity and difficulties gaining new, larger bases to replace multiple smaller ones or gaining expansions with host nations (an exception might be if the United States were establishing bases in new regions and host nations). Beyond this, we did not find systematically higher fixed costs for overseas bases. The exception to this was the Air Force, which we found to have higher fixed costs for overseas bases than for U.S. bases.

Additionally, there are incremental variable recurring costs for stationing personnel overseas in Europe and the Pacific region due to higher allowances related to the cost of living and higher PCS move costs. However, there is wide variation among these incremental costs, such that the cost outcomes of posture changes depend greatly

[81] GAO, *Overseas Master Plans Are Improving, but DOD Needs to Provide Congress Additional Information About the Military Buildup on Guam*, GAO-07-1015, 2007.

on the service and region under consideration. The incremental differences in other regions, were posture to shift, might be different.

Combining the variable cost findings of higher variable costs overseas with the finding that there are fixed costs per base means that consolidating forces at fewer bases would provide more savings when relocating forces from overseas to the United States (i.e., closing overseas bases and moving their forces to U.S. bases) than simply consolidating among bases within overseas regions. The fixed costs would be saved whether consolidating only in the United States or only in an overseas region, but closing an overseas base and relocating to the United States also saves on per-person (i.e., variable) costs due to the higher overseas personnel-related costs (e.g., allowances, PCS costs). However, the United States cannot repurpose overseas bases as it could in the United States, which could produce non-DoD economic benefit. We did not examine the typical benefits gained by the broader economy as the result of U.S. closures.

Finally, the costs of rotational presence present a complex picture. Generally, we found that the savings produced by merely realigning forces while keeping installations open is not sufficient to offset the cost of providing full presence through rotational deployments. In most cases, realignments of permanent forces can underwrite only partial rotational presence in the same location. If an installation is closed, however, this will generally offset the costs of full rotational presence, with some net savings. But the net savings depend greatly on the service, unit type, and location. For ground forces, sealift to move equipment or available equipment for prepositioning is necessary for savings. Further, if a base were to be closed and its forces realigned, another permanent base in that country or region must be maintained to support the rotating forces (including considerations of facilities, space, and other accommodations) or a host nation must agree to provide access to one of its bases.

Second, the illustrative postures protect the ability "to confront and defeat aggression anywhere in the world" by maintaining global power-projection capabilities called for in the 2012 Defense Strategic Guidance.[1] Consequently, these postures maintain global infrastructure that enables U.S. power projection, including key en route ports and airfields as well as key communications facilities.

Third, the illustrative postures do not take into account current war plans, which are in part designed based on current posture. Instead, the assumption is that war plans would have to be modified and that in some situations the United States might be incurring greater risk. In others, it may be possible to develop alternative courses of action to meet national security objectives. The various analyses will identify the strategic and operational trade-offs that the United States would be making if it were to implement any changes that reflect these illustrative postures.

Fourth, these illustrative postures do not consider the current political situation in proposed host nations, including whether U.S. basing rights are at risk or whether it is feasible for the United States to obtain access to military facilities in a particular country. These are important factors, but ones that will be assessed in other parts of the report. These notional choices are made to broadly explore options and understand the potential value to U.S. national security of securing such rights.

Finally, the illustrative postures are all intended to meet the intent of the Defense Strategic Guidance, in particular the decision to rebalance toward Asia[2] while also ensuring that the United States upholds its current security commitments and alliances and maintains presence in the Middle East. But, as called for in the guidance, the postures explore different ways and new approaches by which DoD objectives might be achieved. This is not to say that the illustrative postures eliminate risk or are equally effective. On the contrary, each illustrative posture drives trade-offs among strategic benefits, risk, and cost–trade-offs that can facilitate, but not replace, the deliberations of decisionmakers.

Elements of Current U.S. Posture That Are Held Constant

A number of elements of the United States' current global defense posture are not changed in any of the alternate postures. First, TRANSCOM's en route infrastructure remains largely intact. Currently, it consists of three routes across the Atlantic and two across the Pacific.[3] Each alternate posture keeps in place the most important mobility

[1] DoD, 2012a.

[2] DoD, 2012a.

[3] AMC identified particular locations using the lens, or sweet spot, concept, which is based on the range limitations of current aircraft. Using the 3,500 nautical miles unrefueled planning factor, AMC defined the outermost bound that strategic airlifters could reach from a mid-Atlantic CONUS base, while the other side of the lens

hubs as well as enough mobility bases to maintain the three Atlantic routes and two Pacific routes, although some postures accept more risk by closing some of the backup en route bases.

In addition to the mobility infrastructure, the illustrative postures also maintain other critical enabling capabilities. In particular, the illustrative postures do not change any satellite stations or facilities or realign forces that joint forces rely on for intratheater as well as intertheater communications, including RAF Croughton, USAF satellite stations, Fifth Signal command at Wiesbaden, and Ramstein Air Base. Moreover, none of the postures suggests removing the Army hospital at Landstuhl.[4]

Finally, given that the 2012 Strategic Guidance places the greatest emphasis on the Asia-Pacific and the Middle East, all postures prioritize these regions, keeping substantial bases and forces in Japan and South Korea and access and forces distributed in the Middle East. The illustrative postures also take into account the fact that the United States must fulfill its Article 5 commitment to NATO. Therefore, there are certain capabilities that all of the illustrative postures retain in Europe, including U.S. nuclear weapons, missile defenses, participation in NATO's Heavy Airlift Wing and AWACS component, and contribution to the NATO Response Force (NRF). The United States currently stores nuclear weapons in several European countries.[5] Several of these bases are host-nation facilities and the only U.S. presence consists of a munitions support squadron. Therefore, even if one of the illustrative postures suggests removing the U.S. presence from a base currently containing nuclear weapons, the weapons could be transferred to another base in the same host nation or another NATO member's base and stationed with a U.S. munitions support squadron.

In addition to the nuclear commitment, the illustrative postures do not vary the European Phased Adaptive Approach (EPAA), the architecture to defend Europe from Iranian ballistic missiles. Therefore, all of the illustrative postures contain Ramstein Airbase, which is the command and control center, as well as all missile-defense radar sites and planned Aegis-Ashore systems. Additionally, all of the illustrative postures continue with the planned stationing of four BMD destroyers at Rota Spain.

Finally, the United States will uphold its commitment to important NATO initiatives to improve the alliance's combined capability, most notably the Heavy Airlift Wing in Papa, Hungary, the AWACS component in Geilenkirchen, Germany, and the commitment of a U.S. Army brigade to the NATO NRF. The United States will des-

identified the outermost bound that could be reached from a location in Southwest Asia. Air Mobility Command, 2010, p. 5.

[4] Landstuhl is actually going to be replaced by a new facility at Rhine Ordnance Barracks in 2019. Nevertheless, all of the illustrative postures retain this medical capability in Germany.

[5] DoD, *Nuclear Posture Review*, April 2010d, p. 32.

ignate one brigade stationed in CONUS to contribute to the NRF and elements of the unit will rotate to Europe annually for exercises.[6]

Illustrative Cost-Reduction Posture

The illustrative cost-reduction posture (CRP) aims to minimize the cost of the U.S. global posture while simultaneously maintaining enough forward presence to achieve national security goals, including enabling the United States to project power around the world, tying it to key allies, and ensuring the freedom of the global commons. This posture assumes that these two objectives are not mutually exclusive. In other words, the CRP rests on the assumptions that (1) the United States can meet its national security objectives with a smaller overseas presence because if the need arises, the United States would be able to deploy its forces to and operate in distant theaters, and (2) it can maintain alliances and pursue effective security-cooperation efforts through other means with lower overseas permanent presence. Additionally, this posture is predicated on the notion that closing bases and bringing troops back to the United States would yield significant cost savings. The CRP does not simply close all overseas facilities and bring all forces to U.S. territory. Instead, the illustrative CRP is intended to represent the *minimum* forward military presence that the United States would need to remain a globally responsive military power. Other specific choices could potentially achieve similar strategic benefits at similar costs—this is just one representation.

In designing the illustrative CRP, we developed the following guidelines. The United States would adopt a less-dispersed posture that would rely on larger bases, and it would keep multipurpose bases over those that have only one function. Wherever possible, within the bounds of the guidelines, the illustrative CRP would shrink the size of the U.S. overseas military presence by consolidating its military footprint and reducing the number of forces temporarily deployed, as well as permanently stationed abroad. As a result, the United States would seek to maintain a military presence within a region, rather than in a particular country. If there is room at large multipurpose bases, the United States would consolidate its forces at these locations or a number of facilities proximate to each other to achieve economies of scale. Global transportation and support infrastructure, such as communications nodes and en route locations, are largely retained, although the United States would accept greater risk by closing some backup locations.

[6] Hans Binnendijk, "Testimony Before the Senate Foreign Relations Committee: Defense Issues for the NATO Summit," May 10, 2012.

Notional Cost-Reduction Posture Changes

Based on the guidelines above, this study identified one set of specific changes that could be made to the current U.S. posture if cost reduction were the primary objective, while maintaining the en route basing infrastructure and the presence of U.S. forces in Europe, the Middle East, and Asia. In general, this illustrative CRP is characterized by fewer forces and bases overseas, although there are differences across the services and regions.

The CRP would make significant cuts to the U.S. overseas force posture in Europe. To enable the United States to project power into Europe and other regions in the future, critical en route airfields and ports, such as Rota, Mildenhall, Ramstein, Spangdahlem, and Sigonella, would be retained. The USAF, however, would significantly reduce the size of its combat forces in the theater—particularly those in the UK, which are located far from likely contingencies—by closing RAF Lakenheath and sending the 48th Fighter Wing's three squadrons of F-15s to CONUS. Additionally, all forces from RAF Mildenhall would return to the United States, but a modestly sized air base wing would permanently remain at the base, which would be kept as an expansible location due to its importance for mobility operations. To save money, the USAF also would shutter many of its intelligence and support facilities in the UK, planning instead to rely increasingly on NATO partners to fulfill these missions. RAF Croughton and Fylingdales would be retained because they provide critical communications and satellite capabilities that the United States needs to support intra- and intertheater operations, and Fylingdales hosts the phased array missile defense radar. The USAF would also close less vital en route locations in Lajes, Azores, and Moron, Spain, and significantly downsize the size of its support units at Incirlik Air Base in Turkey. Shrinking the global en route infrastructure, however, would create increased risk. If the United States were to lose access to any of the primary en route locations due, for instance, to bad weather or political denial, it may be unable to quickly respond to all contingencies that might arise. But on the other hand, if a contingency had strong international support, access to new bases might be quickly granted. Aviano Air Base in Italy would be closed, but the two squadrons of F-16s that are a part of the 31st FW would be moved to Spangdahlem, which currently has only one F-16 squadron in the 52nd FW. As a result of these changes, the only permanently stationed USAF forces would be located at the important mobility hubs of Ramstein and Spangdahlem in Germany.

Currently, Germany is also home to the bulk of USAREUR.[7] For the CRP, the Army would withdraw most of its forces from Germany and close most of its garrisons.

---

[7] Because USAREUR is already in the process of consolidating its presence in Germany, further consolidation does not seem possible without also necessitating considerable additional construction, which would be costly. However, ground forces in Germany are not particularly well positioned to respond to contingencies arising in the Middle East. For instance, during OIF, it took U.S. Army units stationed in Germany nearly as long to deploy to the Middle East because a host of nations restricted or denied U.S. forces the right to transit through their

In particular, the Army would close its facilities at Ansbach/Illesheim, Baumholder, Grafenwoehr, Hohenfels, and Vilseck, in addition to the planned closures already in progress at Heidelberg, Schweinfurt, and Bamberg. The 170th and 172nd brigades would deactivate as planned, while elements of the 173rd Airborne BCT would proceed with the planned consolidation in Vicenza, Italy. All additional forces at these bases would return to CONUS. The Army's presence at Kaiserslautern would also shrink as the 21st Theater Sustainment Command, 30th Medical Command (Deployment Support), would be withdrawn, but the Medical Center at Landstuhl and the 10th AAMDC would remain. Additionally, the Army would divest itself of excess infrastructure but keep open the garrisons at Wiesbaden and Stuttgart, with Fifth Signal Command remaining at the former and the EUCOM and component forces headquarters and Special Forces units remaining at the latter. The Army would also close its training locations in Bulgaria, while retaining Mihail Kogalniceanu (MK) Air Base in Romania, which is also a TRANSCOM location, to house the Army units rotating to Europe as a part of the NRF.

The U.S. Navy would also reduce its presence in Europe, particularly in the Mediterranean, by closing its bases at Souda Bay and Larissa, Greece, but retaining its facilities in Naples and Sigonella, Italy. The Sixth Fleet, which is stationed at Naples and Gaeta, would be reduced, as the facility at Gaeta would be closed and the command ship USS *Mount Whitney* would be homeported in the United States. The Navy would go forward with the current plans for the EPAA by stationing four BMD Aegis destroyers in Rota, Spain, by 2015 and placing Aegis-Ashore units in Romania and Poland by 2015 and 2018, respectively. In contrast with the other services, the Marine Corps has a very small presence in EUCOM, consisting only of Marine Corps Forces, Europe (MARFOREUR) HQ and Marine Corps Forces, Africa (MARFORAF) HQ at Stuttgart, prepositioned equipment in caves in Norway, and the Special Purpose MAGTF that deploys to Europe as a part of the Black Sea Rotational Force, primarily in Romania. The only change to the USMC presence in Europe would be to remove the prepositioned equipment from Norway, since it is far from prioritized regions, and to relocate it to the United States.

PACOM is the other region where the United States has traditionally maintained a significant permanent military presence. Given that DoD has identified this region as its focus for the foreseeable future and there is considered to be some risk of major war demanding some in-place forces at the outset, there are fewer reductions made in this theater. U.S. forces would be removed from Misawa Air Base in Japan, recognizing that this step would incur some additional risk. In particular, TRANSCOM would lose an important en route node in the Pacific, while the USAF would reduce its ability to rapidly reinforce its presence on the Korean Peninsula. Moreover, the government of

countries, forcing the Army to sail from northern Germany (when most Army bases are actually located in the south, far from ports) to Iraq.

Japan, which greatly values the presence of U.S. forces, may feel less assured. Despite these concerns, the United States would seek to remove forces from Misawa because the USAF has a significant number of forces stationed at other locations in Japan that are more proximate to likely contingencies. The USN's maritime surveillance aircraft that rotate to Misawa would also be affected by this move, but naval maritime surveillance aircraft would still be able to operate from Kadena Air Base, as they do now. In addition, PACOM would relinquish its rights to some access bases, primarily in Southeast Asia, and reduce its rotations to its remaining access bases in Darwin, Australia; Singapore; and U-Tapao in Thailand. For its part, the Army would consolidate its presence in Japan at Camp Zama by closing its facilities on Okinawa. The forces at these locations would come back to the United States—with the exception of the Army's First Battalion, 1st Special Forces Group, which would move to Camp Zama.

USFK is in the process of consolidating its presence into two large hubs at Pyeongtaek (Camp Humphreys and Osan Air base) and Daegu.[8] Although this process involves all of USFK, the effort centers on the relocation of the Army's Second Infantry Division from numerous dispersed camps along the demilitarized zone and its primary garrison in Yongsan to large hubs located in less-populated areas in southwestern and southeastern Korea. Unlike the consolidation processes in other countries, such as Germany, the Army does not plan to remove any more forces from South Korea, just move them away from the most heavily populated areas.[9] Given the tense security situation on the Korean Peninsula, and the fact that the Army is already consolidating its footprint, the CRP would make no further changes to the U.S. presence on the Korean peninsula beyond completing the planned changes.[10]

Because the Pacific has traditionally been a maritime theater, the Navy has a sizable presence in this region and the Marine Corps has its only significant presence on

[8] The Korean Relocation Plan consists of the Yongsan Relocation Plan and the Land Partnership Plan. Together, these two programs are requiring a $10.3 billion construction program, with South Korea paying roughly 55 percent. See Greg H. Reiff, "Korea Relocation Plan Construction Update," briefing slides, U.S. Army Corps of Engineers, June 15, 2011. The Land Partnership Plan was signed in 2002 and aims to close U.S. Army camps near the demilitarized zone north of the Hahn River and relocate forces to two more southern hubs. The Yongsan Relocation Plan was signed in 2004 and involves closing Yongsan Garrison in downtown Seoul and relocating U.S. forces to Camp Humphreys. In total, USFK facilities will drop from the 104 that it had in 2002 to 48 facilities consolidated in southwestern (Osan/Humphreys) and southeastern (Daegu) hubs. These two enduring hubs will comprise five enduring bases: Osan Air Base, U.S. Army Garrison (USAG) Humphreys, USAG Daegu, Chinhae Naval Base, and Kunsan Air Base. Additionally a Joint Warrior Training Center North of Seoul will be constructed. See Mark E. Manyin, Emma Chanlett-Avery, Mary Beth Nikitin, and Mi Ae Taylor, *U.S.–South Korea Relations*, Washington, D.C.: Congressional Research Service, R41481, November 3, 2010, pp. 12–13.

[9] Ashley Rowland, "Fewer Bases, Same Number of Troops in South Korea, US Ambassador Says," *Stars and Stripes*, February 15, 2012.

[10] One slight change is the decision not to close Camp Casey and relocate the 210th Fires Brigade south. This revision was made because it was questionable whether the brigade would be able to respond quickly enough in the event of a North Korean attack. Jon Rabiroff, "US Artillery Unit May Not Realign South of Seoul," *Stars and Stripes*, June 18, 2012.

foreign soil—in Japan, primarily on the island of Okinawa. In the CRP, the Navy would keep most of the components of its current presence in place, including homeporting the aircraft carrier USS *George Washington* and the Seventh Fleet at Yokosuka, Japan. The Navy would also move ahead with its plan to rotationally deploy four LCSs at Changi in Singapore. The Marine Corps would seek to relocate the majority of its ground and logistics forces and the III MEF Headquarters Group from Okinawa to CONUS, while also choosing not to proceed with two of the three concurrent battalion landing team and aviation detachment UDP rotations of CONUS-based units to Okinawa that are planned for temporary reinstitution. The marines removed from Okinawa would be stationed in CONUS, with plans for increasing stationing in and rotations to Hawaii and Guam eliminated and the size and frequency of planned rotations to Robertson Barracks in Darwin, Australia, reduced. In addition to mitigating Okinawan opposition to the U.S. military presence, this could also yield cost savings. The force elements required to constitute the 31st MEU would remain in Okinawa, supported by UDP deployments of battalion landing teams and aviation detachments, and the four ships assigned to COMPHIBRON 11 that constitute the *Bonhomme Richard* ARG would remain in Sasebo to maintain the responsive ARG/MEU capability. Air elements capable of early strike or self-deployment would also remain in Okinawa and at MCAS Iwakuni. If required for contingency response, the Marine Corps infantry realigned to CONUS would be either airlifted to marry up with prepositioned shipping or embark in ATF shipping in CONUS for use in contingency phases beyond immediate response. With the most mobile, self-deployable, and longer-range strike assets retained in Okinawa, the III MEF and ground force relocation to CONUS would likely not significantly affect contingency responsiveness.

The illustrative CRP would also seek to make some changes to the current U.S. force posture in CENTCOM. In particular, while the Army and USAF would try to keep most of their facilities in Kuwait as expansible bases because of the HNS, they would not maintain a large, continuous rotational presence, but keep only small supporting units at the bases.[11] Therefore, the USAF would move the 386th AEW to CONUS and the Army would reduce its presence to essential support personnel. Similarly, the 380th AEW currently stationed at Al Dhafra Air Base would return to the United States, but the base would retain essential supporting units so that forces could quickly return, if necessary. The primary USAF presence in the region would be the 379th AEW stationed at Al Udeid Air Base, which also contains CENTCOM's CAOC. After 2014, when combat forces will be withdrawn from Afghanistan, the United States would seek to turn all of its current bases over to the Afghan government, with the exception of Bagram Air Base, which it would aim to keep as an expansible base. Removing some ground and air forces from CENTCOM could weaken the U.S. ability to deter Iran and to respond rapidly to contingencies in the Persian Gulf.

---

[11] U.S. Senate, 2012.

Nevertheless, the rationale would be that the United States could save money while mitigating risk by keeping a considerable naval presence in the region, in addition to significant air capabilities, prepositioned equipment, and access to facilities.

The United States has a very small presence in AFRICOM, and would retain its one base on the African continent—Camp Lemonier in Djibouti—as well as the en route and communications node at Ascension Island. Nevertheless, the United States would relinquish its other access bases.

Illustrative Global Responsiveness and Engagement Posture

The global responsiveness and engagement posture (GREP) aims to create an overseas military presence that maximizes the United States' ability to respond rapidly to smaller-scale contingencies in vital regions and to build the military capabilities of allies and partners. U.S. force posture would resemble a regional hub-and-spoke network, in which permanently stationed U.S. forces are consolidated at regional hubs (i.e., one or more primary bases) that can support rotational forces that periodically deploy to the spokes (i.e., access bases) for operations or exercises. This posture rests on the assumption that U.S. forces will be well positioned to react quickly to small-scale contingencies, including humanitarian assistance/disaster relief, peacekeeping operations (PKO), counterterrorism, counterpiracy, limited strikes, counternarcotics, and no-fly zones. Moreover, the GREP also assumes that a regionally focused hub-and-spokes posture would enable U.S. forces to engage in sustained and diversified security force assistance activities with partners in key regions of the world. A regional hub would allow partner nations' militaries to exercise with U.S. forces at advanced training facilities, thereby improving their capacity to respond to contingencies on their own (or as a part of a coalition). At the same time, spokes or access bases would be important because they allow U.S. forces to work with a larger number of partners, some of whom may not be proficient enough or have the funds to travel to the U.S. regional hubs. Rotating American forces to spokes also has the added benefit of familiarizing U.S. forces with partner facilities that the United States hopes to secure access to for future operations.

In most regions, hubs would comprise a mix of facilities, including garrisons, airfields, ports, and ranges. The facilities selected as hubs would be proximate to potential trouble areas where U.S. forces might have to operate (i.e., in Southern Europe, the Middle East, Africa, and the Pacific) to reduce response time. Moreover, facilities designated as hubs would contain or be near mobility infrastructure—such as ports, airfields, and rail lines—to facilitate expeditionary operations. Spokes would be geographically and politically distributed to increase the probability of securing permission to use a facility for any particular operation and to increase the number of partner nations that can train with U.S. forces. Moreover, because it is uncertain where small-scale contingencies would develop, it is prudent for the United States to have access to a number of geographically dispersed access bases in each region.

In the GREP, the United States would station a mixture of forces at each hub, so that it has a wide range of capabilities at its disposal for rapid response and for engagement activities. In other words, because the United States is not certain what type of contingencies might arise or where and when they might occur, it would adopt a capabilities-based approach to force planning that emphasizes versatile forces that can carry out an array of different operations.[12] In general, therefore, the permanently stationed forces would be expeditionary in nature. The Army, for instance, would forward-base light or medium units, such as airborne, infantry, or Stryker brigades, as response forces, while the USMC would station units capable of constituting an MEU or a smaller MAGTF, depending on the mission, at key hubs. The USAF would forward-base composite wings that include a mix of fighters, tankers, ISR platforms, and airlifters at each hub. Similarly, the USN would station a mixture of surface vessels, such as MCM ships, LCSs, destroyers (DDGs), and carriers (CVNs), as well as SSNs and maritime surveillance aircraft (e.g., P-3s, P-8s, or broad area maritime surveillance UAVs). The United States would also want to tailor the forces that it rotates to a region to particular partners' capabilities. For instance, along with Special Forces, the Army would send smaller infantry units to work with less-capable partner ground forces, while it would deploy armored units to exercise with Western European partners because they have heavy forces.

Notional Global Responsiveness and Engagement Posture Changes

In the illustrative GREP, the United States would reshape its current overseas presence into a regional hub-and-spoke construct to facilitate rapid response and engagement. Due to the existing infrastructure, hubs would typically consist of existing primary bases that collectively could host a diverse set of capabilities.

Much of the existing posture in EUCOM is already oriented toward expeditionary operations and engagement. Therefore, there would be fewer changes than in the illustrative CRP. For instance, the Army would complete the consolidation process that is already underway by deactivating the 170th and the 172nd brigades, moving elements of the 173rd Airborne BCT to Vicenza, and closing the garrisons at Bamberg, Heidelberg, and Schweinfurt. The 173rd, which is located in Italy, would serve as the Army's primary quick-response force, while the training facilities at Grafenwoehr and Hohenfels would be used as the engagement hub. The Army would retain its remaining BCT in Germany (2nd Stryker Cavalry Regiment), the 12th Aviation Brigade, and its enabling forces to facilitate expeditionary operations and enable engagement with

[12] For more on capabilities-based planning, see Paul K. Davis, *Analytic Architecture for Capabilities-Based Planning, Mission-System Analysis, and Transformation*, Santa Monica, Calif.: RAND Corporation, MR-1513-OSD, 2002.

NATO countries. Additionally, the Army would rotate Stryker brigades to Novo Selo in Bulgaria and MK in Romania to facilitate expeditionary operations into the Middle East and improve allied capabilities through training activities. Because most of the United States' European allies have heavy armies, the U.S. Army would also regularly rotate an ABCT from CONUS. Additionally, two Special Forces battalions would be stationed at Baumholder to improve the U.S. capability to counter irregular threats, such as terrorism, through rapid response and to build partner capabilities around the periphery of Europe and in Africa.

The USAF would retain the critical mobility hubs of Ramstein and Spangdahlem air bases in Germany, but it would seek to move the 52nd Fighter Wing to Aviano Air Base in Italy, leaving only a small air base wing at Spangdahlem. The United States would seek to move the fighter squadron from Germany to Italy so that it would be closer to areas where small-scale contingencies are likely to arise. In short, the USAF hub would consist of Ramstein and Aviano, with Spangdahlem as an en route base instead of a fighter base. Due to the UK's distance from likely hotspots, the 48th Fighter Wing would return to CONUS, but the USAF would keep open its intelligence and communications facilities to enhance responsiveness and interoperability with NATO partners. Moreover, RAF Mildenhall and its current tenants would remain unchanged due to the fact that the aerial refueling wing and special operations group would provide critical capabilities needed to respond rapidly to contingencies.

With the Sixth Fleet stationed at Naples and Gaeta in Italy, the Navy is fairly well positioned to respond to smaller-scale contingencies in North Africa and the Levant. In the illustrative GREP, the USN would rotate a CSG to the Mediterranean for six months (reducing its CSG presence in CENTCOM), so that it would have an enhanced ability to respond to potential situations, but also could engage with more European partners. For its part, the USMC would keep its prepositioned equipment in Norway, which it would use for its training activities and rapid response. The USMC would increase its rotations as a part of the Black Sea Rotational Force by sending a Special Purpose MAGTF twice a year for six-month deployments.

In the illustrative GREP, the United States would seek to make some adjustments to its forces in PACOM, largely by adding additional access bases or spokes in Southeast Asia and more frequently rotating forces to these locations. The Army would continue with the ongoing consolidation of its presence in South Korea into two hubs, and would retain its current force levels so that it could continue to engage with all levels of the Korean Army. By contrast, because of the limitations on using forces in South Korea for operations beyond the Korean Peninsula, the USAF would return the 8th Fighter Wing to CONUS, but it would keep the 51st Fighter Wing at Osan Air Base to engage with South Korea. The USMC, which does not have a large permanent presence in South Korea, would rotate forces from Japan to Camp Mujuk.

Bases in Japan, therefore, would serve as the regional hub for the Navy, the Marine Corps, the Army, and the Air Force. The USAF would designate Yokota and

Kadena air bases as its regional hubs. Kadena would be used because of its proximity to the South China Sea, a potential flash point, and Yokota is a critical en route location. Because the 18th Wing and the 374th Airlift Wing provide sufficient capability to engage with partners and respond to smaller-scale contingencies, and because Misawa is distant from many trouble spots, Misawa Air Base would be closed and the 35th Fighter Wing would be returned to the United States. Closing Misawa and Kadena, however, could increase U.S. risk if a conflict broke out on the Korean peninsula. Moreover, TRANSCOM would lose an important en route node at Misawa, and the governments of Japan and South Korea may feel less assured of the U.S. commitment to their security. The USMC would increase its capacity to respond to contingencies and to engage partners by fully reinstituting the UDP as planned for the near term, rotating three battalion landing teams and three aviation combat element detachments to Okinawa. Additionally, the USMC would rotate a MAGTF and station elements necessary for rotating and constituting an MEU at Robertson Barracks in Australia. From these hubs in Japan and Australia, all of the services would strive to gain agreement to rotate forces to access bases located in Thailand, the Philippines, Indonesia, and Vietnam. Moreover, the United States would seek to gain an agreement to permanently station small supporting Marine Corps, Navy, and Air Force units at Subic Bay in the Philippines so that the base could frequently host rotational forces. Doing so, however, would require the Philippine Senate to pass legislation authorizing such a move.[13]

In CENTCOM, the USAF bases at Al Udeid and Al Dhafra would become the regional hub, while the Army would seek to rotate Air Defense Artillery battalions to train with the UAE. Moreover, the Army would aim to transform Camp Buehring in Kuwait into a regional training center by having a continuous—although rotational—ABCT presence and improving training ranges and capabilities. CENTCOM would downsize to one access base per country but seek to maximize the number of nations that it could engage with. After combat forces withdraw from Afghanistan, the United States would strive to rotate forces as a part of a modestly sized training mission to Kabul and Bagram. The USN would aim to expand the size of its presence at Bahrain and deploy MCM ships and LCSs from the Fifth Fleet to train with Persian Gulf partners.

[13] In 1999, the Philippine Senate ratified the Visiting Forces Agreement, which authorized the U.S. military to temporarily deploy to the islands, provided that U.S. forces were not engaged in combat operations. See Thomas Lum, *The Republic of the Philippines and U.S. Interests*, Congressional Research Service, RL33233, April 5, 2012, p. 14. Section 25 of The Philippine Constitution states "After the expiration in 1991 of the Agreement between the Republic of the Philippines and the United States of America concerning military bases, foreign military bases, troops, or facilities shall not be allowed in the Philippines except under a treaty duly concurred in by the Senate and, when the Congress so requires, ratified by a majority of the votes cast by the people in a national referendum held for that purpose, and recognized as a treaty by the other contracting State." Office of the President of the Philippines, "The 1987 Constitution of the Republic of the Philippines—Article XVIII," the Official Gazette Online.

In the illustrative GREP, the United States would seek a significant increase its presence in AFRICOM by transforming Camp Lemonier into a joint regional hub. Djibouti would host Army infantry battalion rotations, an expanded composite AEW, and Marine Corps and Navy Special Operations units. The USAF would diversify the capabilities currently stationed in Djibouti by transforming the AEG into an AEW that would include the continuous presence of F-15E Strike Eagles and MQ-9s. These forces would engage in counterterrorism and counterpiracy operations and would rotate to access bases throughout the continent in an effort to improve partner nations' capabilities. The Army would seek to rotate small units to Burkina Faso and Uganda, while the USAF would engage in training activities in South Sudan, Ethiopia, Kenya, Senegal, Mauritania, and Morocco. The USAF would use the base on the Seychelles for UAV operations. Most of the Navy's engagement activities would be conducted through rotations of the Africa Partnership Station around the continent.

Illustrative Major Contingency Posture

In the illustrative major contingency posture (MCP), the United States positions its forces overseas so that they would be situated to deter and, if deterrence fails, conduct large-scale operations against potential adversaries, in particular Iran, North Korea, and China. The United States would place greater forces forward that are capable of conducting major operations against these three potential adversaries.

Because the MCP is focused on three potential adversaries, the character of the U.S. military presence would vary by region. In general, however, the United States would divest itself of bases and forces that would not be useful against one of these three adversaries. Consequently, the United States would retain only the bases in Africa and Europe that provide critical enabling capabilities for intertheater operations or that could be used for operations in the Middle East. European air and naval bases have significant throughput capacity, which would be necessary to transport forces and supplies to the Middle East. Moreover, because of the proximity of some bases in Southern and Eastern Europe to the Middle East, ground forces at these facilities could be rapidly deployed to the Middle East, while aircraft could fly combat missions directly from the air bases. In short, Europe either would support or directly enable power projection into the Middle East. By contrast, the United States would strive to increase the resiliency of its forces in the Middle East and Asia because they could be held at risk by the three potential adversaries' capabilities. Improving the survivability of its forces in these theaters would entail hardening bases and critical infrastructure and dispersing U.S. forces to a greater number of forward operating locations.

The types of forces that the United States should station would also vary by region. In EUCOM and Western Africa, the United States would forward-base light, rapidly deployable ground forces and supporting units. In PACOM and CENTCOM, the United States would need to add combat forces and preposition additional equipment and munitions.

Notional Major Contingency Posture Changes

Because the illustrative major contingency posture is focused on preparing for combat operations against three particular countries, the United States would strive to significantly revise its existing posture by closing bases not useful for these particular scenarios, bolstering the number of forward-stationed combat forces proximate to the likely adversaries, and securing additional forward operating locations.

In EUCOM, there would be notable changes to the current Army and USAF postures. In particular, the Army would close many of its bases in Germany, which would not significantly improve deployment responsiveness for contingencies against Iran compared with deployment from CONUS. Consequently, the Army would retain only its garrisons at Wiesbaden, Stuttgart, and Kaiserslautern because the units (e.g., Fifth Signal Command, hospital) at these locations would directly support operations in the Middle East. With the closure of the bases at Ansbach and Illesheim, the Army would seek to relocate the 12th Combat Aviation Brigade to Camp Buehring in Kuwait. To improve the Army's ability to be prepared for operations in the Middle East, MK Air Base, which is located on the Black Sea in Romania, would be targeted for transformation into a primary base. Vilseck, Grafenwoehr, Hohenfels, and Baumholder would be closed, and the Army would aim to move the Second Cavalry Regiment and key supporting units from Germany to MK.

In contrast, the USAF would seek to augment the 52nd Fighter Wing at Spangdahlem by adding one F-16 squadron from the United States. The expanded 52nd Fighter Wing could quickly respond to crises with Iran by deploying its forces to access bases in the Persian Gulf. The USAF would keep the 31st Fighter Wing at Aviano Air Base for the same reason. Moreover, fighter squadrons from these bases would regularly rotate to airbases in Bulgaria and Romania as a signal to Iran and to improve USAF operators' familiarity with these facilities, which could be used in a contingency against Iran. In the UK, however, the USAF would close RAF Lakenheath, which is far from the Persian Gulf, and seek to relocate the 48th Fighter Wing to PACOM to expand USAF capabilities in that critical theater. The USN and USMC presence in Europe would remain essentially unchanged. In particular, the Navy would continue with the plans for the EPAA by stationing BMD units in Spain, Romania, and Poland.

In PACOM, the United States would seek to posture its forces for major contingencies against two potential adversaries, North Korea and China. As a result, the United States would aim to add combat units to South Korea, in particular capabilities that the South Koreans have limited quantities of, namely airpower and precision-guided missiles. Therefore, the Army would seek to station a second fires brigade to Camp Casey and a combat aviation battalion to Camp Red Cloud. The USAF would also aim to increase its presence on the peninsula by permanently stationing a fighter wing at Suwon Air Base. Finally, marines would regularly rotate from Japan to Camp Mujuk.

On Okinawa, the Army would aim to improve the air defenses at Kadena Air Base by adding an additional Air Defense Artillery battalion (PAC-3). To signal continued U.S. commitment to defend the Philippines and to deter Chinese adventurism, the Army would strive to obtain an agreement to rotate infantry brigades and SF battalions to the islands.[14] The Army would also seek to rotate an Army infantry brigade to Thailand to participate in multinational training exercises.

The USAF would aim to bolster significantly its combat capabilities in the region by adding the three squadrons of F-15s from RAF Lakenheath to Andersen Air Force Base in Guam. Fighter squadrons from Japan and Guam would regularly rotate to air bases, which would be improved, in the Northern Marianas. In addition, the USAF would aim to obtain access to and make infrastructural improvements at air bases in Malaysia, Vietnam, northern Australia, and the Philippines. It would seek to rotate fighter squadrons from CONUS to these Southeast Asian facilities, while bombers and tankers would rotate to Darwin and Tindal in Australia. This network of access bases and rotations would help the USAF to operate in a dispersed manner in the event of a major contingency and would enable larger aircraft to operate from bases beyond the most severe missile threat. Additionally, the USAF would seek to station a detachment of RQ-4s on Australia's Cocos Islands to enhance situational awareness in the Indian Ocean and Southeast Asia.

The USN and USMC would also significantly increase their permanent presence in PACOM by seeking to station the ships assigned to a PHIBRON currently homeported in CONUS with the force elements required to constitute an MEU (which would be supplemented by a UDP deployment of a battalion landing team) at Subic Bay/Cubi Point in the Philippines. In addition, the USMC would aim to base a Marine Special Operations Battalion at Puerto Princesa Palawan to deter a conflict in the South China Sea over the disputed Spratly or Paracel islands. The Navy and the Marine Corps would also increase the number of combat forces that are stationed on U.S. territories in the Pacific, in particular in Hawaii and Guam, while seeking to retain more marines in Okinawa than currently agreed. The Navy would move a CSG from CONUS to Hawaii, while the Marine Corps would station the force elements required to constitute an MEU from CONUS in Guam, with its units filled out by UDP rotations, while seeking to keep its current force levels on Okinawa. In Australia, the USN would seek to homeport an SSN at Perth, while the USMC would aim to station an MEU-sized MAGTF at Robertson Barracks in Australia, with the units provided through UDP rotations. Finally, the Navy would strive to station a detachment of broad area maritime surveillance UAVs at Port Blair airport in the Andaman Islands, to increase surveillance over the Straits of Malacca.

---

[14] This would require the Philippine Senate to pass legislation. See the previous footnote in this chapter.

In CENTCOM, the United States would seek to increase its military presence in the Persian Gulf to deter Iran. In particular, the Army would aim to establish a continuous rotational presence of an armored brigade at Camp Buehring, in addition to relocating the combat aviation brigade from Germany. The Army would also try to preposition a second set of ABCT equipment at Camp Arifjan, Kuwait. The USAF would maintain the airlift AEW stationed at Ali Al Salem Air Base and strive to add a continuous fighter squadron presence at Al Jaber Air Base in Kuwait. In Qatar, the USAF would maintain the 379th AEW and the Army would endeavor to preposition a high mobility artillery rocket system battalion set at Camp As Sayliyah.

The USAF would maintain its continuous rotational presence of tankers and ISR aircraft at Al Dhafra Air Base. Meanwhile, the Army would aim to add a Terminal High Altitude Area Defense (THAAD) battery and a fires brigade to the Fujariah port and air base. Air bases in Afghanistan could potentially be useful for scenarios against Iran, but given the logistical burdens and post-2014 overflight requirements of these land-locked bases, the USAF would seek to turn all air bases over to the Afghan government except for Bagram Air Base, which would be maintained by an air base wing. Finally, the USAF would try to increase fighter and bomber squadron rotations to forward operating locations in Oman.

Summary of Posture Changes

Table 9.2 provides an overview of the notional changes made in the illustrative postures.

The next chapter turns to an assessment of how these illustrative postures affect strategic benefits, risks, and costs.

Table 9.2
Notional Changes Made in Illustrative Postures

Illustrative Posture Type	Major Changes
Baseline posture	• Army in Europe: remove 2 Army BCTs and 2,500 personnel from enabler units • Reinstitute full level of USMC UDP rotations to PACOM
Cost-reduction posture	• Most Army (~16,400) from Europe • About half of Air Force (~6,400) from Europe • III MEF and USMC ground and logistics units from West Pacific (~7,400) along with 2/3 of rotational personnel • Some USAF (~1,500) from West Pacific
Global responsiveness and engagement posture	• Some USAF from West Pacific (~2,900) and UK (~3,500) • Broadly distributed rotations added
Major contingency posture	• Some reductions in Germany and UK (~8,200) • Increases in West Pacific (~11,300) • Increased rotations in Pacific, Middle East, and Eastern Europe

In the Pacific, the relocation of most of the logistics, ground, and HQ elements of III MEF to CONUS would have only a minor effect on deployment responsiveness. In the cost-reduction posture, the 31st MEU and its associated ARG would remain based in Japan, as would Marine Corps aviation assets. These forces would maintain similar presence and response capabilities to those currently available. The realigned units would require transport from CONUS, either by air to marry up with MPSRON Three and the onboard equipment or by sea by means of amphibious lift. This is nearly the same as in the existing posture, with forces positioned in Okinawa, the main difference being flight time from CONUS in the illustrative posture as opposed to flight time from Okinawa. Fixed-wing aircraft would still be in place to provide regional capability and rotary-wing aircraft would retain an in-theater self-deployment advantage as compared with being U.S.-based.

In CENTCOM, by maintaining most prepositioned equipment and several bases in "warm" expansible status, the effects on deployment of removing the airlift AEW would be limited. Not maintaining access to as many bases in Africa could, however, reduce AFRICOM's ability to respond rapidly to humanitarian crises and could hamper efforts to support rotational security cooperation and counterterrorism deployments. Maintaining Camp Lemonier as the sole U.S. facility in AFRICOM would maintain the U.S. military's ability to deploy to the Horn of Africa, but would provide little benefit to the rest of the continent.

Global Responsiveness and Engagement Posture

The GREP would have only a small positive effect on deployment responsiveness, given the ability of U.S. airlift to move initial forces rapidly from CONUS. Its value on this dimension would most come into play if there were multiple and simultaneous large demands on the airlift fleet, since shorter deployments require fewer aircraft to achieve fast response times. The European posture changes are relatively modest and would have little effect on conventional ground force and air force responsiveness compared with the current posture. However, increased Special Operations Forces based overseas could enable faster response to time-sensitive events in the surrounding region that can be handled with these types of forces. Committing to CSG rotational presence in the Mediterranean would reduce the time required to begin conducting air operations in the region, but could have important rotational implications and reduce the availability of CSGs elsewhere. In Asia, changes in the Air Force posture would increase the need to redeploy units in a Korean crisis but would not affect its ability to respond to a crisis in the South China Sea. The increased regional presence of the USMC would not significantly increase deployment responsiveness unless there was also an increase in forward-deployed amphibious lift assets. A rotational ABCT in Kuwait would provide regional responsiveness, but not substantially beyond what the current prepositioned equipment provides. In Africa, the small unit rotations could increase responsiveness for small contingencies.

Major Contingency Posture

Forward-deployed forces are critical for major contingency scenarios, particularly those involving a surprise or short-notice attack by a strong adversary. South Korea is a potential example of this, given Seoul's close proximity to North Korea's heavy artillery and rocket forces, which are often hidden in underground shelters. In addition, the close proximity of Seoul to the demilitarized zone does not allow defending forces to trade space for time. In-place forces allow for an immediate response to a North Korean provocation. The ARG and MEU assigned to the Pacific, fighters at Misawa Air Base, and prepositioned equipment on the Korean Peninsula and in Guam are also in place for rapid deployment. Other USMC units would also be able to meet up relatively rapidly with their afloat equipment, but the advantage of these units being in the Pacific—as opposed to being in United States—is limited given the ability to fly in and marry up with equipment. It is also not clear what benefit deploying the 48th Fighter Wing to Guam would have since the increased responsiveness to regional crises would be marginal because the aircraft would still need to deploy forward to conduct operations.

Other major contingencies often allow for more deliberate buildup of forces. In the case of the first Gulf War, no in-place U.S. forces existed, and it is arguable whether the rapid airlift of the ready brigade of the 82nd Airborne Division deterred Iraq from continuing its offensive into Saudi Arabia. In the second Gulf War, a U.S. Army brigade, as well as Army prepositioned materiel, both afloat and ashore, were in place, though a substantial and deliberate buildup of forces also supported the effort. The Army and Navy/Marine Corps prepositioning programs are also available to support Persian Gulf scenarios on short notice in this posture.

In Europe, moving bases to the Black Sea region would have only a very minor effect on responsiveness for potential conflict with Iran, compared with keeping them in Germany. It is unclear whether the mobility infrastructure in the Black Sea region exists to support rapid-deployment operations for major units.

Increased levels of prepositioned equipment could be an important component of this posture. Added afloat prepositioning of Army, USMC, and USAF equipment would increase flexibility and allow material to swing to where it would be needed in an Asian crisis. The additional prepositioned sets and rotational presence in the Gulf region would enable the rapid buildup of a sizable force in the region.

Availability of Basing for Direct Operational Support

A related responsiveness analysis examined the illustrative postures based on their ability to support requirements and military operations through the basing of air assets in safe havens in each of the scenarios. This analysis assesses the three postures in terms of the degree to which retained bases affect the ability of the postures to enable effective direct fighter, ISR, and sustainment support. This was done using the illustrative, geo-

The CRP, with its emphasis on consolidating the U.S. military footprint through reduction of forces both temporarily deployed and permanently stationed abroad, may contribute to erosion of remaining forces' assurance and deterrence capabilities. If one excludes the en route and other critical enabling infrastructure common to all three postures, the effect on forward presence may be fairly significant in the signaling of credibility and capability. The reduction of costly demonstrations of *intent* could affect adversaries' calculations of U.S. deterrent commitments, leading them to conclude that the United States was less committed to preventing aggression. Also, the reduction of forward presence in turn contributes to adversaries' perceptions, even if not the reality, that the United States has *less capacity* to prevent aggression. This perceived reduction in intent and capability could lead an adversary to calculate that a quick initial victory at relatively low cost was possible.

However, given that many of the overall reductions occur in Europe, with fewer in higher-risk areas such as the Persian Gulf, South Korea, and the Pacific, the overall effects on deterrence should be limited. An adversary could still be deterred by the U.S. ability to project power rapidly from a distance, especially since even the CRP retains the critical overseas enabling infrastructure for transporting and sustaining U.S. forces overseas. Additionally, adversaries are aware of the broader array of U.S. capabilities that can strike from afar, deploy rapidly (e.g., air units), and remain present through maritime presence.

The CRP could have some cost benefits if it led to greater investment by partners in their own defense capabilities to compensate for the change; but the reduced opportunities for ongoing security cooperation may diminish the ability of the United States to shape those decisions. As noted in Chapter Four, reduced presence could also undermine the close military cooperation and willingness of local partners to commit to military operations, heightening the risk of an adversary's calculation of reduced willingness of allies to collectively respond to aggression, although the evidence on this is unclear. The lack of engagement within the region will also decrease U.S. understanding of regional dynamics and make it more challenging to develop close working relationships with local militaries, leading to more uncertainty inherent in deterrence and assurance. Therefore, among the three illustrative posture options, the CRP performs least well based on the estimated deterrence and assurance value of forward presence.

The GREP and MCP both provide a more robust overseas presence, albeit oriented differently. The GREP provides flexibility in responding to a number of contingencies by utilizing a diverse network of facilities—placed within range of potential trouble areas—and mobility infrastructure to reduce response time. Spokes are geographically and politically distributed to reduce the risk of overdependence on a small number of bases that could be denied for a variety of political and operational reasons. Also, with its diverse basing construct, the GREP enables security cooperation in various regions, at least increasing the prospects of partner participation in response to local aggression and U.S. understanding of local political and military dynamics.

While estimating the deterrent value of this is difficult to quantify, adversaries would have to take into account the flexibility and responsiveness of the distributed posture, again increasing the potential risk and uncertainties faced in any planned aggression.

The MCP places heavy emphasis on an overseas presence oriented to countering specific high-level threats: Iran, North Korea, and China. Changes in this posture emphasize the importance of positioning forces to deter major aggression and to assure regional neighbors and security partners. It does so while still retaining the ability—if less so than in the GREP—of responding to smaller-scale contingencies. But the focused emphasis on major contingencies involving these states is designed to maximize deterrence. U.S. overseas presence would continue to be a significant factor in increasing the prospects of partner participation (the most powerful case of course being South Korea, for a North Korean contingency). The MCP performs much more strongly than the CRP along the deterrence criteria, and performs more strongly than the GREP with respect to the three identified countries. However, in implementing such a strategy, the increased vulnerabilities of forward forces to anti-access threats needs to remain a top consideration. Relative to the GREP, there is also an additional risk that U.S. forces will be positioned less favorably for rapid-response contingencies elsewhere that involve significant forces. This could undermine deterrence in these other regions if adversaries there conclude that U.S. force availability and responsiveness is limited as a result of the increased focus on specific threats. However, for smaller-scale contingencies or slowly evolving situations, this is less likely to be a major concern. Finally, the heightened presence for major contingencies could risk increasing tensions and bring negative political pressure on host-nation governments for allowing an expanded, sustained U.S. presence. In this instance, U.S. presence could undermine assurance rather than reinforce it. Political dynamics and context are key, and engagement through presence may help to provide the understanding necessary to craft the proper balance.

Security Cooperation

U.S. government strategic guidance documents maintain that overseas posture improves security cooperation between the United States and partner countries. The 2012 Defense Strategic Guidance emphasizes security cooperation as a means to improve U.S. military capabilities and to build partnership capacity. The analysis in Chapter Four highlights the likely impacts of overseas posture on security cooperation to qualitatively evaluate the effects of posture changes. The two most significant effects are on training to increase partner capabilities and improving U.S. capabilities to operate with partners. Chapter Four also analyzes the value of rotational forces to execute security cooperation activities.

In general, we conclude that almost all of the effects on security cooperation would stem from reductions of permanently stationed forces as opposed to consolidations, force repositioning, or closures of minimally manned access sites. However, where the postures replace these with rotational deployments, the effects would be somewhat mitigated. None of the posture changes would produce meaningful changes in security cooperation benefits produced by the Navy and Marine Corps, given their continued emphasis on maintaining forward presence, engagement, and expeditionary activities.

Effect on Training to Improve Partner Capabilities

The CRP would significantly reduce the U.S. ability to improve partner capabilities and interoperability. In particular, fewer forces in Europe would lead to less security cooperation. For the USAF, the loss of the 48th Fighter Wing and forces from the UK would have moderate to significant impact on USAFE's ability to conduct security cooperation in both Europe and Africa.[6] The U.S. Army would face even more serious constraints on security cooperation under this posture, given the reductions in its footprint in Europe. In addition to the loss of several brigades in Europe, leaving just one BCT and minimal enablers, the CRP eliminates USAREUR's JMTC, which has been the primary training venue for multinational training. Most European partners would receive far less training with U.S. forces without the facilities in Europe, since most partners pay their own way for this training but would not likely pay to go to training sites in the United States. Therefore, closing the JMTC would have a large influence on security cooperation training with European partners.

The GREP would have a positive effect on improving the training of new or developing partners. For example, increased Air Force presence in the Philippines may slightly improve training efforts, while reductions in South Korea and Japan may have a slightly negative effect. Increased rotations in new regions such as Southeast Asia and Africa would result in modest improvements in training. The GREP also retains the JMTC, a cornerstone for multinational training in Europe.

The MCP would have a mixed impact on security cooperation–oriented training. For the Air Force, this posture would improve opportunities for security cooperation, with greater fighter presence in Europe, the Middle East, and Asia. For the Army, the closure of access sites would complicate security cooperation but not exceedingly so, whereas the loss of JMTC and some combat and supporting units would significantly reduce the ability of the Army to conduct extensive training with partners in Europe and Africa. The negative effect on training due to the loss of the JMTC would be similar in scope to those described in the CRP.

---

[6] USAFE is dual-hatted as U.S. Air Force Africa.

Improving U.S. Force Capabilities

Compared with those based in the United States, U.S. forces based overseas often benefit from more frequent training with foreign partners. In particular, U.S. forces learn to operate more effectively in a coalition and adapt to foreign, oftentimes austere, environments. If the United States continues to emphasize the importance of coalitions for future conflicts, especially for complex conventional military operations, interoperability with modern European militaries will remain important. The presence of U.S. forces abroad also provides the exposure time and direct contact for the U.S. military to better understand regional dynamics. By their sustained presence, U.S. forces gain in-depth knowledge of the local conditions that could be invaluable in a contingency.

The CRP, with its more constrained permanent operational locations, will result in reduced exposure of U.S. forces to various operating environments, particularly in Europe and its periphery, where extended engagements with partners in their own regional surroundings would be reduced. By contrast, the MCP provides more enduring presence and the associated operational benefits from in-depth familiarity, but is heavily focused on specific regions pegged to the three states of concern. So while it is superior to the CRP along this dimension, it does have a narrow focus, driven by its emphasis on major contingencies. The GREP provides the most diverse set of operating locations and associated security cooperation venues, and provides the greatest overall range of operational exposure for a much wider array of contingencies.

The Value of Rotational Forces for Security Cooperation

Rotational forces are a component of posture that provides opportunities for security cooperation and, in some cases, can mitigate the consequences of reductions in permanently stationed forces. The Navy's continuous operations at sea mean that much of its force conducting security cooperation overseas is, in effect, rotational or on temporary deployment. Under all three of the illustrative postures we analyzed, the Navy could continue to conduct large, multinational security cooperation events in international waters almost anywhere in the world. The Marine Corps also conducts the majority of its security cooperation outside of PACOM using rotational forces, with some in PACOM also done by forces on UDP rotations.

By avoiding the high fixed costs of overseas bases, rotational forces can cost less than permanent presence, have the flexibility to train partners wherever the need is greatest, and strengthen the expeditionary capabilities of the rotating forces. On the other hand, these forces tend to focus on tactical training, with less ability to build and maintain high-level relationships, and rely on the infrastructure of other countries. This type of rotational approach to security cooperation could prove an effective partial substitute for permanently stationed forces. However, as Chapter Four notes, direct substitution is likely to be most effective with potential partners in preparing for contingencies that will not involve highly complex conventional operations and systems (e.g., FID, counterterrorism, humanitarian assistance and disaster relief). For South

Korea and many countries in Europe that are envisioned as contributors to major contingencies and that have highly capable militaries, rotational forces could only partially mitigate the loss of security cooperation capability that would come from reductions in permanently stationed forces.

As noted in Chapter Eight, any savings realized from using rotational forces will depend on the scale, frequency, and level of host-nation infrastructure for rotating forces to fall in on. Additionally, greatly expanded use of rotational forces could also require significant investments in additional personnel and equipment to ensure that the strain on the force that comes with more frequent deployments stays within personnel tempo goals. We did not determine the limits on rotational presence, given the force structure as of 2012, or how costs would change if additional units had to be added to the force structure to support rotational presence.

Risks to U.S. Installations

Political and Operational Access Risk Comparison

The United States is most likely to gain access to bases within countries that perceive an increasing level of threat from a mutual adversary. In the absence of shared threat, the United States may have to offer incentives to secure access, but many nations that are sensitive to a foreign military presence may not be willing to grant those rights simply for material benefits. Nevertheless, the analysis attempted to identify, at the broadest level, the relative degree of access risk associated with each illustrative posture, as well as to note those factors that made basing in each particular notional posture more or less reliable.

To do so, the following analysis uses two complementary analytic frameworks. The first analysis develops rough metrics to assess each of the key variables—regime type, size of U.S. military presence, and the type of access relationship—that influence peacetime access risk, and then compares them across the illustrative postures. Using the 24 illustrative scenarios from Chapter Two, the second analysis measures the diversity of host nations making operating locations available in each of the three postures to determine access risks.

Analyzing Political Risk: Regime Type, Presence Size, and Shared Threat Relationships

In the first set of analyses, developing metrics to measure the variables that influence basing access was relatively simple for two of the factors. For regime type, we considered the percentage of bases located in authoritarian states compared with consolidated democracies in each of the illustrative postures.[7] To measure the size of the U.S. mili-

---

[7] We used Freedom House's *Freedom in the World 2012* scores to determine whether a nation was a consolidated democracy or not. Only countries that received a "free" score were considered to be consolidated democracies.

tary presence, we compared the percentage of primary bases (which ostensibly increase the risk of access denial) with the percentage of smaller expansible bases and access bases in each posture.

However, there are no easily observable and quantifiable metrics that can accurately identify whether U.S. military presence is based primarily on a shared threat or a transactional dynamic, as noted in Chapter Five. Although the United States frequently provides significant economic and security assistance to countries that face a shared threat, this is not necessarily the reason that a nation provides the United States with access. Rather, the presence of economic assistance and arms sales is an additional symptom to the existence of a common threat.[8] In short, there is no reliable means by which to measure the type of access relationship with great accuracy, but we can assume that a posture focused more on deterrence will have a higher share of relationships based on common interests than one that seeks more regional diversity.

Table 10.2 displays the level of access risk across the illustrative postures based on the identified metrics. When designing the illustrative postures, we did not consider political access-risk indicators. Interestingly, however, at the aggregate level, as shown in Table 10.2, none of the illustrative postures entail significantly different mixes of regime types or base sizes. While there is no significant difference across the illustrative postures in terms of host-nation regime type or portion of primary bases, we assume that the design of the illustrative postures will yield different access relationship profiles—that is, whether the United States obtained basing rights due to a shared perception of threat or in return for some sort of material compensation.

Of course, this observation does not apply to specific bases or specific nations: In individual cases, the possibility of access denial might indeed depend on a country's regime, the size of a particular base, or its location within a host nation. It is also likely that the level of risk associated with regime type and contact varies by region. For instance, most host nations in the Middle East are ruled by hereditary monarchies,

Table 10.2
Measures of Political Risk to Base Access, by Posture

Illustrative Postures	Regime Type (% Consolidated Democracy)	Contact/Base Size (% Primary Bases)	Access Relationship
CRP	46	55	++
GREP	41	46	–
MCP	48	48	+++

NOTE: + = Less political risk; – = greater political risk.

Freedom House, *Freedom in the World 2012: The Arab Uprisings and Their Global Repercussions*, 2012, pp. 14–18.

[8] Harkavy, 2006, concludes that other nations provide the United States with bases in return for arms sales. This, however, is likely a spurious relationship. Pettyjohn, 2012, p. 66.

sary, and U.S. deployments should also be of sufficient size to absorb attack. Geography, resource constraints, and political access limitations, however, can severely constrain the space of feasible postures. Figure 10.6 shows how the three postures, created with these constraints in mind, performed against potential adversary missile threats. Despite differing goals and numbers of bases, each of the three postures has a similar level of exposure across the scenarios. For example, the percentage of bases in each posture that are residing in the threat band most exposed to missile threats (by potential adversary) is similar in both the North Korean and Iranian cases. Also, the China case has a much higher level of exposure to a serious potential threat in all three postures.

Without a large number of relatively secure bases from which to operate, there are two broad options, each with its own strengths and weaknesses. The first approach would be to minimize the number of bases in the most-severe threat band and, therefore, reduce the exposure of U.S. forces to attack. For example, the CRP has the fewest number of bases in the most-severe threat band across the three scenarios, with China and North Korea representing the greatest threats. This approach, however, has three major challenges. Within a given threat band, an adversary has a finite inventory of missiles. If a posture has a small number of bases within the most-severe threat band, then an adversary could allocate more missiles for each base in that threat band. Also, depending on the width of the threat bands, operating from outside the worst threat band might cause serious operational challenges, as tactical aircraft are forced to rely on multiple tankers for each sortie. Finally, host nations that fall within the worst threat band might demand large U.S. force deployments as a condition for access.

The second approach is to maximize the number of bases in the most-severe threat band, which would force the adversary to dilute its missile inventory among a larger number of targets (assuming that all the bases are utilized and the adversary attacks all of the bases). This could motivate one to choose the posture with the most exposure; in this case, the MCP has the most bases in the most-severe threat band across the three scenarios. This approach could result in the adversary's missile inventory being drawn down quickly, setting the conditions for U.S. reinforcements to flow into the newly safe space. This approach faces a major challenge, however, because it might require U.S. forces to operate in small unit deployments at each base to present a diffuse target array for the adversary. The U.S. military would have to develop new techniques, tactics, and procedures to operate in this manner. This approach also requires a new perspective on risk because it accepts that the adversary will launch its missiles at U.S. bases and, thus, focuses on minimizing the effect of those missiles.

Cost Comparisons of Illustrative Postures

In this section, we provide estimated aggregate cost impacts of each illustrative posture and explain the key drivers (e.g., service or region) for each. We first describe the effect of each posture option on the cost of operating and maintaining forces and installations overseas. Second, we describe the investment cost that would be necessary to transition to each posture.

One set of investment costs that is not included in the baseline, nor in any of the illustrative postures, is associated with hardening facilities that are under heavy threat from precision-guided weapons. As is discussed in this and other chapters, hardening is one possible response to the threat of highly accurate precision-guided weapons, but there are other options as well. While it is likely that some investments in hardening will be made in the future, presently we are not able to estimate the scope of such funding.

Effects on Annual Costs

The baseline posture for the cost analysis includes planned, programmed changes. About two-thirds of these changes are to the Army's posture in Europe: three bases closed, three brigades realigned, and two brigades inactivated. The remaining third is split between Army consolidations in PACOM and USAF reductions in CENTCOM. Because the Marine Corps moves in PACOM have been announced but not fully finalized and programmed, we did not include them in our baseline calculation.[12] We did include several rotational missions in our baseline calculation, including three battalion landing team and aviation detachment UDP rotations to Okinawa; an Army ABCT to Camp Buehring, Kuwait, for 12-month rotations; Army deployments in support of the NRF; and Marine Corps rotations to the Black Sea and Australia. Table 10.3 shows the aggregate net-cost effects compared with the baseline posture for the three illustrative posture options for four broad categories:

- *Bases.* This entails closing and completely vacating an overseas installation. The cost savings is the fixed-cost component from our cost analysis in Chapter Eight.
- *Forces.* This includes the variable cost effects of realigning forces, driven by personnel-related costs. Most realignments moved forces from overseas back to the United States, a few increased the U.S. overseas presence, and few involved shifts from one overseas location to another.
- *Training.* This includes the training-cost effect of realigning forces based on unit training costs for different types of units in different regions.
- *Rotations.* This includes the costs associated with providing rotational presence, assuming no changes in force structure.

---

12 Changes at Iwakuni, to include unit relocations, have been programmed, as has initial construction at Guam for the planned relocations there. Final completion of relocation to Guam will be beyond the future years defense program as of the writing of this report, with other adjustments to Marine Corps laydown in the Pacific still in the planning stage with timelines and programming yet to be determined. U.S. Marine Corps, Pacific Division, Plans, Policies and Operations, 2012b.

Table 10.3
Estimated Changes in Annual Recurring Costs ($ billions)

	Posture Option		
Cost Category	**CRP**	**GREP**	**MCP**
Bases	–1.7	–0.2	–0.9
Forces	–0.6	–0.2	0.2
Training	–0.1	0.0	0.0
Rotations (low)	–0.5	0.4	1.0
Rotations (high)	–0.6	0.6	1.2
Net change (low)	**–2.8**	**0.0**	**0.3**
Net change (high)	**–2.9**	**0.2**	**0.5**

NOTE: These are changes compared to the baseline programmed plan as discussed in Chapter One, including Army consolidations in South Korea, reductions in Europe, and the reinstitution of the full level of UDPs to Okinawa.

In the rest of this section, we describe the costs and savings of each posture *beyond* the baseline posture. The CRP would save an estimated $2.8 to $2.9 billion per year beyond the baseline posture. As seen in Table 10.4, about 70 percent—roughly $2.0 billion—of these savings would be due to reductions in Europe. About 55 percent of that is from Army changes, including the realignment of 11,000 soldiers beyond the baseline and the closure of over half of the planned remaining bases in Germany; the rest is from reductions in the USAF presence in Europe. About another 20 percent of the savings for this posture is from reductions in PACOM, split among reductions in marines stationed on Okinawa (which would return to CONUS instead of being distributed elsewhere in the Pacific, as part of the long-term plan agreed to with Japan), reductions in Marine Corps UDP rotations to Okinawa (which would be eliminated rather than being distributed to other Pacific locations as per the long-term plan), and one Air Force base closure in Japan with the realignment of a wing. The remainder of the savings is from reductions in Army and Air Force rotations in CENTCOM.

The GREP could cost about the same as the baseline, or up to $200 million per year, depending on how rotations would be implemented. It produces savings of about $400 million dollars in reductions to forces permanently stationed overseas and base closures, but these are reinvested in increased rotational operations. The largest savings are produced by USAF force reductions (but not base closures) in EUCOM and PACOM (see Table 10.5). There are increased rotational costs for the Army to Eastern Europe, Africa, and Asia, as well as some increases in Air Force rotations.

Finally, the MCP increases costs between $300 million and $500 million per year. This posture does produce savings of about $750 million from reductions in

Table 10.4
Estimated Annual Recurring Cost Changes for the CRP, by Region and Service ($ millions)

		Operating Low	Operating High
Army	AFRICOM	—	—
	CENTCOM	–190	–230
	EUCOM	–1,100	–1,080
	PACOM	–10	–10
	Total	**–1,290–**	**–1,320**
USAF	AFRICOM	–10	–20
	CENTCOM	–130	–160
	EUCOM	–840	–840
	PACOM	–240	–250
	Total	**–1,230**	**–1,270**
USN	AFRICOM	—	—
	CENTCOM	10	10
	EUCOM	–20	–20
	PACOM	—	—
	Total	**–10**	**–10**
USMC	AFRICOM	—	—
	CENTCOM	—	—
	EUCOM	–20	–50
	PACOM	–210	–310
	Total	**–230**	**–310**
Total		**–2,760**	**–2,900**

NOTE: Totals may not add exactly due to rounding.

permanent presence—mostly from reductions in the Army posture in Europe, as well as from some smaller reductions in the Air Force in Europe (see Table 10.6). But this posture entails significant increases in permanent presence and rotations. Much of the increased costs are from a buildup in the Pacific across all four services, including increased FDNF and increased rotational deployments by the Army, Air Force, and Marine Corps to the region. The remainder of the cost increases are mostly driven by Army rotations to Kuwait, followed by a variety of rotational deployments spread

Table 10.5
Estimated Annual Recurring Cost Changes for the GREP, by Region and Service ($ millions)

		Operating Low	Operating High
Army	AFRICOM	10	80
	CENTCOM	–20	–10
	EUCOM	240	300
	PACOM	40	90
	Total	**270**	**450**
USAF	AFRICOM	40	50
	CENTCOM	–20	–20
	EUCOM	–100	–80
	PACOM	–260	–250
	Total	**–340**	**–310**
USN	AFRICOM	10	10
	CENTCOM	20	20
	EUCOM	50	50
	PACOM	10	10
	Total	**80**	**80**
USMC	AFRICOM	—	—
	CENTCOM	—	—
	EUCOM	—	–20
	PACOM	30	30
	Total	**30**	**10**
Total		**40**	**230**

NOTE: Totals may not add exactly due to rounding.

among the services in EUCOM and CENTCOM, with some Army forces increases in South Korea.

Investment Costs to Transition to Posture Options

We now discuss the investment costs necessary to implement the changes proposed in these three postures. These costs are shown in Table 10.7.

Table 10.6
Estimated Annual Recurring Cost Changes for the MCP, by Region and Service ($ millions)

		Operating Low	Operating High
Army	AFRICOM	170	210
	CENTCOM	400	400
	EUCOM	–810	–780
	PACOM	270	310
	Total	**30**	**130**
USAF	AFRICOM	40	50
	CENTCOM	30	50
	EUCOM	–280	–260
	PACOM	190	310
	Total	**–30**	**70**
USN	AFRICOM	—	—
	CENTCOM	—	—
	EUCOM	30	30
	PACOM	220	220
	Total	**240**	**250**
USMC	AFRICOM	—	—
	CENTCOM	—	—
	EUCOM	—	–20
	PACOM	60	60
	Total	**50**	**30**
Total		**300**	**480**

NOTE: Totals may not add exactly due to rounding.

Table 10.7 shows the aggregate upper bound of costs for the baseline posture and the three illustrative posture options for three categories:

- *Move forces.* This includes one-time moves to relocate units from overseas bases to the United States or vice versa.
- *Close bases.* This includes a range of one-time costs to close military installations overseas, such as contract termination, and mothball/shutdown of facilities.

Table 10.7
Range of Posture Option Investment Costs ($ millions)

	CRP		GREP		MCP	
Cost Category	**Low**	**High**	**Low**	**High**	**Low**	**High**
Move forces	90	370	40	180	70	280
Close bases	—	410	—	40	—	200
MILCON	3,780	6,300	2,240	2,940	2,730	4,230
Total	**3,870**	**7,080**	**2,280**	**3,160**	**2,800**	**4,710**

- *MILCON.* This includes the one-time construction costs to expand U.S. or overseas facilities to accommodate realigned forces. To the extent there would be existing capacity that could be used, this spending would be unnecessary. In particular, this cost could be offset by the freeing of capacity from the planned downsizing of the Army and the Marine Corps. For example, the Army has announced a reduction of eight BCTs, including the two inactivating in Europe, plus there will be a reduction in support brigades as it reduces its size by 80,000.[13]

For all options, costs scale linearly with the number of forces realigned and bases closed, and MILCON makes up about 90 percent of the costs. The CRP could cost between $3.9 billion and $7.1 billion over the baseline to implement, but more likely toward the lower end given the planned force structure reductions, which would limit the need for realignment from overseas or free up sufficient CONUS capacity. This posture realigns nearly 50,000 personnel from all four services. At steady-state savings, this produces a payback in the range of from about 1.5 to three years. Note, though, that the paybacks for the individual changes within the posture vary substantially. For example, the payback period for the changes in EUCOM in the CRP is estimated at one to 2.5 years, depending on the availability of infrastructure for the Army forces that would be realigned. For the restationing of Marine Corps units from Okinawa to CONUS, the payback would be six to seven years because the variable cost differential between marines in Japan and CONUS is lower, and, without a full base closure, there are no fixed-cost savings. The Army and Air Force changes in the Pacific would have paybacks of about one year, given higher incremental variable costs per person and base closures.

The GREP realigns over 21,000 personnel beyond the baseline changes, but about 12,000 are returning to the United States and over 9,000 are moving from the United States to overseas locations, mostly in PACOM. This posture could cost about

[13] Michelle Tan, "2 Europe-Based BCTs Pack to Move Out," *Army Times*, October 13, 2012; Paul McLeary, "U.S. Army's Uncertain Future: Generals Study Battalion, Vehicle, Equipment Mix," *Defense News*, October 21, 2012.

$2.3–3.2 billion above the baseline for these trades, which, combined with increased rotations, would not produce recurring savings. Finally, the MCP could cost $2.8–4.7 billion to implement in addition to the increased annual recurring costs. These costs are driven by two factors: a reduction in Army and Air Force forces and bases in Europe, and an increase in Air Force, Navy, and Marine Corps forces in the Pacific.

Comparing Marine Corps CRP Cost Changes to the Long-Term Plan

As discussed in Chapter One and reflected in the April 2012 Joint Statement between the United States and Japan, the United States will be relocating about 9,000 marines from Okinawa. It will also establish a 5,000-person presence in Guam, increase its presence in Hawaii by a little more than 2,500, and establish a rotational presence in Australia of about 2,500 personnel. The MEF command element and HQ Group, the 31st MEU, and elements of the MAW and MLG will remain in Okinawa, with one UDP rotation to fully constitute the MEU remaining. Two other battalion landing team and aviation detachment UDP sets in Okinawa that are reinstituting in the near term will later cease rotating to Okinawa. Guam will have the 3rd MEB Headquarters; the 4th Marine Regiment; and elements of aviation, ground, and support units from III MEF, with roughly 60 percent of the marines to be assigned permanently and the remainder maintaining presence through continuous UDP rotations. Increases in Hawaii will be through permanent assignments, consisting of aviation, ground, and logistics elements. Australia will have an MEU-sized MAGTF formed through UDP rotations.[14]

The total reduction posited in the illustrative CRP is a little larger than the current agreement and plan, at over 11,000 marines, consisting of about 7,400 permanently assigned personnel and 3,700 UDP rotational personnel. In this illustrative posture, however, they would return to CONUS, without establishing a presence in Guam, increasing presence in Hawaii, and building rotations in Australia to envisioned size. This is posited to save a little more than $200 million per year with an investment cost of close to $1.5 billion, producing a six- to seven-year payback for less of a relative financial return than many of the CRP changes. The limited return is due the fact that Camp Butler and Futenma would remain open and the relatively low cost of living and housing allowances for marines there, resulting in allowances just a little higher per person than at Camp Pendleton, for example. For reference, closing Okinawa and returning all forces to CONUS would produce roughly $300 million in savings per year, with a required investment of roughly $2.5–5 billion.[15]

---

[14] DoD, Office of the Assistant Secretary of Defense (Public Affairs), 2012; U.S. Marine Corps, Pacific Division, Plans, Policies and Operations, 2012b.

[15] This includes eliminating two of the three concurrent UDP rotations, leaving one for the 31st MEU, which could still continue rotational deployments from CONUS.

We did not directly compare the estimated costs of the CRP and the current plan; nor did we develop models with sufficient precision to accurately compare very specific options. For that, detailed, base-level analysis should be done. However, to provide some insight into how the costs might compare and how different possible options for the Marine Corps in the Pacific might impact costs, we provide rough estimates of effects on annual variable cost and investment cost for different levels of force increases in different locations, in comparison to Okinawa. We do note, however, that the agreed-upon Marine Corps plan for the Pacific was not motivated by a desire to reduce costs; rather, it was designed with the intent of improving deterrence and contingency responsiveness while assuring Japan and other allies of the U.S. commitment to regional stability and defense and to alleviate political pressures on U.S. stationing in Japan.[16] This section does not address how well the plan might meet these goals. Instead, it only examines the potential cost impacts.

Table 10.8 provides estimates of the annual recurring costs for stationing varying numbers of marines in Japan, CONUS, Hawaii, Guam (with low and high estimates), Camp Pendleton, Camp Lejeune, and a combination of the last two. As per our cost modeling, we estimate constant installation-support costs. MILCON for recapitalization of facilities varies with the area cost factor (ACF), with Japan lower due to HNS contributions. DoD schools are accounted for in places that have them or would need to have them, as are overseas logistics costs. Allowances are based on actual averages for marines by location, as explained in Chapter Eight, with the exception of Guam, because of the lack of historical Marine Corps data for the location. The high Guam rate (Guam-H) is the average of allowances for Air Force and Navy personnel stationed there. The low Guam rate (Guam-L) is the product of the relative Air Force/Navy rates for Japan versus Guam (Guam is 95 percent of the Japan level) and the Marine Corps Japan rate, assuming the dependent and housing mix for marines in Guam would be similar to those currently in Okinawa. PCS costs per person are estimated by service for CONUS and OCONUS, as described in Chapter Eight. The table shows that the total per-person rate would be lower in CONUS, slightly higher in Hawaii, and slightly to moderately higher in Guam as compared with Japan. This translates into the annual cost differences, based on varying force sizes, shown in the bottom table. The relative differences between CONUS and the other locations can also be seen by comparing the respective columns.

Increasing Marine Corps presence in each location in Table 10.8 would require new facilities and infrastructure investment. Rough estimates of the costs are shown in Table 10.9, which applies two different methodologies and resulting models. The RAND columns reflect the one developed and employed for this study, and the USMC columns reflect a Marine Corps methodology produced to translate detailed Guam and Hawaii facility investment requirements, from the time when there was a plan to have

[16] DoD, Office of the Assisstant Secretary of Defense (Public Affairs), 2012.

Table 10.8
Comparison of Estimated Recurring Variable Costs for Marine Corps in Japan Versus Other Locations

	Japan	CONUS	Hawaii	Guam-L	Guam-H	Camp Pendleton	Camp Lejeune	Camp Pendleton/ Camp Lejeune
Installation support	8,501	8,501	8,501	8,501	8,501	8,501	8,501	
MILCON	1,138	2,275	4,869	5,028	5,028	2,548	2,252	
Schools	4,900	—	—	4,900	4,900	—	1,100	
Logistics	734	—	670	1,636	1,636	—	—	
Allowances	20,433	15,390	24,223	19,288	27,733	18,967	12,387	
PCS	6,542	2,687	6,542	6,542	6,542	2,687	2,687	
Total cost per person	42,248	28,853	44,805	45,895	54,340	32,703	26,927	
Difference per person vs. Japan	—	–13,395	2,556	3,646	12,091	–9,546	–15,321	
Relative annual recurring costs								
5,000 personnel ($ millions)		–67.0	12.8	18.2	60.5	–47.7	–76.6	–62.2
7,400 personnel ($ millions)		–99.1	18.9	27.0	89.5	–70.6	–113.4	–92.0
11,000 personnel ($ millions)		–147.3	28.1	40.1	133.0	–105.0	–168.5	–136.8
Area cost factor		1.00	2.14	2.21	2.21	1.12	0.99	1.06

NOTE: L = low estimate; H = high estimate.

Table 10.9
Comparison of Estimated MILCON Investment Costs for Different Locations ($ millions, except MILCON per person)

		Relocate 5,000 Marines		Relocate 7,400 Marines		Relocate 11,000 Marines	
	2011 ACF	MILCON-RAND	MILCON-USMC	MILCON-RAND	MILCON-USMC	MILCON-RAND	MILCON-USMC
Guam	2.21	4,392	856	5,405	2,755	6,925	5,603
Hawaii	2.14	2,044	3,831	3,025	5,669	4,496	8,427
CONUS	1.00	955	1,790	1,413	2,649	2,101	3,938
Camp Pendleton	1.12	1,070	2,005	1,583	2,967	2,353	4,411
Camp Lejeune	0.99	945	1,772	1,399	2,623	2,080	3,899
Camp Pendleton/ Camp Lejeune		1,008	1,888	1,491	2,795	2,217	4,155
MILCON per person		$191,000	$358,000	$191,000	$358,000	$191,000	$358,000

11,000 or more marines relocated there (mostly Guam), as per the 2006 agreement with Japan, to estimate requirements for a more distributed laydown as planned in the 2012 agreement. The Marine Corps model is extremely accurate for the Guam/Hawaii 11,000-marine option, may underestimate for the lower Guam option in the table, and may overestimate for additions at larger, more established bases, because it does not have fixed- and variable-cost components, which we discuss in Chapter Eight. The RAND model employs fixed- and variable-cost components, with a fixed cost of about $2.4 billion (at an ACF of 1) added to the Guam estimates only, because the other locations already have substantial infrastructure, only some of which would have to be expanded or duplicated. Since it employs fixed- and variable-cost components, we believe the RAND model would more accurately estimate the MILCON requirements for an expansion of an existing base and for a new, small base (the Marine Corps model would be much more accurate for a situation analogous to the earlier Guam/Hawaii plan), but we show the cost estimates produced by both models here. A more detailed comparison of the two methodologies and the rationale for this conclusion is discussed in Appendix A. All of the Guam estimates also subtract the agreed-upon $3.1 billion Japanese contribution for facilities there.[17] Finally, the last important element of the comparison is the ACF, which varies significantly, with similarly high factors for Guam and Hawaii.

Using both methods, stationing larger numbers of marines in Guam would be estimated to require more investment than in CONUS, with the difference depending on the CONUS location and estimating method. At the middle force level in the table,

[17] U.S. Marine Corps, Pacific Division, Plans, Policies and Operations, 2012b.

whether investment in Guam would be similar to CONUS or higher depends upon the cost estimating method. At lower force levels, the Marine Corps model estimates that establishing a presence on Guam could require less investment than in CONUS (due to the Japanese contribution), but the RAND model suggests otherwise because of the fixed-cost component in the model. In all cases, costs would be higher for facility expansion in Hawaii than in CONUS.

Overall, due to the higher operating costs in Hawaii and Guam, the CRP's annual recurring savings would likely be just a little higher, roughly $25 million more per year, if compared with the agreed-upon plan, rather than comparing it with the status quo on Okinawa. Using the same methods as for Table 10.8, investment for about 2,500 marines in Hawaii and 5,000 in Guam would come to $3–5 billion versus about $1.5–3.5 billion for all being assigned in CONUS. Thus, the CRP's annual savings roughly hold, or would be a little larger, in comparison to the agreed-upon plan for the Marine Corps in the Pacific, and would likely require less investment. The GREP and MCP would not require the investments because they both seek to sustain status quo levels on Okinawa, but are likely not politically feasible.

Implications for Determining Actual Postures

The three postures were used as analytic tools rather than prescriptive options to help illuminate the types of major trade-offs that decisionmakers face in any real-world choices on overseas posture. There is no "ideal" or "optimized" overseas posture, but rather a set of preferences that decisionmakers must make, corresponding choices related to those preferences, and recognized trade-offs or consequences of those choices. The three illustrative postures share many features (most importantly the en route infrastructure), and all are designed to meet core U.S. security and military objectives. The primary difference is in emphasis (cost savings, global responsiveness and engagement, major contingencies) and the varying levels of risk and cost associated with each. As such, the postures can be useful in both making clear the essential minimum basing arrangements needed to support U.S. objectives and in thinking through potentially valuable but more "discretionary" options.

The insights from analyzing the three illustrative postures can be summarized as follows:

- The CRP is the only option that reduces overall costs, illustrating a rough limit of about $3 billion per year in savings, with a majority coming in Europe, after an initial investment. It does so through a reduced overseas presence (stationed and rotational). This, in turn, reduces the level of security cooperation that could be conducted, along with producing risk with respect to assurance of allies. It also does not position forces forward for deterrence of major contingencies to the same degree as the MCP, although it does preserve them in key locations, relying more on maritime presence and capabilities with global reach. However, given the retention of en route infrastructure, the CRP would still provide responsiveness for smaller-scale contingencies and the ability to reinforce in-place forces that would be involved in major contingencies.

- The GREP provides a more extensive network of access bases and rotations that expands opportunities for security cooperation, preserves current efforts, and provides improved responsiveness for many smaller-scale contingencies. Through its more geographically diverse and distributed basing arrangement, the GREP also provides the potential for more robust military and political access. Deterrence with respect to potential threats from Iran, North Korea, and China would be similar to current levels. Annual recurring costs for the GREP would be similar to current levels, but there would be meaningful transition cost to realign a small number of forces to provide the recurring savings to reinvest in rotations in new areas.
- The MCP provides the highest level of deterrence and assurance of allies and partners for the three principal contingencies of concern. This comes at the expense of reducing security cooperation with allies in Europe, while enhancing it elsewhere. This could also result in some additional risk to assurance and deterrence in other regions of concern. The MCP also risks increased levels of exposure for forward-stationed forces to anti-access threats. The MCP would add annual recurring costs as well as require significant investment.

Now let us examine benefits and costs across the three postures. Only by substantially reducing forces and bases in one or more regions, with limited to moderate replacement by rotations, could posture changes yield meaningful savings. This would force one or more trade-offs in strategic benefits. Conversely, it appears infeasible to substantially increase engagement with new partners while meeting other national security goals and also reducing overall costs; rather, doing so is likely to require some investment to realign forces to produce neutrality in longer-term recurring operating cost. Similarly, efforts to increase presence for specific major threats could require substantial investments.

The contrasts between the CRP and the other two postures suggest that implications for security cooperation, deterrence, and assurance are likely to be greater than for global responsiveness and access risk when considering posture options, as long as they protect global en route infrastructure and the United States sustains emphasis on maintaining geographically distributed access to bases.

The postures differ substantially in the level of effort that the United States would put into building security cooperation, both with traditional allies and new partners, with limits on both of these dimensions for the CRP and increases in the latter for the GREP. Future interoperability levels with partners would differ among the postures, as would building the capabilities of new partners, although the implications are not completely clear. The implications are even less clear for partner willingness to work with the United States in future operations, which might still also be enabled by mutual security interests.

While global capabilities might not result in substantial differences in warfighting capabilities for major contingencies among the postures, differences in presence in some regions could affect perceptions, influencing both deterrence and assurance, and possibly influencing relationships with allies. A possible second-order effect of the relationship aspects of security cooperation and assurance is whether there would be effects on gaining peacetime access to bases. So while a cost-reduction–like posture does not directly affect these dimensions, it is possible that they could be affected as well.

In short, our illustrative CRP offers an example that seeks to reduce costs while retaining power-projection capabilities and maintaining some presence in Europe, the Middle East, and Asia. However, the CRP involves more than simply finding efficiencies but rather forces a significant trade-off in forgone strategic benefits in exchange for cost savings. Therefore, one might view the CRP as essentially the foundation or "core" of a U.S. global-response posture with options to build on this foundation in different policy directions—represented by the GREP and MCP. Policymakers will need to make those determinations based on the importance that they attach to various strategic objectives and the level of risk—including that of financial cost—they are willing to incur. Answers to some key questions will be needed to guide future posture decisions: How much does U.S. presence in Europe assure allies, and how much security cooperation in Europe is needed? How much is security cooperation and assurance in the Asia-Pacific region valued, and how much are these two considerations affected by posture? Will increased presence in the Pacific strengthen deterrence, and how should posture address the missile threat? How should the needs for responsiveness and deterrence in the Persian Gulf be weighed against the potential for political tensions and risks? Informed value judgments bolstered by analysis will be required to address these important questions. We focus on these questions in Chapter Eleven.

CHAPTER ELEVEN

Conclusions

This report responds to legislation that called for an independent assessment of the overseas basing presence of U.S. forces. It specifically asked for an assessment of the location and number of forces needed overseas to execute the national military strategy, as well as an assessment of the advisability of changes to overseas basing in light of potential fiscal constraints and the changing strategic environment. We conclude that there are some enduring minimum overseas posture needs that are necessary to execute the current national security strategy. The location and minimum number of forces needed are generated by formal commitments, critical deterrence needs, and the basic premise in U.S. national strategies and plans that the ability to sustain global power projection is essential. We also conclude that it is advisable to maintain an overseas posture that goes beyond these minimums, where the security value-added outweighs the financial costs and risks. Both how much additional posture and the nature of that posture that is advisable depend on several key questions that do not have definitive answers and that require the application of decisionmaker judgment to weigh the benefits and risks. We identify these questions and the implications of different answers for the advisability of posture changes in this chapter. Advisability could also depend on how decisionmakers prioritize the benefits of forward posture versus the value that could be produced by using the same resources for other defense capabilities.

Minimum Essential Posture Needs

There are several posture elements clearly needed to achieve U.S. strategic goals. The ability to be globally responsive is at the core of the U.S. national security strategy. This requires a global network of ports and air bases—infrastructure and access—and the air and sea lift assets to move forces through that global network. Forward naval forces also provide response flexibility, as well as playing a role in protecting freedom of movement for commerce. In addition, there are other global enablers, such as communications capabilities, that forward forces rely on and that link them to national capabilities.

Where the United States has identified major threats, in-place forces are essential for deterrence and to respond immediately to aggression to prevent defeat. This mandates having presence in South Korea, across East Asia, and in the Middle East, with the level and nature of advisable presence depending on answers to the key questions we pose. Similarly, treaty commitments and other agreements to protect partners and regions against missile and nuclear threats set forth minimum needs in Europe, in addition to creating the need for greater presence in South Korea, East Asia, and the Middle East. While these needs establish clear essential elements of overseas posture, the best set of specific capabilities and locations required within these broad elements depends on the judgment of policymakers; there may be equally effective substitute sets. Beyond these foundations, U.S. strategy also seeks to pursue security cooperation in all regions. Whether that is more cost-effective through permanent presence or rotational presence—and how much of either might be needed—rests on judgments regarding the level and nature of engagement necessary to achieve U.S. goals.

Table 11.1 summarizes the essential elements of posture. In the remainder of this chapter, we discuss the judgments that should inform the specific choices within these essential elements, particularly with respect to configuring posture to meet deterrence, assurance, and security cooperation goals.

Changes Advisable to Consider Depending on Strategic Judgments

Barring a dramatic shift in the U.S. national security strategy, the current overseas posture can continue to serve U.S. interests into the future. DoD has already made significant reductions in its European posture. U.S. posture in Asia is shifting to address the changing security environment in that region and new strategic priorities. Enduring posture in the Middle East and Central Asia relies primarily on rotational forces, base access, and prepositioned equipment and supplies. U.S. security goals in Africa and Latin America are met through engagement, primarily by small, specialized forces through deployments, rather than permanent presence. There are, however, posture changes beyond those currently planned that could be reasonable to consider. Table 11.2 summarizes existing posture elements that could merit more in-depth examination for potential realignment depending on decisionmaker evaluations of the strategic environment and judgments on priorities. Table 11.3 summarizes new elements of posture that could merit pursuit depending on strategic evaluations and priorities.

If decisionmakers especially value deterrence and assurance, concentrating in areas that position U.S. forces to respond to potential high-threat adversaries would be most valuable, if these forces are not overly exposed and vulnerable to large and accurate missile forces.

If decisionmakers especially value broad-based, intensive security cooperation, then significant increases in rotations—force structure permitting—to Southeast Asia, Eastern Europe, and Africa could be merited. Any significant reductions of permanent forces and locations overseas would incur risk.

Table 11.1
Essential Elements of Overseas Posture

Posture Element	Key Bases or Regions	Key Unit Types/Assets
Global mobility infrastructure	• Primary and backup bases to cover at least three East and two West routes (see Figures 2.3 and 2.4)	• Airlift wings • Ground support squadrons
Other intertheater/ global enablers	• Communications and satellite facilities in Europe and Northeast Asia	• Signal Commands in Europe and South Korea
NATO and other European commitments	• Radar sites for missile defense • Rota Naval base • Nuclear weapon storage sites	• Land- and sea-based ballistic missile defense platforms • Nuclear weapons • Stationed or rotating BCT for the NRF • AWACS
South Korea deterrence and assurance	• Planned bases in South Korea • Some air bases in Japan	• Army units in South Korea • Air and missile defense units • Prepositioned Army and Marine Corps equipment in South Korea/afloat • Some air units in Northeast Asia
East Asia deterrence and assurance	• Some air bases in Japan and South Korea • Navy bases	• Most air units in the Pacific • Air and missile defense units • Prepositioned afloat equipment • Naval forces in Asia
Middle East response	• Distributed expansible aircraft bases • Prepositioned equipment sites • Command and control sites	• Rotational Air Force units • Naval presence • Prepositioned equipment sets for ground forces • Air and missile defense units
Operational support flexibility	• Access bases in every region, notably in Southern Europe, Southeast Asia, and the Middle East	
Security cooperation		• Rotations in Southeast Asia • Rotations to Europe or some of the current presence

If reducing cost is the overriding priority, there are some options for realignment for the United States to consider against forgone benefits: much of the remaining Army presence in Europe; much of the Air Force presence in the UK and some of its forces in Central Europe; the level of rotational presence in the Middle East; and Army, Air Force, and, to a lesser degree, Marine Corps presence in Japan. These would entail significant strategic trade-offs, particularly security cooperation in Europe and its periphery and assurance in the Asia-Pacific region.

Table 11.2
Elements of Overseas Posture That Could Be Evaluated for Realignment/Closure

Shift in Priority or Evaluation of Needs	Potential Realignment/Closure
Less security cooperation in Europe	• Most Army units and bases in Europe • Some Air Force units and bases in Europe (some need to be retained for global mobility and bases from which to execute operations)
High anti-access/area-denial missile threat in Asia	• Some reduction in air units and bases in Japan or South Korea • III MEF HQ and ground forces in the West Pacific (retain MEU)
Limited assurance and deterrence value	• III MEF HQ and ground forces in the West Pacific (retain MEU)
Limited deterrence benefit in the Middle East	• Reduced rotations in the Middle East

Table 11.3
New Elements of Posture That Could Be Considered

Shift in Priority or Evaluation of Needs	Potential Addition
More security cooperation emphasis with new partners	• Increased rotations to Southeast Asia, Africa, and Eastern Europe • Additional ARG in the West Pacific
Increased risk of Iranian aggression	• Increased rotations to the Middle East—all services • Increased air and missile defense assets • Increased armor prepositioning
High anti-access/area-denial missile threat in Asia	• Hardening of bases • Increased access to partner bases across the Asia-Pacific region
Increased need for assurance of Asian partners	• Increased air and naval presence

The following section assesses the three main overseas regions of U.S. military presence to highlight the key policy questions for overseas posture, based on enduring and evolving national security concerns.

Regional Considerations

Europe

In Europe, forces and bases have been reduced substantially, with planned cuts taking USAREUR forces 90 percent below their Cold War peak and USAFE forces more than 75 percent below their Cold War peak. The deterrence needs in Europe have largely waned, with Europe projected to remain relatively stable. Forces in Europe, though, help to maintain NATO cohesion and still likely provide some assurance to partners in

Eastern Europe and the Caucasus, given their apprehension about the long-term trajectory of Russia. There is some deployment-responsiveness advantage to be gained from stationing forces in Europe for operations outside of Europe, but it is limited. However, mobility bases, key communications nodes, and regional medical capabilities all support intertheater responsiveness and sustainment and, thus, are critical. Bases in Central and Southern Europe can support air operations around the periphery of Europe, providing immediate response value, as do naval forces in the Mediterranean. However, the ground forces based in Europe do not provide a significant deployment benefit to other theaters. For instance, deploying an ABCT from Germany to Kuwait takes about 18 days, only about four days quicker than a comparable force coming from the east coast of the United States—assuming no delay waiting for an LMSR in Europe—and, if a situation is time-sensitive, lighter units can be airlifted from CONUS in a time span comparable to those for European-based units. The exception would be response times for special operations forces or other small units conducting missions in which mere hours matter. BMD capabilities fulfill commitments and provide assurance to regional partners. Beyond these capabilities, U.S. forces play a significant role in enabling security cooperation, but the amount and nature of security cooperation in Europe that is necessary or advisable, particularly given the need to consider competing requirements and resource constraints, presents a policy choice.

In short, for forces in Europe, the central question is: How much does U.S. presence assure allies, and how much is security cooperation valued? Each service member based in Europe costs about an additional $15,000–40,000 per year, depending on the service, with fixed costs of each European installation costing between $115 million and $210 million per year, due to factors such as the high cost of living, accompanied tours, and lower levels of HNS than key Asian allies. Can substantial cuts beyond current plans be made in Europe to reduce costs while

- maintaining sufficient alliance cohesion and interoperability with NATO partners?
- adequately developing capabilities of new partners?
- maintaining partner willingness to conduct operations outside of Europe when the United States sees vital interests?

Three aspects of European-based posture contribute to security cooperation. The first is advanced multinational training center capacity, exemplified by the JMTC in Germany, which supports advanced military training with NATO partners and training with emerging partners to improve their capabilities. For ground forces, training facilities in Eastern Europe have also facilitated opportunities for the latter. For air forces, there is the Warrior Preparation Center near Ramstein Air Force Base.

The second aspcct is operational forces, both combat units and support or enabling units. Fewer forces in Europe would lead to fewer security cooperation activities, as the marginal cost of such events is very low for forces stationed in Europe. Without continuous presence, the opportunities for a broad range of security cooperation engagements would decrease, with planning more difficult and greater lead times required. Additionally, the direct expenses would be larger and more transparent, potentially depressing the willingness to execute rotational deployments or small-scale security cooperation activities. At the higher end, some of the unit-oriented tactical training could be compensated for through rotational training, particularly if JMTC were retained. But the ability to combine lower-level training events through the course of the year would lessen. This second effect might be greater with enabling units that focus security cooperation efforts on building specialized capabilities, such as logistics, medical, air-ground operations, and intelligence units. The reduction would occur not only in countries that host U.S. bases, but also in regions around Europe's periphery, such as Eastern Europe, Africa, and Central Asia.

The third aspect is forces that focus on the strategic and operational level of engagement, such as headquarters units. A sizable reduction in forces would likely lead to a downgrade of headquarters levels that would lessen senior leader engagement activity. It would be possible to keep the command structure intact, though, to preserve this capability, with forces being shifted to EUCOM and its components when operational needs merit, much like CENTCOM.

We now turn to the spectrum of possibilities. The minimum permanent presence in Europe must preserve bases to support air-based contingency response around the periphery, mobility hubs, air and missile defense, communications capabilities, and limited quick-response capability. As Figure 11.1 illustrates, to the extent that security cooperation is less valued, further reductions could be made.

The option that would go the furthest would restation most Army forces and about half of the remaining Air Force units, while closing a number of bases—including training facilities. The middle change option would reduce force structure somewhat, removing an Army BCT and some enablers along with some Air Force consolidation. A limited change option would consolidate a couple of Air Force bases. The final option would be to make no changes beyond those already planned by the Army and Air Force. In parallel with any of these, rotations could be increased or decreased from those planned. High levels of rotations could make up for much of the forgone unit-level tactical training, but with a significant reduction in the savings from reduced permanent presence. It would also depend on keeping JMTC open or working out agreements with allies to use their bases and training facilities.

The option shown in the bottom box of Figure 11.1 would greatly reduce security cooperation activities and could affect assurance, but would reduce costs by about $2 billion per year. Adding JMTC back in with increased Army and Air Force rotations could mitigate the tactical-level reduction in interoperability and building new partner capability effects, but would offset up to half of the savings. Savings for the second and

In Japan, the U.S. homeports one aircraft carrier and numerous surface ships at Yokosuka, with another major naval base at Sasebo, where an ARG resides. F-18 fighter aircraft are stationed at Atsuki and Iwakuni, with F-16s and P-3s at Misawa in northern Japan. Finally, an airlift wing operates from Yokota airbase in Tokyo. In Okinawa, Kadena Air Base is home to a composite wing that includes F-15 fighter aircraft and a second air base used by the USMC, Futenma. Key elements of the 31st MEU reside in Okinawa, with rotational units filling it out. Okinawa is also home to the III MEF HQ Group, most of the First Marine Air Wing, the Third Marine Logistics Group, and elements of the Third Marine Division, with additional infantry and aviation detachment rotations from the United States.

For those who believe it necessary to maintain a forward presence at these locations to deter and assure, despite threats from precision-guided weapons, there are a few broad options. These facilities could be hardened and protected with missile defenses or the number and mix of aircraft and ships could be reduced—either aiming to restation them elsewhere in the Pacific or back to the United States, striking a balance between assurance and operational resilience.[1] Such measures would make these bases less vulnerable—though certainly not invulnerable—to attack. Furthermore, in the context of efforts to add capabilities to the Pacific region, it may not make sense to add them to facilities in the highest threat zones. While clearly the intent of keeping forces in a potentially vulnerable location is to bolster assurance and deterrence, if doing so leaves U.S. forces open to a crippling attack, it might have the opposite effect. Alternatively, the number and kinds of aircraft and ships that are based at these facilities could be changed, again so as not to present such a lucrative target while yet retaining military utility. For fighter aircraft, this problem could become more acute when Joint Strike Fighters begin to enter the inventory, because the number of fighters in the entire U.S. inventory will decline, potentially reducing the tolerance for accepting risks to aircraft. Ships that homeport in high-threat zones are vulnerable in port, but with sufficient attack warning can head to sea and reduce this vulnerability. So for ships, assumptions about attack warnings are important in choosing among options. A final option, dispersal to many bases, may be challenging given limited options for gaining access to enough facilities to effectively dilute the inventory of attacking missiles. Moreover, fighters require special facilities to handle weapons, which would need to be provided at dispersed locations. There are more dispersal options with regard to large support aircraft, which can operate from dual-use civilian airfields—often without any special modifications to existing infrastructure. If such a strategy is pursued, it may have implications for costs and U.S. requests for host-nation burden sharing, because it

---

[1] While not the driving factor of such an action, moving a wing to CONUS and closing a base would also reduce costs by about $250 million per year.

would involve a shift of resources to a larger number of facilities possibly located away from close allies who have provided substantial HNS.[2]

For those who believe that the implications for deterrence and assurance would be limited were there to be major changes to forces in Japan, the U.S. footprint could be reduced. The United States has already developed plans and begun to shift forces to Guam and other locations in the Pacific through restationing and the shifting of rotational deployment locations. While some of these locations are still potentially under some threat from Chinese missiles, the threat is qualitatively different from the threat faced by forces in Japan or South Korea.

The Navy has moved some of its submarine force forward to Guam to increase undersea presence in the Pacific. Similar moves may be warranted for the surface fleet, particularly the carrier fleet, but also amphibious vessels and cruisers/destroyers. Forward stationing of vessels is only one essential element to increase naval forward presence; additional homeporting in the Pacific would also have to be tied to changes in personnel and maintenance policies and schedules to achieve the maximum presence benefit of overseas homeporting. So while forward stationing of assets can improve availability to an extent, naval presence is more influenced by the size of the fleet than the level of overseas basing. Were one to consider substantially increasing FDNF, the benefits of increased naval presence and responsiveness would need to be weighed against the financial costs and feasibility of aligning personnel, training, and maintenance policies. So in the near term, increasing presence in the Pacific is likely to demand reducing presence elsewhere.

Currently, the USMC posture in the Pacific is in transition. As agreed to with Japan and Australia, the Marine Corps plans to reduce some forces in Okinawa, maintain a rotational presence in northern Australia, establish a substantial presence in Guam, and increase forces in Hawaii. DoD has set a policy goal of having 22,000 Marines based west of Hawaii.

For humanitarian response and security cooperation, the dedicated MEU/ARG in the Pacific provides unique capabilities. Beyond the value of keeping the 31st MEU in the Pacific, there seem to be several open questions regarding the USMC posture in the Pacific. In addition to the 2,200-strong 31st MEU, over half of which is constituted through rotations, there are substantial numbers of other ground, logistics, and MEF headquarters group marines stationed or rotationally deployed to Okinawa, comprising a substantial number of the 13,000 permanently stationed and 5,600 rotationally deployed (when the UDPs are fully reinstituted) marines in Okinawa. However, the absence of dedicated lift for the other ground and logistics forces makes their forward position less of an advantage. If the forces are to fly to a contingency location to marry up with equipment, their presence in Okinawa or elsewhere in the Western Pacific

2 The exact number of bases required depends on the assumptions about the effectiveness of the missiles and the types of warheads used.

saves limited travel time. If they require maritime lift, that will have to come from the United States, erasing the responsiveness advantage of forward presence, or from the addition of dedicated, collocated maritime lift. In contrast, fixed-wing strike aircraft in Japan can operate from home bases in the event of contingencies in the region (although they could also deploy rapidly to expansible bases from the United States), and rotary-wing aircraft can self-deploy from Okinawa within the region, but they could not from CONUS. However, the Marine Corps may not want to station aircraft units apart from ground forces.

The Marine Corps pays more to have forces forward stationed on Pacific islands than in CONUS. The annual costs of keeping marines in Hawaii are a little higher than in Japan, and costs in Guam are likely to be higher as well. In addition, construction costs in Guam are the highest, although Japan has agreed to make a substantial contribution to construction there. We estimate that restationing all but the MEU elements and aviation units to CONUS (instead of potentially elsewhere in the Pacific) and not reinstituting two of three UDP rotations (leaving just the one for the battalion landing team and aviation detachment for the MEU) would save roughly $200 million per year. In addition, construction costs are likely to be similar or lower in CONUS, even considering the substantial contribution to construction in Guam for marines that Japan has agreed to make.[3]

Thus, depending on how decisionmakers assess the benefit that additional Marine Corps forces beyond the 31st MEU play in Okinawa or elsewhere in the Pacific, with respect to assurance, security cooperation, and responsiveness, keeping them there merits weighing against the somewhat higher costs, the lack of dedicated lift beyond the one ARG, the potential threats to Okinawa from China, the opposition in some quarters in Okinawa to a continued U.S. presence there, and the training limitations for marines stationed there. Among these considerations, the biggest is likely to be how it would affect Japanese and other regional nations' perceptions of U.S. commitments to the region. One possibility to ameliorate this risk would be to retain the III MEF Command Element in the region to preserve frequent high-level engagement activities. The broader decision to keep these forces in the Pacific also merits linkage to the Navy's force-structure considerations, particularly if the intention is to increase the presence of MEU/ARGs or otherwise mobile MEU-sized MAGTFs in the Pacific, which might be one, but not the only, objective. If so, the Navy might have to shift some of its amphibious fleet to the Pacific. For training purposes, ideally USMC and USN forces preparing to embark on a MEU/ARG would be located close enough to train together, but as of the writing of this report, there are no plans to locate more amphibious ships near Marine Corps forces in the western Pacific.

To guide these decisions, it may be helpful to consider what factors are being used to guide and judge the Marine Corps posture in the Pacific. Currently, one guideline is

---

[3] See Tables 10.8 and 10.9 for details of variable and MILCON cost differences by location.

the number of marines based west of the International Date Line. A more operational metric might be considered to identify the operational capabilities and response times desired, while still providing assurance.[4] The expeditionary orientation of the Marine Corps might make meeting presence, or at least responsiveness, goals through other means than forward basing possible. For example, it may be that if greater Marine Corps presence in the Pacific is desired, alternative deployment concepts, such as flying marines from the United States for rotations on which they employ prepositioned equipment stored in Southeast Asia, could meet policy intentions. Similarly, engagement goals may be met by using units much smaller than MEU-sized MAGTFs, with less-frequent large exercises with partners handled through deployments from the United States. For example, if a country in Southeast Asia were to agree to allow joint high-speed vessels port access, these might be used to transport smaller numbers of marines within the region, who could fly from their home station to conduct regional engagement rotations aboard these vessels.

In the Pacific, the United States is trying to improve responsiveness in Southeast Asia with rotational forces in Singapore, the Philippines, and Australia. In addition, this increases the number and frequency of security cooperation activities with Southeast Asian partners. Whether policymakers view this as sufficient could have implications for the overseas posture. If greater responsiveness or security cooperation is desired, the Navy or Marine Corps presence in the region could be increased. If host nations were willing, the USAF rotational presence in the region could be increased. Such a rotational presence could help to increase the response capabilities of U.S. forces, improve training benefits to regional partners, and help to better align security interests in the region, but will come with increased costs. Finally, in South Korea, the U.S. and South Korean governments are in the process of implanting a bilateral agreement to consolidate U.S. forces. This consolidation will increase efficiency and reduce costs. In the future, this consolidation could lead to vulnerability if North Korea invests substantially in an accurate missile force, but to date it has not done so. The United States has managed to continue to assure South Korea of its commitment to its security while drawing down the number of forces stationed on the Korean Peninsula.

The U.S. has far-reaching goals in Asia, but its forces are concentrated in Northeast Asia, where they are now within range of large numbers of precision-guided weapons. In Northeast Asia, the question is how the United States can continue to both deter potential foes and assure allies while making adjustments to reduce the chances of catastrophic failure from attacks by precision-guided weapons. The problem is complex enough that no one solution is likely to meet all the competing needs and fit

[4] Such a policy change would require discussions with allies so that they understand the intention of the policy changes and the capabilities the USMC can provide, not only from forward-based forces but the entire force. For instance, marines based in North Carolina already participate in the rotation of forces with the 31st MEU in the Pacific, and marines based in the United States could deploy to the Pacific to join prepositioned equipment in less than a day.

within resource constraints. Therefore, if the United States still wants to maintain forces in Northeast Asia, some combination of hardening, active defenses, dispersal, and changes to the mix of forces that are exposed to the highest threats will likely be in order. These adjustments would likely require substantial investments in infrastructure. At the same time, the United States wants to strengthen the security partnerships it maintains in South and Southeast Asia, but for now that will have to be accomplished through rotational presence. Rotational presence can be more cost-effective than permanent presence if rotations provide partial-year presence or are of long duration (assuming sufficient force structure to support such presence), but can be more expensive under some conditions. At present in this area, only rotational presence is viable in the near term; so in this case, there is no cost trade-off to consider. If greater security cooperation is desired, some permanent presence in Southeast Asia in the long term may be the most cost-effective means to meet that need, but that would depend on finding a willing host nation, which the United States could pursue through relationship-building steps. Alternatively, more arrangements like the recent agreements with Australia, Singapore, and the Philippines, which allow regular access but not permanent presence, may meet security cooperation goals for the region.

Overall, modest reductions in the Asia-Pacific Region could produce some savings while preserving in-place forces in South Korea and some additional capabilities in Japan for broader regional security. These reductions would include some of the Marine Corps forces and an Air Force base and wing, contributing roughly equal amounts of up to roughly $450 million per year. However, the nature and size of these reductions would depend on how decisionmakers judge the likely impact of modest force reductions in Asia on regional perceptions of the U.S. commitment to the region, how critical they believe large in-place forces are to deterrence, and the degree to which forces should be kept in higher-threat zones. But these modest reductions would reflect the call for pursuing new approaches to defense in the face of resource constraints. Any of these steps, though, might appear incompatible with the U.S. government's stated intention to rebalance toward the Asia-Pacific region, even if alternative approaches could provide similar capabilities. Concerted efforts to explain to allies how security could still be provided would have to be made, with some risk of not fully assuring key U.S. allies in the region.

Alternatively, emphasizing different aspects of the 2012 Defense Strategic Guidance could lead to increased presence in Asia and the Pacific. If increased security cooperation in South and Southeast Asia is highly valued and increases in rotational presence are pursued, costs would increase. If done, however, in combination with modest reductions in Northeast Asia, costs in the greater region might be held relatively steady. If such rotations were added while maintaining or even increasing presence oriented toward meeting perceived needs to increase assurance and deterrence, then annual recurring costs in the region would increase, potentially substantially. Any costs for hardening of facilities would be in addition to the cost estimates in the report.

The region presents a complex set of judgments and trade-offs regarding assurance, deterrence, security cooperation, and risks, with a range of options corresponding to different judgments on how different posture choices are likely to affect these factors.

Middle East

As noted in the 2012 Defense Strategic Guidance, U.S. defense efforts in the Middle East will be aimed at countering violent extremists, upholding commitments to allies and partner states, and addressing the enduring concerns of the proliferation of ballistic missiles and nuclear weapons. In supporting these objectives "the United States will continue to place a premium on U.S. and allied military presence in—and support of—partner nations in and around this region."[5] The United States has a diverse network of air bases that provide a range of capabilities in the Persian Gulf region and are complemented by maritime capabilities that also provide enduring regional presence. In the Persian Gulf, the United States has an interest in preventing Iran from disrupting commerce in the region, seeking to politically pressure or destabilize neighboring states, or (in the future) threatening regional states with nuclear coercion. U.S. facilities in the Persian Gulf region tend to be located in wealthy states that have overlapping interests with the United States, particularly regarding the free flow of natural resources. In other areas, U.S. interests and policies do not overlap with those of the host nation, particularly regarding U.S. relations with Israel, and potentially over human rights and political representation. There is public information that several host nations in the region provide burden sharing; as shown in Chapter Seven, Kuwait and Qatar have a history of helping to offset basing costs.

Basing in the region traditionally has involved trade-offs between the need to deter adversaries and reassure partners while also weighing political sensitivities to the U.S. military presence in host nations. While the wind-down of the conflicts in Iraq and Afghanistan have reduced the need for major combat and support assets in the region, enduring concerns over Iran continue to heavily motivate U.S. force presence. The Arab Spring has added another element of uncertainty to the internal political dynamics in many states, which could create a perceived need for presence to respond to new crises, but could lead to difficulties in maintaining U.S. presence. So in many respects the United States continues to face the balance between adequate force presence to secure national objectives and political risk to sustain those forces.

Overall, the central question for this region is how responsiveness and deterrence needs in the Persian Gulf should be weighed against the potential for political tensions and risks.

More specifically, the posture in the region, particularly in the Persian Gulf, involves judgments on some key aspects of that presence, representing the tradespace for consideration. The United States currently maintains substantial forces on a rota-

[5] DoD, 2012a, p. 2.

nonstate actors in the region. This is likely to be a function of many factors, including the degree to which partners and the United States share common concerns over the threats to be countered.

The middle tier represents a reduction in presence and cost relative to the top-tier option. Army rotations to Kuwait would be reduced or eliminated, with continued or increased emphasis on prepositioning. The planned Navy expansion in Bahrain would be deferred. Under this option, responsiveness and forward deterrence would be reduced, although this would be a function of the threats to be countered. If major contingencies, including Iran, were no longer viewed as primary concerns, then a reduced presence as illustrated in this mid-tier option would not necessarily erode responsiveness and deterrence. It could also serve to reduce sources of political opposition in host states to a sizable U.S. military presence in the region.

The third tier would further reduce the routine presence in the Persian Gulf region and instead focus on maintaining access to host-nation facilities as needed. This would result in additional cost savings and possible political benefits from further reducing routine U.S. presence. Again, these potential benefits would have to be weighed against the threats and the corresponding impact on deterrence and assurance. A reduced presence of this scale could risk both increasing the chances of adversary aggression and undermining the confidence of regional partners in the U.S. commitment to their security. To the extent that latent state threats become acute on short notice, the United States would be less well positioned to respond quickly.

All of the tiered options involve trade-offs and are highly contingent on regional developments and dynamics. The degree of risk assumed in each is likewise contingent. Depending on how decisionmakers assess deterrence needs in the region and the role of forward posture in meeting them versus how they assess the risks of creating political tensions or a backlash in the region would lead to choices on different points along the spectrum. Depending on the weight given to the two competing sets of factors, decisionmakers could elect to selectively reduce rotations in the region, maintain the status quo, or seek to increase rotations to the region across the services.

Value Perceptions and Priorities Are Critical to Posture Decisions

There are some clear limits to how far consolidation in the U.S. overseas posture could be pursued, beyond which achieving national security goals and executing the 2012 Defense Strategic Guidance would become untenable. There is a minimum threshold of foreign posture that the United States must retain. Beyond that, there is additional posture that is almost certainly advisable to retain or even add. As described in this chapter, there are a number of choices specific to each region, where different judgments could lead to differing calculations of the advisability of reductions, additions, or changes in the nature of posture. Again, the three illustrative postures presented in this report represent policy options, not right or wrong choices, because only the cost side of the equation can be determined with precision. Decisions will reflect judgments based on the perceived values of the competing goals—how they should be prioritized—and the degree to which overseas posture is perceived to advance those goals.

APPENDIX A

Cost Analysis Appendix

This appendix explains the cost analysis approach discussed in Chapter Eight and explores its results in more detail. The bulk of the appendix discusses the recurring installation-support costs and PRV-based estimated recapitalization MILCON cost needs regression analyses.[1] Thus, the first section describes our overall approach to regression analyses. The second and third sections specifically describe our installation-support cost and PRV regression analyses, respectively. Both sections are structured in the following order by service: Air Force, Army, Navy, and Marine Corps. The remaining sections are shorter, adding some details excluded from the main chapter. These include rotational deployment costs, DoD dependents' education, how we derived Army prepositioning costs, how we utilize FCM costs, and some comparisons of personnel and PCS costs.

Installation-Support Cost Analytic Approach Overview

First we reiterate our approach to give context to the detailed discussion in this appendix. Our overall analytic approach was as follows.

- *Process personnel data.* For all regression analyses, we used DMDC-provided personnel data on the number of people assigned by UIC. By installation, we classified these personnel, first by unit, then as operational, institutional, and installation support, to determine independent (operational and institutional) and dependent variables (installation support).
- *Process cost data.* We aligned cost categories across the services to ensure we included comparable costs for all. Then we aggregated costs by installation.
- *Match personnel and cost data by installation.* In this step, we mapped the installation names from each data set to develop a final regression data set

[1] The PRV analysis was used to estimate both recurring and one-time (for a unit relocation) MILCON expenditures.

- *Assign variables and categories to inform regression analysis.* We classified all bases by service and region and whether they have assigned operational forces.
- *Run regression analysis and interpret results.* We ran many regression models for each data set, testing the significance of different variables individually and together to arrive at statistically significant mathematical models that best explain the costs.
- *Construct cost models.* We assembled the findings from each regression model into a total cost model for each service and region to compute the cost effects of posture changes for the posture-options analysis in this report. These models could be used to estimate the cost effects of other options as well.

General Approach to Installation-Support Cost Regression Analyses

The goal of our regression analyses was to produce reliable cost models to estimate the cost effects of realigning forces and closing installations, whether for our analysis of options or in general. If one wanted to look only at the installation-support cost impacts of closing bases and realigning all forces in a region, a model to determine the changes in costs for that region would not be needed—it would simply be the actual budget amount for each base. But if forces were realigned, this would not reflect the change in total DoD costs. A model would still be needed to understand the impact of adding those personnel to another base or of opening a new base in another region. Likewise, a model is needed to estimate the cost impact if only some people on a base were to be realigned, with the base itself staying open, to determine how much the costs would change, since budget data are not neatly binned into fixed and variable cost categories. Rather, many budget or PE categories have both fixed and variable costs associated with them. Note also that this is just the installation-support cost portion, which is only one of several cost categories affected by base closures and force realignments.

In performing these analyses, we faced tension due to competing objectives. We sought to capture cost drivers based on installation attributes to the maximum extent possible to accurately represent how these attributes relate to installation-support costs and personnel counts, as well as how these variables relate. At the same time, a multiplicity of terms in a regression analysis can decrease statistical significance and confidence in the model. As total sample sizes overall and for given subsets get smaller, typically fewer and fewer explanatory terms can be used in a model while still maintaining statistical significance. If only one or a small number of data points have a certain attribute, for example, one cannot model the effects of that attribute. Since the number of overseas U.S. military installations is already small (for the purposes of regression analysis), further subdividing the data exacerbates the challenge of identifying statistically significant models. As we conducted our regression analyses, we kept this tension in mind as we tested different models and determined the most appropriate models to use.

Because the aim of this part of the analysis was to construct cost models for the analysis of the illustrative postures—as well as to quickly estimate the costs effects of other options, our cost models were shaped to serve this intended use. In our case, the service, region, base population, and presence of operational forces (and aircraft for the Air Force) were the key explanatory variables that could be employed from the changes in posture options. Other potential explanatory variables could be the types of units, but this would lead to very few or even one base in some categories (e.g., a base with a certain unit type in Europe), preventing the use of regression modeling. We therefore need cost models that operate with these inputs, and with few others, to make the cost models and posture options compatible.

Additionally, when it comes to the actual dynamics of installation-support costs, many, but not all, costs are population-driven, making population a reasonable variable for determining variable costs. Every base has some unique characteristics that prevent its costs from following the same mathematical formula perfectly. For bases within a certain range of base population, some have extensive training facilities, some have unique tenant units, some have communications towers and other peculiar equipment, and some serve as hubs of one kind or another. Differences specific to the local region, such as price levels, local resources, or proximity to other bases, could influence installation-support requirements and costs. While these attributes can be captured with additional variables, we run into the problem of sample size or data availability when trying to use them.

In the actual conduct of our regression analysis, we started with some assumptions about cost relationships, but explored the actual relationships in the data by iteratively testing a range of regression models, including and excluding different terms and aggregating and disaggregating data to see which combinations and levels of data aggregation have the best explanatory power. We also believe that some of the terms should have different impacts on costs (e.g., military versus civilian personnel), but also correlate, or co-vary, with one another. Including these collinear terms would require different models that would be difficult to handle with the sample sizes. For example, there may be a relationship between the number of military and civilian personnel at installations. Incorporating every potential explanatory variable and installation attribute creates more complicated models but may decrease fidelity because of the small data sets described above. These are all factors we took into account as we analyzed the data and tried to arrive at the most meaningful cost models feasible given the available data and sample sizes.

Because the purpose of our overall model is to predict cost, we view it as a descriptive model rather than a causal model. The estimated model statistically summarizes the cost patterns with respect to region, base population, and other variables where possible, such as whether operational forces are on the base. In a larger framework—for example, from the viewpoint of making decisions about basing—explanatory variables such as region and base population are endogenous variables determined by policy, and with this in mind, we also considered these variables to be endogenous in our data set. As a result, the estimated coefficients on these variables should not be interpreted as unbiased causal effects of the variables, but they can be interpreted as providing a statistical description of the relationship between the explanatory variable and the dependent variable (cost). The descriptive models can be used for predicting cost, which is our purpose. The predictions we make are typically within the range of our sample observations, and our estimated cost model is, in effect, a way of interpolating from the existing data what costs are likely to be present under alternative basing arrangements.

The two subsequent sections, on recurring annual installation-support costs and PRV, which was modeled using the same general regression analysis approach, follow the same organization:

- *Data sources, caveats, and limitations.* We discuss known caveats for the data and what categories or bases we excluded up-front.
- *Raw data plots and initial observations.* We show some of the raw data in tables and plots and make high-level observations about them. The reader should be careful not to jump to too many conclusions from the data plots alone, as the two-dimensional plots do not capture all the data elements and dimensions/attributes that we analyzed, and the analytic software we use is able to discern and quantify relationships that may not be obvious to the naked eye. We include the plots for reference.
- *Regression analysis final results.* We show the final regression model at which we arrived and describe our analytic process, including important findings and the final values for the cost model.

Installation-Support Costs

Air Force Installation Support

Data Sources, Caveats, and Limitations

The Air Force provided us with comprehensive data from the AFTOC database.[2] AFTOC data include cost categories irrelevant to our analysis, as well as PEs for all operations and sustainment costs, not just installation support. Thus, we had to filter out much of the AFTOC data to ensure that we appropriately modeled USAF installation-support costs.

[2] U.S. Air Force, Air Force Total Operating Cost database, FYs 2009–2011.

Two categories of spending captured in AFTOC that we excluded are overseas contingency operations (OCO) funds and transportation working capital funding. This portion of our cost analysis focuses on the *base budget* impacts of *permanently* stationed personnel and activities, so it was necessary to exclude OCO funds. Transportation working capital funding is for reimbursable transportation services provided to DoD customers, not local installation support, so we excluded it. Below, we show the proportions of costs for these categories.

Because AFTOC includes PEs relevant to installation support beyond the PEs common to all four services, we made two estimates for USAF installation costs. The first estimate employed only categories common to all of the services, ensuring that we could make cost calculations comparable across the services. We refer to this data set and its cost model as "Common PEs." The second estimate enabled us to capture the full installation-support cost the United States bears for overseas USAF forces and bases. In addition to the 14 PEs listed in Table 8.5 in Chapter Eight, we included in our second estimate a category of PEs the Air Force calls "Combat Support," which contains military personnel who support both deployed and home-station missions.[3] From a home-station perspective, these forces provide installation-support capabilities, rather than, for example, relying on civilian personnel and/or contract services; so we included their military pay (in the Combat Support PEs) in our installation-support costs. We refer to this second data set and its cost model as "Combat Support." For our entire data set of active duty bases, the Common PE data set comprises about $10.5 billion in annual support costs. The additional Combat Support PEs alone account for about $4.8 billion in annual support costs. Thus, the entire data set for the Combat Support estimate accounts for about $15.1 billion in annual costs. Ultimately, we chose to use the Combat Support data set for the overall cost model used to estimate the costs effects of different posture options. We explain why in detail below. We include the "Common PEs" model in the report as well, for comparison with the other services.

In addition to cost categories, we also excluded some bases from our cost model, for two reasons. Some of these we excluded up-front, because we could identify that they were not good analogs for the types of bases in the United States and overseas in which we were interested. Others we excluded because we observed during our analysis that they had an outsized effect on the regression model values and fit, but were not vital to our posture options. This outsized effect likely indicated that we were missing a key explanatory variable related to how the base is different than others. We ultimately excluded the following types of USAF bases from our regression analysis:

- *Guard- and Reserve-only bases.* These are located only in the United States and are not good analogs to overseas bases. They have very different cost profiles from

[3] Combat Support represents a category that comprises a number of PEs. Whereas the other services often utilize only a single PE for a category, the Air Force utilizes multiple PEs, usually to track costs for a category across its different major commands, as these major commands manage their own portfolios of resources.

active duty bases, often because they have different support concepts, facilities, and footprints. Also, while the idea of shifting forces from active duty to the reserve component is a topic of discussion, this analysis did not consider that policy option.
- *Major command headquarters*, e.g., Ramstein Air Base. These bases usually had installation-support costs far higher than bases of similar populations, whether in the United States or overseas. These bases tended to have an outsized effect on the regression model. The only major command headquarters located on foreign soil is Ramstein Air Base. It also likely has higher costs in recent years due to extensive operational support of CENTCOM operations and an associated high transient population. Because none of our posture options included the closure of Ramstein (due to its importance as a regional hub), we excluded it from our cost model. We had no reason to believe the forces stationed there would have higher per-person costs than those stationed at other German bases.
- *Air logistics centers*, e.g., Hill Air Force Base. These bases, while having some active duty units, have large industrial operations, which drive their installation-support costs. While it is theoretically possible to account for these bases with additional variables, in practice that tends to dilute the fit of the regression models. We had no reason to believe the forces stationed there would have higher per-person costs than those stationed at other U.S. bases. Because those air logistics centers are not of interest to our broader posture analysis and because there are no overseas analogs to these bases, we excluded them.
- *Army installations with Air Force personnel*, e.g., Ft. Rucker. AFTOC does track USAF expenses for USAF forces stationed at several Army installations in the United States. Because the Army provides installation support to USAF forces at Army installations, those personnel would be supported in a way consistent with Army cost profiles, not Air Force ones. Furthermore, because there are USAF personnel hosted by Army installations, the Army would bear the fixed-cost component of installation support, so the costs for USAF personnel would look relatively low. Finally, those Army installations might not be appropriate for the realignment of USAF forces from overseas to the United States. For these reasons, we excluded these bases from our USAF cost models.

Regression Data

Before discussing the details of the regression analysis, we show the raw data that fed our regression analysis and make some general observations about these data. Table A.1 shows the raw cost and personnel data for U.S. locations with permanently stationed aircraft. The military personnel column shows the total operational and institutional personnel, i.e., the explanatory variables. The annual support cost shows the average support costs for FYs 2009–2011. Table A.2 shows the U.S. bases with no permanently stationed aircraft. Tables A.1 and A.2 do not list the bases we excluded. Finally,

Table A.1
Regression Data for Air Force Bases, U.S. Locations with Aircraft

Bases	Military Operations Personnel	Military Institutional Personnel	Military Support Personnel	Civilian Personnel	Annual Cost, No Combat Support ($ millions)	Annual Cost, with Combat Support ($ millions)
Eielson AFB (AK)	732	87	1,099	555	107.6	160.7
Elmendorf AFB (AK)	2,720	70	2,455	1,842	306.7	387.9
Altus AFB (OK)	305	167	915	1,081	66.0	110.5
Andrews AFB (MD)	1,427	467	2,606	2,607	198.7	236.6
Barksdale AFB (LA)	3,015	791	1,728	1,039	124.1	220.9
Beale AFB (CA)	2,710	57	1,279	672	93.7	148.1
Buckley AFB (CO)	959	74	460	923	98.8	111.0
Cannon AFB (NM)	2,901	112	1,380	434	102.8	166.1
Charleston AFB (SC)	2,447	196	1,377	1,706	90.0	149.8
Columbus AFB (MS)	5	468	395	426	55.0	69.2
Davis-Monthan AFB (AZ)	4,366	284	1,669	1,363	96.8	206.1
Dover AFB (DE)	2,166	65	1,249	1,275	97.3	160.6
Dyess AFB (TX)	3,221	111	1,333	407	74.7	146.1
Edwards AFB (CA)	9	1,390	700	3,991	158.8	191.2
Eglin AFB (FL)	1,297	2,052	2,463	4,502	216.8	320.4
Ellsworth AFB (SD)	2,054	25	1,336	589	77.6	147.5
Fairchild AFB (WA)	1,242	477	1,148	880	85.6	142.5
Francis E Warren AFB (WY)	2,096	130	898	598	80.6	121.0
Grand Forks AFB (ND)	201	1	1,073	325	66.3	120.5
Hill AFB (UT)	1,833	830	941	10,178	156.7	226.1
Holloman AFB (NM)	2,111	118	1,471	859	97.1	176.6
Hurlburt Field (FL)	4,825	1,276	1,562	1,312	107.1	182.9
Keesler AFB (MS)	285	3,018	1,655	1,470	105.0	134.7
Kirtland AFB (NM)	1,179	1,064	627	2,395	132.9	178.7
Laughlin AFB (TX)	134	300	389	807	58.7	74.7
Little Rock AFB (AR)	3,673	400	1,132	817	93.7	165.3
Luke AFB (AZ)	2,143	217	1,319	997	100.5	161.0

Table A.1—Continued

Bases	Military Operations Personnel	Military Institutional Personnel	Military Support Personnel	Civilian Personnel	Annual Cost, No Combat Support ($ millions)	Annual Cost, with Combat Support ($ millions)
Macdill AFB (FL)	1,036	875	1,570	2,102	143.6	202.5
Malmstrom AFB (MT)	2,284	22	947	858	73.7	115.3
Maxwell AFB (AL)	89	1,456	563	2,207	137.6	161.3
Mcchord AFB (WA)	2,271	69	1,050	661	69.6	130.9
Mcconnell AFB (KS)	1,806	24	1,028	881	93.1	147.2
Mcguire AFB (NJ)	2,751	105	1,491	1,807	189.9	266.1
Minot AFB (ND)	3,818	16	1,632	551	92.3	163.3
Moody AFB (GA)	3,263	48	1,197	475	83.5	185.2
Mountain Home AFB (ID)	2,186	16	1,310	441	76.7	145.0
Nellis AFB (NV)	5,055	1,330	3,160	1,117	171.0	284.5
Offutt AFB (NE)	2,563	1,095	1,504	2,624	114.7	173.9
Patrick AFB (FL)	733	47	604	1,977	170.7	201.3
Peterson AFB (CO)	980	1,079	1,342	3,000	205.0	253.1
Scott AFB (IL)	1,595	1,177	1,609	3,573	141.3	240.1
Seymour Johnson AFB (NC)	3,184	64	1,390	670	86.9	164.8
Shaw AFB (SC)	3,028	36	1,525	605	104.7	203.3
Sheppard AFB (TX)	8	2,169	603	1,102	119.3	141.2
Tinker AFB (OK)	4,163	693	844	14,211	173.0	208.2
Travis AFB (CA)	3,547	80	2,826	1,640	131.5	211.8
Tyndall AFB (FL)	1,247	438	819	877	293.5	325.8
Vance AFB (OK)	137	266	310	175	51.9	64.2
Whiteman AFB (MO)	2,001	106	1,610	850	98.7	173.9

SOURCE: Authors' analysis of DMDC and AFTOC data.

Table A.3 shows the same data for overseas locations, including all the bases for which we received data, including those that we ultimately excluded from our regression modeling. We denote bases we excluded with an asterisk next to the base name.

Table A.4 summarizes the data points in each category for our regression analysis. This table shows the total bases, military personnel (operations and institutional only), and average FY 2009–2011 support costs. The next to the last column, labeled "No

Table A.2
Regression Data for Air Force Bases, U.S. Locations without Aircraft

Bases	Military Operations Personnel	Military Institutional Personnel	Military Support Personnel	Civilian Personnel	Annual Cost, No Combat Support ($ millions)	Annual Cost, with Combat Support ($ millions)
Arnold AFB (TN)	2	48	—	242	34.1	34.2
Cape Cod Air Force STN (MA)	—	—	—	—	4.7	4.7
Cavalier AFS (ND)	—	—	—	—	4.7	4.7
Cheyenne Mountain (CO)	—	—	—	—	0.1	0.4
Clear AFS (AK)	—	—	—	—	12.1	12.1
Goodfellow AFB (TX)	5	535	487	654	52.8	69.8
Hanscom AFB (MA)	62	565	321	1,702	89.5	108.9
Kaena Point (HI)	—	—	—	—	3.5	3.5
Kelly AFB (TX)	—	10	—	—	0.0	0.0
Los Angeles AFB (CA)	42	920	222	1,334	61.9	70.6
Maui Island (HI)	—	—	—	—	0.0	0.0
New Boston Af STN (NH)	—	—	—	—	5.7	5.8
Newark AFB (OH)	—	1	—	—	0.0	0.0
Onizuka AFS (CA)	—	10	—	4	0.0	0.1
San Antonio (TX)	—	—	—	—	2.0	2.2
Schriever AFB (CO)	984	249	359	681	73.2	86.0
Socorro (NM)	—	—	—	—	0.1	0.1
Vandenberg AFB (CA)	590	646	1,162	1,172	185.5	223.0
Washington (DC)	—	—	—	—	0.0	0.0

SOURCE: Authors' analysis of DMDC and AFTOC data.

Combat Support," includes only Common PEs and excludes Combat Support PEs. The rightmost column includes Combat Support PEs.

Data Plots and Regression Analysis Findings for Common PEs

Figure A.1 shows the raw data by installation for USAF installation-support costs. The x-axis shows the total of operational and institutional personnel (both explanatory variables). The y-axis shows annual installation-support costs (FYs 2009–2011 averaged) for the Common PEs data set. Each point on the graph represents a single

Table A.3
Regression Data for Air Force Bases, Overseas Locations

Region	Bases	Military Operations Personnel	Military Institutional Personnel	Military Support Personnel	Civilian Personnel	Annual Cost, No Combat Support ($ millions)	Annual Cost, with Combat Support ($ millions)
PACAF	Kadena Air Base	4,394	41	2,186	382	127.1	230.7
	Kunsan Air Base	1,350	7	1,104	25	61.2	114.0
	Misawa Air Base	1,585	21	1,304	145	78.8	149.2
	Osan Air Base	2,734	12	2,237	130	121.7	207.6
	Yokota Air Base	1,390	159	1,345	245	87.4	156.2
	*Andersen AFB	1,087	2	926	118	59.3	116.6
	*Wonju	7	—	—	—	0.1	0.1
USAFE	Aviano Air Base	1,946	35	1,860	—	104.7	193.3
	RAF Lakenheath Air Base	2,646	26	1,949	—	124.9	201.1
	RAF Mildenhall Air Base	1,858	36	1,000	174	156.5	222.3
	*Ramstein Air Base	3,976	1,270	2,960	959	358.9	584.9
	Spangdahlem Air Base	1,978	21	1,702	200	117.5	209.9
	*Geilenkirchen Air Base	—	377	106	26	0.5	1.2
	*Incirlik Air Base	347	19	986	103	110.2	148.2
	*Kapaun	96	28	40	—	0.0	0.0
	Lajes Field	102	13	558	65	54.3	90.1
	*RAF Alconbury Air Base	69	111	291	61	0.0	0.0
	RAF Croughton Air STN	50	1	342	43	22.2	27.6
	RAF Menwith Hill	107	—	69	48	42.4	42.4
	*Rota	221	1	—	5	0.0	0.0
	*Stuttgart	10	311	22	4	0.2	0.2
	*Thule Air Base	27	—	119	—	19.4	25.5

SOURCE: Authors' analysis of DMDC and AFTOC data.

* Base excluded from regression analysis.

Table A.5
Air Force Installation Support Regression Results, Common PEs

Regression Term	Parameter Estimate	Standard Error	Lower Bound	Upper Bound
United States, with aircraft	52,300,000	15,100,000	37,200,000	67,500,000
United States, no aircraft	–43,900,000	16,700,000	–60,600,000	-27,300,000
USAFE	47,500,000	16,500,000	31,000,000	64,000,000
PACAF	34,800,000	18,100,000	16,700,000	52,900,000
Military – Operations	8,300	5,000	3,300	13,300
Military – Institutional	41,000	10,100	30,900	51,100
Civilian – Operations	79,200	22,400	56,800	101,600

SOURCE: Authors' analysis of DMDC and AFTOC data.

c_{ir} is the total installation-support cost of installation i in region r, a_R is the fixed-cost component of an installation in region r, and $b_r p_{ir}$ is the product of the variable cost per person in region r and the number of personnel at installation i in region r.

The fixed cost for a U.S. base a_R is simply a_{US}. The total fixed cost for an overseas base is $a_R = a_{US} + a_r$, where a_R is the total fixed cost for an installation in region R, and a_r is the regression component additive cost for a base in the same region. As we can see from Table A.5, a_{US} = 52,300,000 and a_r = 47,500,000. Thus, for USAFE, a_R = 52,300,000 + 47,500,000. This means that the regression model estimated that, on the whole, USAFE bases have a fixed-cost component that is about $47.5 million higher than comparably sized U.S. bases. The parameter estimate for U.S. bases with no forces is negative, so the net fixed-cost component is lower than for bases with aircraft. Both the overseas parameter estimates are positive, so the regression model estimated their fixed-cost components to be systematically higher than U.S. bases.

The USAFE and PACAF terms add about $48 million and $35 million beyond U.S. bases, respectively. The USAFE data points in Figure A.1 are not obviously higher than comparably sized U.S. bases, but the p-value of the regression term (0.005) gives us reasonable statistical confidence that USAFE base's costs are systematically higher than comparable U.S. bases. The p-value for PACAF was much higher (0.05), but still at a level generally considered statistically significant, and, as one can see from Table A.5, the standard error is much higher in relation to the parameter estimate.

The net result of these individual parameter estimates, following the equations shown above, is that the fixed costs (i.e., the amount saved only when closing a base) we attribute to USAF bases are as follows:

- U.S. bases with aircraft: $52 million
- U.S. bases without aircraft: $8 million
- USAFE bases with aircraft: $100 million

- USAFE bases without aircraft: $56 million
- PACAF bases with aircraft: $87 million.

This fixed cost for installation support forms the core of our total fixed-cost calculation. Later in this appendix we also discuss MILCON recapitalization. Not discussed in this appendix, but covered in Chapter Eight, is overseas DoD dependent schools.

We arrived at three significant explanatory variables that define the variable cost component, i.e. the per-person installation-support costs. Operational personnel are estimated to have a variable cost of about $8,000 per year. This means that at any base, the addition of one operational person is associated with an increase in total installation-support costs (for the narrow set of installation support PEs included in this regression model) of $8,000 per year. For institutional military personnel (of which there are few at overseas bases), it is about $41,000 per person, and for civilians supporting operational activities, about $79,000 per person.

The estimates with respect to operational versus institutional personnel are for the baseline case of U.S. bases. Given the sparseness of the data for overseas bases (only five bases with aircraft each in EUCOM and PACOM), doing a separate regression analysis was not feasible.

Data Plots and Regression Analysis Findings for Combat Support

We now show comparable data and results for our second estimate, Combat Support, which includes military personnel providing installation support. This is the data set we use for our overall cost model.

Figure A.2 shows the raw data by installation for USAF installation-support costs for the Combat Support data set. As in Figure A.1, the x-axis shows the total of operational and institutional personnel (both explanatory variables). The y-axis shows annual installation-support costs (FYs 2009–2011 averaged), including the Combat Support PEs. Each point on the graph represents a single active-duty Air Force base. The colors and symbols differentiate the region and type of base. U.S. bases with permanently stationed aircraft are shown as blue diamonds; those without permanently stationed aircraft are hollow blue diamonds. PACAF bases are shown as red circles. USAFE bases are shown as yellow triangles for bases with aircraft, and brown triangles for those without. U.S. bases that we excluded are shown with a blue X. We do not show the overseas bases we excluded.

In Figure A.2, we see that the overseas bases are somewhat higher relative to U.S. bases than they appeared to be in Figure A.1, without Combat Support. Table A.6 shows the results for our estimate including Combat Support, and it follows all the same formatting and organization as Table A.5.

This regression model has an adjusted R-squared value of 0.63. Here, the fixed-cost component of a U.S. base with aircraft is about $73 million per year, while the same value for U.S. bases with no aircraft is about $68 million per year lower than the baseline, for a net fixed cost of about $5 million. This fixed-cost estimate for U.S. bases

operations and institutional personnel to support personnel. The x-axis shows the total of operational and institutional personnel. The y-axis shows military support personnel. Each point on the graph represents a single active duty Air Force installation. The colors and symbols differentiate the region and type of base. Bases with permanently stationed aircraft are shown as blue diamonds for the United States (excluding territories), red circles for PACAF, and yellow triangles for USAFE. U.S. bases with no permanently stationed operational units are shown as hollow blue diamonds. We excluded Ramstein and USAFE bases with no aircraft for this plot.

The first thing to notice about Figure A.3 is that it shows the same characteristics as Figure A.2 (and even Figure A.1, which excludes the military pay in the Combat Support PEs). The plot of personnel by categories suggests a fixed and variable component. Second, overseas bases appear to have more support personnel relative to supported personnel than most of their U.S. counterparts. To determine how many more support personnel (in relative terms) this tends to be, we did a regression analysis of these data. Table A.7 shows the results.

The adjusted R-squared value for this model is 0.80. Table A.7 shows that the fixed-cost component for military personnel providing installation support is about 315 support personnel for U.S. bases with operational forces. USAFE and PACAF bases are estimated to have 319 and 374 support personnel beyond the U.S. baseline, respectively.[5] This results in a total of 634 and 689 support personnel total for USAFE and PACAF bases, respectively. The standard costs of these personnel are in the fixed-cost component of the installation-support cost model. We say standard costs because AFTOC uses a standard cost per person (referred to as a standard composite rate) irrespective of base location. These costs would be saved if an overseas base were closed.

At this point, we also note that there are additional costs associated with these fixed support personnel when overseas that are not accounted for in the installation-support costs and, thus, not in the fixed installation-support cost with Combat Sup-

Table A.7
Air Force Installation Support Personnel Regression Results

Regression Term	Parameter Estimate	Standard Error	Lower Bound	Upper Bound
United States	315	112	202	427
USAFE	319	120	199	439
PACAF	374	132	242	507
Operations personnel	0.39	0.04	0.36	0.43
Institutional personnel	0.41	0.07	0.34	0.47

SOURCE: Authors' analysis of DMDC data.

[5] As discussed earlier, these numbers are generalizations. RAF Mildenhall does not exhibit these higher support personnel numbers, yet does have noticeably higher support costs excluding the Combat Support PE.

port parameters shown earlier (Table A.6). This is because the overseas personnel cost differential (allowances, PCS costs, schools, and overseas logistics) for those personnel, as well as the standard costs of the personnel (e.g., basic pay) that are already included would be saved if an overseas base were closed. These additional overseas costs are not captured in our earlier regression of AFTOC spending data. We return to this additional cost later, computing how much it would be and adjusting the fixed-cost component of installation-support cost accordingly.

The last two rows in Table A.7 show the variable component for the number of support personnel relative to supported personnel. The basic concept is that operational personnel are supported by a proportional number of support personnel, while the workload of support personnel is driven, at least in part, by the number of operational personnel. This relationship is shown in Figure A.3. If, for example, a squadron of aircraft moved from one base to another, some number of support personnel would move to provide installation support. The question is: How many support personnel would move? The parameter estimate in Table A.7 for operations personnel (leftmost column, fourth row) is shown as 0.39. This means that in general, as the population of operational personnel increases or decreases by one, the population of support personnel changes by 0.39. Thus, if 100 operational personnel move from base A to base B, roughly 39 support personnel ought to move from base A to base B, also, to provide enough installation support for the increased population at base B. The variable component estimate for institutional personnel is virtually the same as operational personnel, at 0.41 support persons.

So what does this mean for the question of higher installation-support costs (even excluding the Combat Support PEs) from Table A.5? This means that for comparable operational units (e.g., a single base with two fighter squadrons), USAFE and PACAF bases have significantly more military personnel in support roles than their U.S. counterparts. Why might that be? To better understand that, we dug deeper into the personnel data and did a base-by-base comparison of overseas bases with U.S. counterparts.

Figure A.4 shows the type and number of support personnel at U.S. and overseas bases of comparable size. For each base along the x-axis, the stacked column shows the total number of manpower positions in support functions, by function. The black line shows the total number of operations and institutional personnel. Red arrows denote overseas bases.

Given the correlation of operational and institutional personnel to support personnel demonstrated above, we would expect to see, in general, the support personnel numbers tracking with the nonsupport personnel. However, several bases, such as the bases in Japan, Spangdahlem, and Aviano, appear to have high support personnel relative to nonsupport personnel.[6]

---

[6] These personnel results not include heavy construction, which are in a different PE.

either from the United States or partner nations in those regions, or for the various mobility and throughput missions at these overseas bases. While some of these facilities could persist from past drawdowns (overseas bases may have had greater facility space to accommodate deploying forces), this seems unlikely given the significant consolidation and cost-cutting, at least in Europe. Whatever the reason, overall greater relative facility space is another likely driver of higher overseas support costs.

We conclude this section with one final point about the inclusion of Combat Support costs (pay for military personnel providing installation support) in our overall cost analysis. Because the Air Force chooses to use military personnel for installation support as a policy, and the high correlation we observed between operational and support personnel in the section above, if the operational population at a base were reduced, the military support population would be reduced as well. Those personnel are thus a part of the variable costs of USAF installation support. Thus, because of the nature of military support personnel at USAF bases, we conclude that their military pay (and other costs associated with having personnel overseas) should be included when tallying the total cost the USAF and DoD bear to maintain USAF presence in the United States and overseas.

Accounting for the Overseas Variable Costs of Combat Support Personnel

Up to this point, we have discussed the installation-support cost model that results from our analysis of AFTOC data. The standard composite rates for military pay for these military support personnel are captured in AFTOC, so the regression results that include the Combat Support PEs reflect the cost of military pay as a component of Air Force installation support. However, because AFTOC uses standard rates for all personnel irrespective of region, it does not reflect the regional cost differentials discussed in Chapter Eight. Therefore, AFTOC understates the cost of overseas installation support by using standard composite rates for those military personnel that do not reflect the higher allowances received by personnel overseas or other costs associated with being assigned overseas, such as higher PCS move costs.

To better reflect the real cost to DoD of providing Air Force installation support overseas, and the cost differential between the United States and overseas, we supplemented our Air Force cost model to adjust for these with additional overseas costs. We do this with a twofold approach. The first step was to derive the relationship between military personnel demanding installation support (i.e., operational and institutional personnel, our explanatory variables) and the personnel providing installation support, which we showed in Table A.7.

The second step is to apply the total overseas per-person variable cost differentials (i.e., all categories, not just installation support) to the support personnel ratios developed in the first step. The result will be the additional amount that could be saved if operational (or institutional) forces were relocated to the United States from overseas. In other words, when we model the movement of a unit from overseas to the United States, we explicitly account for only the number of people in the unit to be moved, not the additional support personnel required to support it; we are examining the potential realignment of units/operational activities. But, as shown here, additional support personnel would also shift and incur the lower U.S. personnel costs. Table A.8 shows the results of the calculations: the fixed and variable cost component adjustments for USAFE and PACAF.

In Table A.8, the left two columns—with fixed cost calculations—function essentially independently of the right two columns, which adjust the variable costs. We start with the left side, fixed costs. The first two entries in the first row show the original fixed cost parameter estimates from our regression analysis shown earlier in this appendix (with Combat Support). This is taken directly from Table A.6 and the surrounding discussion above. The second row shows the fixed component for the number of support personnel, drawn from Table A.7. The fifth row, still on the left side, shows the number of support personnel multiplied by the incremental overseas variable cost. What this means is that the baseline cost (first row) reflects the fixed-cost component as shown in AFTOC. The fifth row adds to that the missing cost per person to reflect regional cost differences. The last row totals these to show the sum of these installation-support costs. The bottom section of the table adds to these installation-support costs the other components of cost: MILCON for recapitalization and DoD schools. These costs are drawn from the respective sections of Chapter Eight.

Table A.8
Calculated Values of Additional Air Force Installation-Support Costs

	Fixed		Variable	
	USAFE	PACAF	USAFE	PACAF
Baseline cost	$146,100,000	$123,000,000	$27,700	$19,500
Support personnel	634	689		
Operational personnel parameter			0.39	0.39
Incremental overseas cost per person			$27,700	$19,500
Additional overseas cost	$14,800,000	$10,400,000	$10,800	$7,600
Total overseas cost, installation support	$160,900,000	$133,400,000	$38,500	$27,200
MILCON fixed cost	$28,300,000	$14,200,000		
DoD schools fixed cost	$21,700,000	$21,700,000		
Total fixed cost	$211,000,000	$169,400,000		

SOURCE: Authors' analysis of DMDC and AFTOC data.

Now we address the right side of the table. The first row shows the total incremental variable cost (all categories), as calculated in Chapter Eight. The PACAF value here is a composite of Japan and South Korea. The third row shows the number of support personnel per operational person, taken from Table A.7. The fourth row repeats the incremental cost per person.

For the fifth row, the two variable-cost columns show the additional per-person cost we must add to each operational person when applying posture changes: the overseas incremental variable costs for the additional Combat Support personnel needed as the number of supported personnel increase (i.e., 0.39 more support personnel per supported person multiplied by $27,700 or $19,500 per additional support person).[8] The two right columns in the last row show the new total overseas incremental variable cost per operational/institutional supported USAF person in USAFE and PACAF—their direct cost plus the variable cost associated with the additional support people they require.

When we assess the cost changes resulting from posture changes, we do not count the military support personnel separately, but only as they follow the operational personnel who are realigning. In effect, each operational person realigned is counted as 1.39 people in determining the variable cost shift to account for the "0.39 support people" that would also move with them.

To illustrate how this all works together, we provide an example for the closure of one base and the realignment of its operational units to CONUS. In Table A.9, we show example calculations for three USAFE bases in our posture options: Lakenheath, Aviano, and Mildenhall. As an example, our cost-reduction posture posits the closure of Lakenheath, with all forces realigned to the United States; the closure of Aviano, with forces relocated to Spangdahlem; and the relocation of some forces from Mildenhall to the United States, but leaving the base open with the air base wing staying in place.

In Table A.9, the first two rows show the actual installation-support costs shown in AFTOC for these three bases. The first shows the total for PEs we included, excluding the Combat Support PEs. The second row includes Combat Support. The next two rows show the fixed-cost components our regression analysis modeled for those two categorizations. To explain the table, we walk through the Lakenheath column. For Lakenheath, of the $125 million in installation-support costs with Combat Support, we attributed $100 million to fixed costs that would be saved if the base were closed; of the $201 million for installation support with Combat Support, we attributed $146 million to fixed costs that would be saved if the base were closed. The next row shows that there is an additional $14.8 million to be saved from the incremental variable cost for the additional fixed number of Combat Support personnel at overseas bases. This

---

8 The PACAF value for incremental overseas cost per person is the personnel-weighted average of incremental per person costs for Japan and South Korea.

Table A.9
Example of Overseas Air Force Base Closure Costs

	Annual Cost ($ millions)		
	Lakenheath	**Aviano**	**Mildenhall**
Actual cost, no Combat Support	124.9	104.7	156.5
Actual cost, with Combat Support	201.1	193.3	222.3
Fixed cost installation support, no Combat Support	99.8	99.8	
Fixed cost installation support, with Combat Support	146.1	146.1	
Additional fixed cost for Combat Support personnel	14.8	14.8	
MILCON recapitalization fixed cost	28.3	28.3	
Schools fixed cost	21.7	21.7	
Total modeled fixed cost, with Combat Support	211.0	211.0	
Incremental variable costs	73.3		51.7
Additional incremental variable costs for Combat Support	29.3		20.7
Total saved, with Combat Support	313.6	211.0	72.4
Variable cost that transfers	73.6	44.5	52.0

SOURCE: Authors' analysis of DMDC and AFTOC data.

figure is drawn from Table A.8. The next two rows show the fixed-cost components for MILCON recapitalization and DoD dependents schools. These are discussed in Chapter Eight. The final row shows the total modeled fixed cost for Lakenheath, with Combat Support costs, a total of $211 million per year that would be saved were the base closed, not including the savings from moving operational personnel in the fighter squadrons. That $211 million is not directly comparable with the original installation-support cost in the second row; our total modeled cost includes additional fixed cost categories beyond just installation support, as walked through in Chapter Eight and Table A.9. For reference, our total modeled installation-support cost (just fixed installation-support costs plus variable installation-support costs, equivalent to the categories in the "Lakenheath actual, with Combat Support" row) for Lakenheath is $214 million compared with an actual $201 million, our modeled installation-support cost for Aviano is $187 million compared with an actual $193 million, and our modeled installation-support cost for Mildenhall is $194 million compared with an actual $222 million. Thus, some modeled costs are higher than the actual values, and some are lower. This is to be expected given the natural variation in spending.

Continuing the Lakenheath column explanation, the next two rows show the additional per person costs that would be saved by moving operational personnel from

Lakenheath to the United States. The first row shows the costs for the operational personnel; the second row shows the additional amount for the Combat Support personnel that would move from Lakenheath to a U.S. base. That factor is calculated from per-person costs found in Table A.8. The next to the last row in the table shows the estimated total of $313 million that would be saved according to our calculations. For Lakenheath, this includes the closure of Lakenheath ($211 million) with all its operational forces relocated to the U.S ($73 million plus $29 million). For Aviano, this includes only the fixed cost saved from closing the installation, $211 million, because in the illustrative CRP option, Aviano's forces are relocated to Spangdahlem, still in USAFE, so the regional incremental variable costs are not saved. Finally, Mildenhall stays open in the illustrative CRP option, but its operational forces are relocated to the United States. Thus, only the incremental variable costs per person costs are saved; a total of $72 million.

The last row in the table shows the variable (i.e., population-driven) installation-support cost burden that would simply transfer from base to base. For Lakenheath and Mildenhall, that cost burden would transfer to the United States, but would still be borne within the Air Force writ large. For Aviano, because the forces are moving within USAFE, the cost burden stays within USAFE, albeit at a different base.

Army Installation Support

Data Sources, Caveats, and Limitations

The Army provided the RAND research team with two separate data sets to derive installation-support costs: Installation Status Report for Services (ISR-S) and environmental data.[9] As described by Army G-8, Program Analysis and Evaluation, "ISR-S is essentially self-reported execution from Garrison Resource Managers to track service costs and performance. As such the ISR-S data cannot be audited directly by J-Book task or activity group on a dollar-for-dollar basis."

The ISR-S data included CONUS Regional Headquarters and OCONUS headquarters service overhead costs. Army personnel included these cost data "to provide a more complete regional assessment," but informed the team that these costs could not reasonably be assigned to individual installations.[10] The ISR-S data did not include OCO-funded services.

Environmental program data are not tracked in ISR-S, so additional spreadsheets were provided. These environmental cost data included data from the Defense Finance and Accounting System (DFAS) and General Fund Enterprise Business System (GFEBS) for FY 2009 and FY 2010, and FY 2011 for GFEBS only.[11] These environ-

[9] Data provided by Army G-8, Program Analysis and Evaluation, on June 8, 2012.

[10] Email communication with Army G-8, Program Analysis and Evaluation, CIPAD, on June 8, 2012.

[11] According to the GFEBS website, GFEBS is "the Army's new web-enabled financial, asset and accounting management system."

mental data did not include reimbursable costs because environmental programs are mostly accounts where funds are earned and then expended. We were informed this might cause some double accounting, but as a rough comparison, the average annual dollar value of costs in the ISR-S data was $11.3 billion, while the same costs for the environmental data (sum of DFAS and GFEBS) were $333.5 million. So the relatively small level of environmental cost data mitigates this potential issue.

Unlike the USAF data described above, we did not see the need to exclude any particular bases up-front, but left that question for the regression analysis. However, about $2 billion out of the original $11.3 billion in the ISR-S data fell into a category labeled as "HQ Central funds," which could not be assigned to specific bases.

Regression Data and General Observations

Table A.10 shows how we mapped installations from the Army cost data to the DMDC personnel data.[12] Total active duty Army personnel numbers are shown. Because the Army cost data provided to us (shown later) are rolled-up costs by parent (direct report) installation, those costs reflect an aggregation of a number of bases or sub-sites of the parent, direct reporting installation. The base mapping reflects the direct report garrisons of the time period of the data, FYs 2009–2011. They have since changed in some cases due to planned closures.

Table A.11 shows the raw cost and personnel data for overseas locations.

Table A.12 summarizes the data points in each category for our regression analysis. This table shows the total bases, military personnel (operations and institutional only), and average FY 2009–2011 support costs. This table excludes about $363 million in overseas costs from records we excluded from our regression. Most of these entries were HQ Central Funds that could not be allocated to individual locations.

Figure A.6 shows the raw regression data for Army installation-support costs. The x-axis shows the total of operational and institutional personnel (both explanatory variables). The y-axis shows annual installation-support costs (FYs 2009–2011 averaged). Each point on the graph represents a single active duty Army installation. The colors and symbols differentiate the region and type of base. Bases with operational forces are shown as blue diamonds for the United States (excluding territories), red squares for USAREUR, and yellow triangles for USARPAC. U.S. bases with no permanently stationed operational units are shown by hollow blue diamonds.

We make a few observations about Figure A.6:

- Overseas bases are significantly smaller than U.S. bases. Nearly all overseas bases are smaller than those with forces in the United States.

[12] Installation Management Command (IMCOM) Installations Map Feb 12.pdf provided to RAND by HQ IMCOM personnel in November, 2012.

Table A.10
USAREUR Parent Installation Mapping (FY 2009–2011 Time Period)

Base from Cost Data	Bases from Personnel Data	Active Duty Army	Total Personnel
Ansbach	Katterbach Kaserne (Ansbach)	1,503	
	Shipton Kaserne Ansbach	236	
	Ansbach Barton Barracks	35	
	Illesheim Germany	1,054	2,828
Bamberg	Bamberg Warner Barracks	3,206	3,206
Grafenwoehr	Grafenwohr Germany	2,199	
	Viseck	4,635	
	Hohenfels Germany	1,436	
	Garmisch	41	8,311
Stuttgart	Stuttgart Germany	377	
	Boblingen Panzer Kaserne	632	
	Vaihingen – Patch Barracks	377	
	Mohringen Kelley Barracks	201	1,587
Vicenza	Vicenza Italy	2,817	2,817
Baden Wuerttemberg	Heidelberg Patton Barracks	1,008	
	Heidelberg Campbell Barracks	887	
	Kaiserslautern	1,823	
	Baumholder H.D. Smith Barracks	4,794	
	Landstuhl Medical Center	1,092	
	Sembach	449	
	Seckenheim	434	
	Mannheim	274	
	Sandhofen	274	
	Kaefertal Germany	270	
	Schwetzingen	218	
	Schwetzingen Tompkin Barracks	200	
	Worms	182	11,905
Benelux	N/A		
EURO HQ Central Funds	N/A		

Table A.10—Continued

Base from Cost Data	Bases from Personnel Data	Active Duty Army	Total Personnel
Schweinfurt	Schweinfurt Ledward Barracks	3,371	
	Schweinfurt Conn Barracks	1,367	
	Giebelstadt	385	5,123
Weisbaden	Wiesbaden Germany	1,880	
	Miesau Army Depot	457	2,337

SOURCE: Authors' analysis of DMDC data and IMCOM Installations Map.

Table A.11
Army Overseas Regression Analysis Data Set

Region	Base	Operational and Institutional Military Personnel	Annual Support Cost
USARPAC	Camp Humphreys	2,904	$77,544,689
	Camp Red Cloud	1,464	$115,127,105
	Zama/Sagamihara	640	$94,765,121
	Camp Henry/Walker	876	$66,756,747
	*KORO HQ Central Funds		$57,032,797
	*Pacific HQ Central Funds		$20,205,166
	Yongsan	3,151	$113,223,212
USAREUR	Ansbach	2,729	$103,829,756
	Baden Wuerttemberg	11,495	$331,956,774
	Bamberg	3,136	$60,868,208
	*Benelux	—	$89,349,544
	*EURO HQ Central Funds	—	$196,373,875
	Grafenwoehr	8,047	$323,064,742
	Schweinfurt	5,016	$76,865,647
	Stuttgart	1,452	$132,616,625
	Vicenza	2,647	$130,658,965
	Wiesbaden	1,936	$241,454,903

SOURCE: Authors' analysis of DMDC, ISR-S, and Army Environmental data.

NOTE: USARPAC = U.S. Army, Pacific.

* Base excluded from regression analysis.

with the USAF regression, operational and institutional personnel proved to be statistically significant drivers of the variable-cost component, i.e., the per-person installation-support costs. We tested a model with the number of civilians as a separate variable, but this did not improve fit.

Operations personnel are associated with a variable cost of about $5,700 per person per year. This means that at any base, the addition of one operational person is estimated to increase total installation-support costs by $5,700 per year. For institutional personnel, it is roughly five times that, or $27,600. Those figures are for the baseline case of U.S. bases. But we also found no statistical difference for installation-support variable costs overseas.

This brings us back to the USAREUR outlier bases. With the fixed- and variable-cost components, we can now estimate what those USAREUR base-support costs would be if predicted perfectly by the regression model. The total support costs would be expressed by the equations below.

The basic model for cost has the form $c_{ir} = a_R + b_r p_{ir}$. Here, c_{ir} is the total installation-support cost of installation i in region r, a_R is the fixed-cost component of an installation in region r, and $b_r p_{ir}$ is the product of the variable cost per person in region r and the number of personnel at installation i in region r. The total fixed cost for an overseas base is $a_R = a_{US} + a_r$, where fixed cost for a U.S. base is a_{US}, a_R is the total fixed cost for an installation in region r, and a_r is the regression component cost for a base in the same region.

The variable cost can be expressed as $bp_{ir} = b_O p_{iO} + b_I p_{iI}$, where $b_O p_{iO}$ is the variable cost of operational personnel multiplied by the number of operational personnel at base i, and $b_I p_{iI}$ is the variable cost of institutional personnel multiplied by the number of institutional personnel at base i.

Table A.14 shows the results of these calculations. This table shows the calculated support costs c_{ir} for the three USAREUR outlier bases using the equations shown above, the actual cost (average FYs 2009–2011), and the cost difference between the two. Given the data at hand, then, this cost difference represents the fixed-cost component for each base above and beyond U.S. bases of comparable base population. This is the estimated cost associated with their unique activities that are independent of support to permanently assigned personnel at these bases.

In our cost analysis, when a posture option contemplates the closure of one of these bases, we apply this cost difference in addition to the $65 million baseline. In our overall posture analysis, Grafenwoehr is the only base in this set that is closed in one of our posture options. Heidelberg, which is technically accounted for under Baden Wurttemberg, is also closed in one of our posture options, but because it accounts for such a small portion of the parent installation's facilities, personnel, and costs, we do not apply the additional fixed-cost component. The entire Baden Wurttemberg complex as mapped in our analysis from the FY 2009–2011 timeframe would have to be closed to realize the full savings indicated in Table A.14.

Table A.14
Calculated Values of USAREUR Outlier Base Costs

Base	Calculated Support Cost	Actual Support Cost	Cost Difference
Baden Wuerttemberg	$139,367,013	$331,956,774	$192,589,761
Grafenwoehr	$130,316,507	$323,064,742	$192,748,761
Weisbaden	$77,519,887	$241,454,903	$163,935,016

SOURCE: Authors' analysis of DMDC, ISR-S, and Army Environmental data.

Navy Installation Support

Data Sources, Caveats, and Limitations

For installation-support cost analysis, the Navy provided data from the Claimant Financial Management System (CFMS). We used the Navy's Installation Management Accounting Project (IMAP) Model to map spending to PE categories.[13] The Navy's PE categories were all comparable to the other services, with the exception of port operations, which we included since it is key to naval operations.

The naval base cost data were extremely comprehensive, with nearly 400 separate installations included (ranging in annual support costs from $1,000 to over $500 million), so we excluded many bases that were not relevant to our analysis. We excluded bases with one or more of the following characteristics to simplify and streamline our cost analysis:

- bases with fewer than 100 military personnel
- locations with only a personnel support detachment
- bases with less than $1 million per year in support costs.

For our regression analysis, we excluded several more bases in an effort to develop an accurate cost model. We discuss these exclusions in a subsequent section.

Regression Data and General Observations

Table A.15 shows the raw cost and personnel data for overseas locations. We include Souda Bay in this table, even though it has less than 100 operational and military personnel, because of its support cost and for completeness.

Table A.16 summarizes the data that fed our regression analysis.

Figure A.7 shows the raw regression data for Navy installation-support costs. The x-axis shows the total of operational and institutional personnel (both explanatory variables). The y-axis shows annual installation-support costs (FYs 2009–2011 averaged). Each point on the graph represents a single active duty Navy installation. The

---

[13] Installation Management Accounting Project (IMAP) Model 2012, Commander, Navy Installations Command (NV52) provided Thursday June 7, 2012, via OSD/Policy from Directorate of Strategy and Policy (N5), Office of the Chief of Naval Operations.

Table A.17
Navy Installation Support Regression Results

Regression Term	Parameter Estimate	Standard Error	Lower Bound	Upper Bound
United States, with forces	55,900,000	8,100,000	47,800,000	64,000,000
United States, no forces	–46,980,000	8,843,200	–55,823,200	–38,136,800
Overseas base	17,925,100	9,904,900	8,020,100	27,830,000
Military – Operations	12,700	3,900	8,900	16,600

SOURCE: Authors' analysis of DMDC and CFMS data.

threshold for term inclusion), in our best cost model. The table shows, for each variable, the parameter estimate, the standard error, and the resulting lower and upper bounds.

This final regression model had an adjusted R-squared value of 0.44. The y-intercept (fixed cost) of the baseline, a U.S. base with forces, is about $56 million per year. If the base has no forces, the fixed-cost component is $47 million lower, or $9 million. We found it necessary to exclude Pearl Harbor because it had a large influence on the results and is not representative of the other cost relationships.[14]

We then assessed the cost differences for USPACFLT and USNAVEUR bases. We found it necessary to exclude both Guam and NSA Andersen for the same reason we excluded Pearl Harbor from the U.S. data set. None of the posture options we consider close either base, so it is not necessary for us to separately assess their unique fixed costs. We found it necessary to combine the USPACFLT and USNAVEUR data points to achieve statistical significance. The regression analysis did produce a parameter estimate of $18 million for this combined overseas data set. However, we saw no discernible pattern in the remaining overseas data points. They are sufficiently spread out to make any generalization essentially meaningless.[15]

Our posture options only consider the closure of one of the bases in this set of overseas bases (the posture options do call for the closure/relinquishment of several smaller access bases and support locations): Souda Bay, Greece. Thus, we took an approach similar to that used with the USAREUR outliers discussed in the previous section. We applied the equation resulting from the fixed and variable components of the regression model for U.S. bases and then calculated the fixed-cost premium for Souda Bay.

---

[14] It is possible for a data point to have an outsized influence on the model fit, but still be in line with the cost relationships of other data points if it is far to the upper right of the other data points. This can greatly increase the R-squared value but might not change the overall cost function.

[15] We tested EUCOM and PACOM data points separately, but their parameter estimates, though both positive, were not statistically significant. We were able to achieve statistical significance by combining the two data sets into one overseas set, but the meaning of this additional parameter (about $18 million above U.S. bases) is questionable given the obvious spread in the data points.

For variable costs, only operations personnel were found to have a statistically significant effect on installation-support costs and are included in our model. Operations personnel are associated with a variable cost of about $12,700 per person per year, regardless of location. We found that there is still a positive correlation between costs and military institutional personnel after accounting for military operational personnel, but the standard error on that coefficient is very high. We also tested the effect of civilian personnel on costs but found a lack of correlation between the two.

Because of the way Navy forces are bedded down in the United States, with several very large bases, these regression results are fairly sensitive to the inclusion of these data points. We felt it was important to include these data points because such large portions of naval forces are based there, but we express caution in extrapolating these results too far. In our cost analysis of posture options, the incremental variable costs per person are wholly driven by costs other than installation support, and only one significant overseas base is closed. This brings us to our last topic in this section.

The support costs associated with Souda Bay, Greece, are clearly lower than the fixed-cost component estimated for both U.S. and overseas Navy bases. Because the use of our generalized parameter estimate for a single base would, in this case, overestimate its costs, we estimate the fixed-cost component of this base individually. We use the same equation in the previous section that was used to calculate the same values for Army bases. According to our data, Souda Bay has a total of 315 Navy military personnel, only 55 of whom we classified as operational. Thus, the component costs are as follows:

- Calculated support costs: $56,570,528
- Actual support costs: $29,510,789
- Cost difference: $27,059,739
- Fixed cost used for base closure: $28,800,000.

We arrived at the last figure by simply subtracting the sum of variable costs attributable to operational military personnel. The remaining cost would be saved if the base were closed.

Marine Corps Installation Support

Data Sources, Caveats, and Limitations

For installation-support costs, the Marine Corps data came from the Standard Accounting, Budget, and Reporting System (SABRS), the official accounting system for the U.S. Marine Corps.[16] Given the size of the Marine Corps and the way it bases

[16] SABRS was designed to meet fiduciary standards established by the General Accounting Office, Office of Management and Budget, United States Treasury Department, and DoD.

Table A.20
Marine Corps Installation Support Regression Results

Regression Term	Parameter Estimate	Standard Error	Lower Bound	Upper Bound
United States, with forces	49,200,000	10,700,000	38,500,000	59,900,000
Military – Operations	8,500	800	7,700	9,300

SOURCE: Authors' analysis of DMDC and SABRS data

Table A.21
Marine Corps Regression Model Summary

	Parameter Estimates, $		
Model inputs	Adjused R-squared	Fixed component	Variable component
With OCO, with large bases	0.88	50,500,000	8,500
With OCO, no large bases	0.47	45,100,000	9,900
No OCO, with large bases	0.90	49,200,000	8,500
No OCO, no large bases[a]	0.54	28,600,000	10,500

SOURCE: Authors' analysis of DMDC and SABRS data.
[a] Fixed component parameter estimate not statistically significant.

Lejeune. One can see that the first three models do not show a significant difference in their parameter estimates, but the final one, no OCO and excluding large bases, had a much lower fixed cost and much higher variable cost. This means that the two large bases essentially hold the variable cost down and attribute more of the cost to the fixed component when included in the model. This makes sense, because in reality the fixed cost may vary in steps or be different for completely different classes of bases. The results are obviously very sensitive to this inclusion. One can also see how much those two bases drive the R-squared value.

Ultimately, though, the choice of models matters little for our overall cost analysis. We utilize the same variable cost for U.S. and overseas Marine Corps personnel (i.e., zero incremental cost difference). The one calculation for which it would matter is the estimation of an additional fixed component for Camp Butler, if it were found to be an outlier. If one uses the upper-bound estimates for each model (not shown here), the two models that include large bases show Camp Butler to be slightly higher than the upper end of the range that would be expected given its number of assigned personnel (by about $15 million), and the two models that exclude large bases show Camp Butler to be within the range produced by the standard error. The main reason is that the models with lower R-squared values have much higher standard error values, and, thus, result in higher ranges for upper and lower bound estimates. If the outcome of these calculations mattered more for our analysis, we would devote more attention to settling the question. As it is, one should take this particular regression model with a grain of salt, given how small the sample size and how sensitive it is to the inclusion of these data points.

Our illustrative posture options do not include the complete closure of Camp Butler, so none of the aggregate estimated cost effects from the illustrative postures are affected by any of the modeling for the Marine Corps discussed in this section. In the few calculations in the report where we do need to attribute a fixed-cost component to Camp Butler to discuss other possible options or to frame the discussion of Okinawa, we use the cost model selected above and attribute a $40 million per-year fixed-cost differential (i.e., a total fixed cost for installation support for Okinawa of about $90 million per year).

Plant Replacement Value

The purpose of performing a regression analysis of PRV data was to estimate two different costs: recurring MILCON cost requirements for modernization and restoration and one-time costs associated with implementing changes to the posture. As with installation support, our regression modeling determined a fixed and variable component to PRV per base for each service. For recurring costs, we applied the 67-year recapitalization factor mentioned in Chapter Eight to both the fixed and variable components, applying only the fixed element to base closures and the variable element to changes in base population. For one-time investment for posture transition, we used the per-person variable cost only, using it as-is for the relevant region since realignments include adding personnel to bases, not opening new ones, in our modeling and posture analysis, Thus, this accounts for the one-time costs of building new facilities to handle the new units and some expansion in general base facilities but does not include the fixed cost of facilities and infrastructure for standing up a whole new installation, which would involve some facilities for which only one is needed within a wide band of installation size. We derived installation-level PRV values from RPAD. While RPAD does have known problems with facility conditions (as documented in Chapter Seven), these same caveats do not apply to PRV, and we were not advised of any other significant, systematic issues with the PRV data. We excluded from our calculations all closed or disposed facilities, all land-only facilities, and any Guard- or Reserve-only installations.

Air Force PRV Analysis

Regression Data and General Observations

We excluded most of the same base types as for the installation-support cost data, plus two additional ones:

Corps base. The colors and symbols differentiate the region and type of base. Bases are shown as blue diamonds for the United States (excluding territories) and yellow triangles for MARFORPAC.

We now make a few observations about Figure A.13:

- There are simply very few data points. There are a small number of Marine Corps bases to begin with, but there were installation-level PRV data available for only a subset of them.
- We combine Camp Butler and Futenma, and include a data point that also includes the 5,600 UDP personnel mentioned in the installation support section. Both of these data points appear to be outliers relative to U.S. bases.

Regression Analysis Findings and Final Results

Given the data inputs shown above, we iteratively tested a range of models that specified different independent and control variables to identify the primary drivers and best predictors of support costs. We selected the final regression model shown in Table A.25. This table shows each of the variables, all of which have a p-value ≤ 0.01 (0.1 was the threshold for term inclusion), in our best cost model. The table shows, for each variable, the parameter estimate, the standard error, and the resulting lower and upper bounds.

This model has an adjusted R-squared of 0.58. The y-intercept of the base case, U.S. bases, is $2.4 billion. The variable PRV is about $191,000 per operational military person. For recurring costs, this translates to about $36.3 million per base and $2,844 per person for fixed and variable components, respectively.

For investment, we use only the per-person costs. So for each operational person moved from overseas to the United States, the PRV value would translate to an upper bound of about $191,000 in one-time MILCON to accommodate the arriving personnel.

This regression model includes only U.S. data points. There were not enough data points to distinguish overseas costs from U.S. bases: only one for Iwakuni and one for Okinawa. If we estimate the difference between Camp Butler, including UDPs, and the upper bound of U.S. costs, we find that Camp Butler has a PRV roughly $2.0 billion higher than comparable U.S. bases. Thus, it is well outside the upper bound of

Table A.25
Marine Corps PRV Regression Results

Regression Term	Parameter Estimate	Standard Error	Lower Bound	Upper Bound
United States, with forces	2,435,000,000	704,100,000	1,731,100,000	3,139,000,000
Military – Operations	190,600	45,700	144,800	236,300

SOURCE: Authors' analysis of DMDC and RPAD data.

U.S. PRVs. However, when we factor in Japanese HNS contributions to MILCON (as discussed in our Chapter Eight discussion of MILCON recapitalization), the net cost for MILCON recapitalization is estimated to be very close to that in the United States. The U.S. value is about $29 million per year, and the Okinawa value is about $27 million per year. However, none of the illustrative posture options completely close Camp Butler, so this does not affect the aggregate cost estimates in Chapter Ten, only the Okinawa-specific discussion in Chapter Ten in the section "Comparing Marine Corps CRP Cost Changes to the Long-Term Plan."

Marine Corps MILCON Requirement Estimating Methodology

To develop a budget-quality estimate of the MILCON requirement for relocating marines to Guam and Hawaii associated with a reduction in Okinawa, the Marine Corps determined the precise facility laydown that would be required, building by building. It basically represented a complete, detailed installation plan, as would need to be done to stand up any new installation and develop a budget-quality estimate of needed funding. This produced an extremely accurate estimate of the MILCON requirements. When this analysis was performed, it was based on the 2006 agreement with Japan and a plan to relocate a substantially larger number of marines to Guam than currently agreed to and to add a modest number of marines to Hawaii in what was called the "preferred laydown." The facility requirements also were based on the specific needs of the specific units that would have relocated to Guam and Hawaii.

In the 2012 agreement with Japan, it was agreed to pursue a more distributed laydown of marines in the Pacific associated with the reduced presence of marines in Okinawa, which would involve a smaller number of marines ultimately being located in Guam, with some additions still planned in Hawaii, rotational presence planned in Australia, and possibly some in CONUS. To quickly develop a modified MILCON cost estimate for a smaller presence in Guam than envisioned in the preferred laydown and to develop cost estimates for the revised presence additions in these other locations to develop an overall cost estimate for the distributed laydown, the Marine Corps developed an estimating methodology based on the detailed Guam and Hawaii facility requirements analysis, with about 90 percent of the requirement being for facilities in Guam. The preferred laydown MILCON requirement (not including training projects), including the portion that Japan would have paid for to determine the full construction requirement for the specified number of marines, was translated to a per-person cost. Then the ACFs for Guam and Hawaii, which are almost the same, were backed out to produce a MILCON requirement per person at an ACF of 1. This produced an estimate of $358,000 per person. Then this was applied to estimate MILCON requirements at all four locations based on planned stationing levels and specific ACFs.[20]

---

[20] U.S. Marine Corps, Pacific Division, Plans, Policies and Operations, 2012b.

As noted, the base MILCON requirement was as accurate as possible for a new Marine Corps base in Guam with the specified number of personnel and units and for the expansion in Hawaii. It would likely produce a very good ACF-adjusted MILCON requirement estimate for any other new bases with somewhat similar populations or instances in which the increase represented a large relative expansion of an existing base. However, by taking a per-person average MILCON requirement, this estimating methodology does not decompose MILCON requirements into some level of fixed or base "startup" costs versus variable MILCON requirements associated with increasing numbers of marines and units per the approach we discuss in Chapter Eight. The facility requirements for Guam were categorized mostly into the four categories listed in Table A.26, with small portions also for land acquisition, planning and design, environmental mitigation, and defense access roads. If a new base were being established, all of these facilities would be needed; with some types of facilities not changing in size for a range of base size. If a modest increase was to occur at a large base, say Camp Pendleton or Camp Lejeune, or Marine Corps Air Station, such as Miramar or Cherry Point, then not all of the same facilities would likely be needed. There would likely be some facilities that would not need to be expanded or duplicated, while others would be needed—what we term fixed versus variable components in the PRV analysis.

However, determining which of the facilities would typically need to be expanded or duplicated is not completely feasible—it depends to a great degree on the facilities at the receiving base, the relative and absolute size of the increase, and the nature of the increase. Clearly, the operational facilities would be in the variable category, as likely would be quarters, and perhaps facilities such as fitness centers. Only a few facilities would clearly be in the fixed category for moderate changes in base population. For example, this might be a main exchange or a commissary, for which capacity could be increased to a degree through labor. With a large enough increase, first new satellite facilities and then major ones might be needed. Similar categories might be base warehouses and the like. In the middle between these examples, schools might be able to absorb some increase in population, depending on class sizes and utilization, with first

Table A.26
Facility Categories and Examples from the Guam Facility Requirements Analysis

Category	Description/Examples
Operational facilities	Facilities for units: armories, maintenance shops, aircraft hangers, administrative, HQ, unit warehouse
Quality of life facilities	Fitness centers, outdoor playing fields, main exchange, schools
Upgrades and site improvements	Utilities, parking
Base facilities	Post office, enlisted dining, fire stations, medical clinics, public works/base maintenance, base warehouses, bachelors enlisted and officers quarters, commissary, central issue facility, religious facilities

some increases in teachers. But more quickly, more schools—or school expansions—might be needed, compared with a main exchange. On the other hand, some locations may have public schools that dependents attend, with the base not using DoD schools. Other types of base facilities and quality-of-life facilities likely would require some expansions as well, but not always one-for-one with the change in population.

So on one end, assuming the MILCON requirement per person from the Guam preferred laydown expansion would likely overestimate the requirement for a modest expansion at major U.S. installations. It would likely also underestimate the requirement for a new, smaller base for which the fixed facility portion or the portion less variant with population would be a greater percentage. It might also overestimate the requirement for a very large new base. In short, the further away it gets from the preferred laydown case, the more likely it is to produce error because of the assumption of linearity with no fixed cost element. In fact, the analysts who extrapolated from the preferred laydown MILCON per person estimate noted that there could be inelastic effects that could cause estimating errors. This inelasticity is akin to our definition of some level of fixed cost to MILCON per base. But we cannot just assume that the variable requirement would only be operational facilities either, and without a detailed base analysis knowing how much more of each of the other facility categories would be needed is impossible.

Our methodology for CONUS PRV requirements is based solely on existing U.S. facilities, and changes we consider in the illustrative postures would be modest with respect to the larger CONUS installations. As seen in Figure A.13, there is a visually strong relationship between CONUS installation PRV and base population for the Marine Corps, with an apparent fixed portion or a portion invariant with population. The regression analysis enables us to decompose PRV into fixed and variable components. The y-intercept of $2.4 billion represents the fixed component with an increase in PRV of $191,000 per person beyond that. For a large base such as Camp Pendleton, with nearly $10 billion in PRV, this puts about three-quarters of the PRV into the variable component, and one can see that the fixed component becomes relatively larger the smaller the base. Given the strong relationship, we believe this method to be reasonable at estimating the relationship between base population increases and PRV, in CONUS. However, we also recognize it could underestimate requirements if not all requirements have been fully met. The PRV represents what has been built, not what is necessarily required. One thing to note is that the estimates per person are somewhat similar, particularly for the Army, Marine Corps, and Air Force.

Figure A.14 compares the MILCON cost estimates our model would produce against the estimates the Marine Corps model would produce varying the base population from zero to 20,000 for CONUS and Guam, using the respective ACFs. This would be a comparison for a new base. One can see that for populations below about 15,000, the RAND model produces a higher estimate and the Marine Corps model produces lower ones, because the RAND model has an initial fixed cost while the

Table A.27—Continued

How RAND Uses. . .	Cost Element	Armored Brigade Combat Team — U.S. Cost ($ millions)	EUCOM Cost ($ millions)	PACOM Cost ($ millions)
	Training Ammunition & Missiles	$9.8	$9.8	$9.8
Training Cost–Table 8.2	Post Production Software Support	$4.3	$4.3	$4.3
Applied to unit type	Annual Maintenance Cost	$0.4	$0.4	$0.4
	Modernization Amortized Cost	$3.8	$3.8	$3.8
Training Cost–Table 8.2	Indirect Support Cost	$10.9	$32.4	$9.4
Applied to unit type	Transportation of Things	$0.6	$2.7	$0.5
	Supplies and Equipment	$3.7	$4.3	$2.7
	Contractual Services – Field	$0.5	$1.7	$0.8
	Mission Travel	$1.3	$4.0	$1.6
	Equipment Leases	$0.3	$0.6	$0.1
	Contractual Services	$2.6	$13.4	$1.7
	ADP	$0.3	$6.0	$0.1
	Other	$2.3	$7.4	$1.6
	Purchased Equipment	$1.6	$2.3	$1.8
	Admin Travel	$0.2	$1.0	$0.1
	Civilian Labor	$0.0	$0.9	$0.0
	Other	$0.1	$1.4	$0.0
Not accounted for	Personnel	$262.7	$338.7	$361.2
	Replacement Personnel Training	$4.2	$3.9	$3.9
	Training Through Initial MOS	$3.7	$3.4	$3.4
	Military Pay Funded	$1.4	$1.3	$1.3
	O&M Funded	$2.1	$2.0	$2.0
	Other Funded	$0.2	$0.2	$0.2
	Clothing Initial Issue	$0.5	$0.5	$0.5
Not used (see Chapter Eight for sources of pay/allowances/PCS data)	PCS Travel: Military & Dependents	$18.2	$24.4	$24.1
	Military Personnel	$240.3	$310.4	$333.3
	Basic Pay and Allowances	$187.6	$187.6	$187.6

Table A.27—Continued

		Armored Brigade Combat Team		
		U.S.	EUCOM	PACOM
How RAND Uses. . .	Cost Element	Cost ($ millions)	Cost ($ millions)	Cost ($ millions)
	BAH/OHA	$46.3	$95.3	$100.1
	COLA	$0.0	$21.1	$39.2
	Special/Incentive/Hazardous Duty Pay	$6.4	$6.4	$6.4
Not used (modeled separately with regression analysis)	Other Unit Support	$53.2	$83.0	$70.0
	Base Operations	$27.7	$57.5	$44.5
	Acquisition	$0.0	$0.0	$0.0
	Army Family Housing O&M	$0.7	$9.0	$4.2
	Command and Control	$1.5	$3.0	$2.1
	Engineering	$16.5	$31.1	$27.4
	Information Technology	$1.7	$1.9	$2.2
	Logistics	$4.2	$5.7	$4.0
	Operations	$0.1	$0.3	$0.1
	Personnel	$2.7	$5.8	$3.8
	Resource Management	$0.3	$0.8	$0.6
Not used (alternative data source described in Chapter Eight)	Defense Health Program	$25.5	$25.5	$25.5
	Total annual operations	$353.1	$480.1	$465.6

SOURCE: Authors' analysis of FCM 2012.

NOTES: POL = petroleum, oils, and lubricants; OSMIS = Operating and Support Management Information System; O&M = operation and maintenance; MOS = military occupational specialty.

APPENDIX B

Detailed Cost Analysis Results

This appendix contains five tables that present the results of our detailed cost analysis.

Table B.1
Rotations by Posture with Estimated Costs Using Sealift, Airlift, and Prepositioning for Equipment

Alternative	Service	Region	Location	Unit/Unit Type	Sealift ($ millions)	Airlift ($ millions)	Prepositioning ($ millions)
Baseline	Army	CENTCOM	Camp Buehring, Kuwait	1x12 month ABCT	$209	$558	$232
	Air Force	AFRICOM	Djibouti	CJTF-HOA, 449 Air Expeditionary Group (C-130), rotational F-15Es, UAVs, 2x6 months	$17	$20	$17
	Air Force	CENTCOM	Al Udeid, Qatar	379 AEW (90 AC including B-1, F-15E, E-8C, C-130, RC-135), 609 CAOC, 2x6 months	$307	$354	$311
	Air Force	CENTCOM	Al Dhafra, UAE	380th AEW (KC-10, U-2), 2x6 months	$97	$116	$98
	Air Force	CENTCOM	Ali Al Salem, Kuwait	386th AEW (C-130), 2x6 months	$17	$20	$17
	Air Force	CENTCOM	Manas, Kyrgyzstan	376th AEW (KC-135), 2x6 months	$25	$28	$26
	Air Force	PACOM	Andersen, U.S.	6xB52, 2x6 months	$19	$21	$19
	Marine Corps	PACOM	Makiminato Okinawa, Camp Butler, Japan	3 BLT (Inf), 3,837 personnel, 2x6 months, 3 ACE (6xAV-8), 2x6 months	$194	$356	$166
	Marine Corps	PACOM	Robertson Barracks, Australia	1 MEU, 2500 pers, 2x6 months	$65	$65	$71
	Marine Corps	EUCOM	MK, Romania	1 SPMAGTF, 400 pers, 2x6 months	$24	$43	$48
CRP	Army	EUCOM	Livorno (Camp Darby), Italy	1x12 months IBCT (-)	$72	$127	$71
	Air Force	CENTCOM	Al Udeid, Qatar	379th AEW (90 AC including B-1, F-15E, E-8C, C-130, RC-135), 609 CAOC, 2x6 months	$308	$354	$311
	Marine Corps	PACOM	Makiminato Okinawa, Camp Butler, Japan	1 BLT (Inf), 1,279 personnel, 2x6 months 1 ACE (6xAV-8), 2x6 months	$79	$133	$69

Table B.1—Continued

Alternative	Service	Region	Location	Unit/Unit Type	Sealift ($ millions)	Airlift ($ millions)	Prepositioning ($ millions)
GREP	Army	AFRICOM	Djibouti (proposed)	2x6 months I BN	$38	$65	$54
	Army	AFRICOM	Entebbe, Uganda	1x3 months 1 BN	$20	$34	$26
	Army	PACOM	Nakhon Ratchasima (Korat), Thailand	1x3 months IBCT	$65	$150	$59
	Army	PACOM	Zamboanaga, Philippines	2x6 months SF BN	$23	$42	$21
	Army	EUCOM	Novo Selo, Bulgaria	1x2 months SBCT from FRG	$78	$249	$57
	Army	EUCOM	Ansbach/Illesheim, Germany	1x6 month AVN BDE (-)	$66	$157	$67
	Army	EUCOM	Grafenwoehr, Germany	2x2 month ABCT(-)	$84	$562	$76
	Army	EUCOM	Shinnen, Netherlands	3x4 months SBCT BN	$143	$458	$124
	Army	CENTCOM	Camp Buehring, Kuwait	1x12 months ABCT	$202	$551	$209
	Army	CENTCOM	Fujariah, UAE	1x3 months THAAD, 2x2 months PAC-3 BTY	$11	$14	$12
	Air Force	AFRICOM	Djibouti	CJTF-HOA, 449th Air Expeditionary Group (C-130), rotational F-15Es, UAVs, 2x6 months	$56	$66	$57
	Air Force	CENTCOM	Al Udeid, Qatar	379th AEW (90 AC including B-1, F-15E, E-8C, C-130, RC-135), 609 CAOCm 2x6 months	$308	$354	$311
	Air Force	CENTCOM	Al Dhafra, UAE	380th AEW (KC-10, U-2), 2x6 months	$97	$116	$98
	Air Force	CENTCOM	Manas, Kyrgyzstan	376 AEW (KC-135), 2x6 months	$25	$28	$26

Table B.1—Continued

Alternative	Service	Region	Location	Unit/Unit Type	Sealift ($ millions)	Airlift ($ millions)	Prepositioning ($ millions)
	Air Force	EUCOM	Lakenheath, UK	24xF15x1x3	$24	$29	$25
	Air Force	EUCOM	MK, Romania	24xF16x1x3	$21	$25	$22
	Air Force	EUCOM	Graft Ignatievo, Bulgaria	24xF16x1x1	$14	$18	$15
	Air Force	PACOM	Subic Bay, Philippines	24xF16x1x2	$19	$24	$20
	Marine Corps	PACOM	Makiminato Okinawa Camp Butler, Japan	3 BLT (Inf), 3837 personnel, 2x6 months 3 ACE (6xAV,-8), 2x6 months	$194	$356	$166
	Marine Corps	PACOM	Camp Mujuk, South Korea	1 MEU, 2,200 personnel, 1x3 months	$28	$28	$34
	Marine Corps	PACOM	Robertson Barracks, Australia	1 MEU, 2,500 personnel, 2x6 months	$65	$65	$71
	Marine Corps	EUCOM	MK, Romania	1 SPMAGTF, 400 personnel, 2x6 months	$24	$43	$22
MCP	Army	AFRICOM	Djibouti (proposed)	2 x 6 months IBCT(-)	$208	$358	$191
	Army	PACOM	Nakhon Ratchasima (Korat), Thailand	1x3 months IBCT	$65	$150	$59
	Army	PACOM	Misawa, Japan	2x6 months IBCT(-), w/ENG, ADA additional	$154	$274	$144
	Army	EUCOM	Novo Selo, Bulgaria	1x6 months SBCT	$10	$253	$78
	Army	CENTCOM	Camp Buehring, Kuwait	1x12 months ABCT + AVN BDE	$382	$870	$394

Table B.1—Continued

Alternative	Service	Region	Location	Unit/Unit Type	Sealift ($ millions)	Airlift ($ millions)	Prepositioning ($ millions)
	Army	CENTCOM	Fujariah, UAE	1x12 months ADA BN, Fires BDE(-)	$230	$372	$233
	Air Force	AFRICOM	Djibouti	CJTF-HOA, 449th Air Expeditionary Group (C-130), rotational F-15Es, UAVs, 2x6 months	$56	$66	$57
	Air Force	CENTCOM	Al Udeid, Qatar	379th AEW (90 AC including B-1, F-15E, E-8C, C-130, RC-135), 609 CAOC, 2x6 months	$307	$354	$311
	Air Force	CENTCOM	Al Dhafra, UAE	380th AEW (KC-10, U-2), 2x6 months	$97	$116	$98
	Air Force	CENTCOm	Al Musanah, Oman	24xF16x1x3	$28	$36	$29
	Air Force	CENTCOM	Thumrait, Oman	12xB1x1x1	$9	$11	$10
	Air Force	CENTCOM	Al Jaber, Kuwait	24xF16x1x2	$23	$28	$24
	Air Force	CENTCOM	Ali Al Salem, Kuwait	12xC-130 2x6	$17	$20	$17
	Air Force	EUCOM	MK, Romania	24xF16x2x1	$18	$26	$19
	Air Force	EUCOM	Graft Ignatievo, Bulgaria	24xF16x1x3	$21	$25	$22
	Air Force	PACOM	U-Tapao, Thailand	24xF16x1x3	$20	$25	$21
	Air Force	PACOM	Subic Bay, Philippines	24xF16x1x3	$20	$25	$21
	Air Force	PACOM	Labuan, Malaysia	24xF16x1x3	$21	$25	$22
	Air Force	PACOM	Darwin, Australia	12xB1x1x1	$9	$10	$10
	Air Force	PACOM	Tindal, Australia	12xKC135x2x2	$8	$10	$8
	Air Force	PACOM	Saipan, Mariana Islands	24xF16x1x1	$19	$24	$20

Table B.1—Continued

Alternative	Service	Region	Location	Unit/Unit Type	Sealift ($ millions)	Airlift ($ millions)	Prepositioning ($ millions)
	Air Force	PACOM	Tinian, Mariana Islands	24xF16x1x1	$19	$24	$20
	Air Force	PACOM	Rota, Mariana Islands	24xF16x1x2	$19	$24	$20
	Marine Corps	PACOM	Camp Butler, Japan	3 BLT (Inf), 3837 pers, 2x6 months 3 ACE (6xAV-8), 2x6 months	$194	$356	$166
	Marine Corps	PACOM	Camp Mujuk, South Korea	SPMAGTF, 500 pers, 2x6 months	$36	$68	$35
	Marine Corps	PACOM	Robertson Barracks, Australia	1 MEU, 2500 pers, 2x6 months	$65	$65	$71
	Marine Corps	EUCOM	MK, Romania	1 SPMAGTF, 400 pers, 2x6 months	$24	$43	$22

Table B.2
Estimated Costs of Baseline Posture Changes

Service	Region	Base	Changes	Operational personnel to/ from CONUS	Close base?	Personnel ($ millions)	Base ($ millions)	Personnel to/ from OCONUS locations	Inactivate (Personnel)
Army	EUCOM	Bamberg, Germany	173rd ABN to Italy, Others to CONUS	−866	Yes	−35	−115		−2,298
Army	EUCOM	Heidelberg, Germany	All to CONUS	−2,243	Yes	−34	−115		
Army	EUCOM	Schweinfurt, Germany	172 inactivates (total 3,800), 173 elements to Vicenza	−2,217	Yes	−70	−115	−70	−2,384
Army	PACOM	Camp Casey, South Korea	AVN BDE and 1 BCT to Humphreys, Fires remains			—	—	−4,254	
Army	PACOM	Camp Humphreys, South Korea	Continue with planned consolidation					9,455	
Army	PACOM	Camp Red Cloud, South Korea	2 ID HQ to Humphreys		Yes	—	−79	−1,465	
Army	PACOM	Yongsan, South Korea	Relocate to Humphreys		Yes	—	−79	−3,736	
Air Force	EUCOM	Lajes, Portugal	Move elements of 65th ABW to CONUS	−400	No	−16	—		

NOTE: All personnel are operational personnel. ABW = air base wing.

Table B.3
Estimated Costs of Cost Reduction Posture Changes

Service	Region	Base	Changes	Operational Personnel to/ from CONUS	Close Base?	Personnel ($ millions)	Base ($ millions)	Personnel to/from OCONUS Locations	Inactivate (personnel)
Army	EUCOM	Ansbach/ Illesheim, Germany	12 AVN to U.S.	–2,759	Yes	–42	–115		
Army	EUCOM	Bamberg, Germany	173 ABN to Italy , Others to CONUS	–866	Yes	–35	–115		–2,298
Army	EUCOM	Baumholder	170 inactivates, return other units to CONUS		Yes	–57	–115		–3,758
Army	EUCOM	Garmisch, FRG			Yes	—	–115		
Army	EUCOM	Grafenwoehr	172 inactivates, other units to CONUS	–785	Yes	–28	–308		–1,073
Army	EUCOM	Heidelberg	All to CONUS	–2,243	Yes	–34	–115		
Army	EUCOM	Hohenfels	Close return 1-4 to CONUS	–917	Yes	–14	–115		
Army	EUCOM	Kaiserslautern	Move 21 TSC, 212 CSH and 30 MEDCOM to CONUS but keep hospital and AAMDC	–1,170	No	–18	—		
Army	EUCOM	Schweinfurt	172 inactivates (total 3,800), 173 elements to Vicenza	–2,217	Yes	–70	–115	–70	–2,384
Army	EUCOM	Vicenza, IT			No	—	—	936	
Army	EUCOM	Vilseck	Units return to CONUS	–4,420	Yes	–67	–115		
Army	EUCOM	Wiesbaden	V Corps HQ inactivates, 66 MI to CONUS 5th SIG Com & 2 SIG BDE remain	–1,109	No	–17	—		
Army	PACOM	Camp Zama, Japan	Add 1-1 SF BN		No	—	—	405	

Table B.3—Continued

Service	Region	Base	Changes	Operational Personnel to/ from CONUS	Close Base?	Personnel ($ millions)	Base ($ millions)	Personnel to/from OCONUS Locations	Inactivate (personnel)
Army	PACOM	Torii Station, Okinawa	Return to CONUS, except 1-1 SF BN to Zama	–287	Yes	–7	—	–405	
Army	PACOM	Camp Casey, Korea	AVN BDE and 1 BCT to Humphreys, Fires remains		No	—	—	–4,254	
Army	PACOM	Camp Humphreys, Korea	Continue with planned consolidation		No	—	—	9,455	
Army	PACOM	Camp Red Cloud, Korea	2 ID HQ to Humphreys		Yes	—	–79	–1,465	
Army	PACOM	Yongsan, Korea	Relocate to Humphreys		Yes	—	–79	–3,736	
Air Force	EUCOM	Alconbury	Move 501 Combat Support wing and 423 ABG to CONUS	–472	Yes	–18	–78		
Air Force	EUCOM	Aviano, IT	Move 2 FS to Spang, other units to CONUS	–1,600	Yes	–62	–210		
Air Force	EUCOM	Fairford, UK	Move 420 ABS to CONUS	–30	Yes	–1	—		
Air Force	EUCOM	Incirlik	Move part of 39th ABW to CONUS	–347	No	–13	—		
Air Force	EUCOM	Lajes, PT	Move 65th ABW to CONUS	–673	Yes	–26	–78		
Air Force	EUCOM	Lakenheath, UK	Move 3 FS (F-15s) to CONUS	–2,646	Yes	–103	–211		
Air Force	EUCOM	Menwith Hill	Return units to CONUS	–176	No	–7	—		
Air Force	EUCOM	Mildenhall, UK	Move all units to CONUS	–1,868	No	–72	—		
Air Force	EUCOM	Moron, ES		–109	Yes	–4			

Table B.3—Continued

Service	Region	Base	Changes	Operational Personnel to/ from CONUS	Close Base?	Personnel ($ millions)	Base ($ millions)	Personnel to/from OCONUS Locations	Inactivate (personnel)
Air Force	EUCOM	Sigonella, Italy	Move RQ-4 back to CONUS	–70	No	–3	—		
Air Force	EUCOM	Spangdahlem, Germany	Add 2 FS from EUCOM	1,600	No	62	—		
Air Force	PACOM	Misawa, Japan	35 FW to CONUS	–1,585	Yes	–56	–169		
Navy	CENTCOM	NSA Bahrain	Add 8 LCS	400	No	11	—		
Navy	EUCOM	Alconbury, UK RAF Molesworth		–108	Yes	–3	—		
Navy	EUCOM	Deveselu AB, Romania	No Change	35	No	1	—		
Navy	EUCOM	NSA Gaeta, Italy	Close Send LCC to CONUS	–175	Yes	–5	—		
Navy	EUCOM	Larissa, Greece	Close	–100	Yes	–3	—		
Navy	EUCOM	Menwith Hill RAF, UK		–55	Yes	–2	—		
Navy	EUCOM	NSA Naples, Italy	Reduce Support	–100	No	–3	—		
Navy	EUCOM	Naval Station Rota, Spain	Add planned 4 Destroyers	1,080	No	31	—		
Navy	EUCOM	Souda Bay Crete, Greece	Close	–259	Yes	–7	–29		
Navy	EUCOM	TBD, Poland	Unchanged	35	No	1	—		
Navy	PACOM	Misawa AB, Japan	No Change	–379	Yes	–8	—		

Table B.3—Continued

Service	Region	Base	Changes	Operational Personnel to/ from CONUS	Close Base?	Personnel ($ millions)	Base ($ millions)	Personnel to/from OCONUS Locations	Inactivate (personnel)
Navy	NORTHCOM	Naval Station Mayport, Jacksonville, Fl	Disestablish 4th fleet	–100	Yes	–3	—		
Navy	PACOM	Changi Naval Base, Singapore	4 LCS crews +shore support	300	No	–8	—		
Marine Corps	PACOM	Camp Butler, Okinawa, Japan	Relocate some units to CA	–7,428	No	–73	—		

NOTE: All personnel are operational personnel. AB = Air Base; TSC = theater sustainment command; CSH = combat support hospital.

Table B.4
Estimated Costs of Global Responsiveness and Engagement Posture Changes

Service	Region	Base	Changes	Operational Personnel to/ from CONUS	Close Base?	Personnel ($ millions)	Base ($ millions)	Personnel to/from OCONUS Locations	Inactivate (personnel)
Army	EUCOM	Bamberg, Germany	173 elements to Vicenza	–866	Yes	–48	–115		–2,298
Army	EUCOM	Baumholder, Germany	170 inactivates, Add 2 SF battalions	600	No	–48	—		–3,758
Army	EUCOM	Grafenwoehr, Germany	172 inactivates		No	–16	—		–1,073
Army	EUCOM	Heidelberg, Germany	Relocate to Wiesbaden	–2,243	Yes	–34	–115		
Army	EUCOM	Schweinfurt, Germany	172 inactivates, 173 elements to Vicenza	–2,217	Yes	–70	–115	–70	–2,384
Army	EUCOM	Vicenza, Italy			No	—	—	936	
Army	EUCOM	Wiesbaden, Germany			No	–2	—		–164
Army	PACOM	Camp Casey, South Korea	AVN BDE to Humphreys, Fires remain		No	—	—	–4,254	
Army	PACOM	Camp Humphreys, South Korea	8 Army HQ & units; SBCT + Fires BN		No	—	—	9,455	
Army	PACOM	Camp Red Cloud, South Korea	As planned		Yes	—	–79	–1,465	
Army	PACOM	Yongsan, South Korea			Yes	—	–79	–3,736	
Air Force	EUCOM	Alconbury, UK		–30	No	–1	—		

Table B.4—Continued

Service	Region	Base	Changes	Operational Personnel to/ from CONUS	Close Base?	Personnel ($ millions)	Base ($ millions)	Personnel to/from OCONUS Locations	Inactivate (personnel)
Air Force	EUCOM	Aviano, Italy	Add 52 FW from Spang	1,981	No	77	—		
Air Force	EUCOM	Incirlik, Turkey	Shrink ABW	–347	No	–13	—		
Air Force	EUCOM	Lajes, Portugal		–473	No	–18	—		
Air Force	EUCOM	Lakenheath, UK	Move 48th FW to CONUS, rotate 1x3 months 1 fighter squadron	–2,646	No	–103	—		
Air Force	EUCOM	Spangdahlem, Germany	Move 52 FW to Aviano, keep ABW for ENR	–1,981	No	–77	—		
Air Force	PACOM	Misawa, Japan	Move 35th FW to CONUS	–1,585	Yes	–53	–169		
Air Force	PACOM	Kunsan, South Korea	Move 51st FW to CONUS	–1,351	No	–24	—		
Navy	AFRICOM	Camp Lemonnier, Djibouti	Establish regional support facility and add Seal Team from CONUS	210	No	6	—		
Navy	CENTCOM	NSA Bahrain	Expand rotational forces	600	No	17	—		
Navy	EUCOM	Deveselu AB, Romania	Unchanged	35	No	1	—		
Navy	EUCOM	Larissa, Greece	Close	–100	Yes	–3	—		
Navy	EUCOM	NSA Naples, Italy	Expand	300	No	8	—		
Navy	EUCOM	NAS Sigonella, Italy	Regional hub/Expand	250	No	8	—		
Navy	EUCOM	Naval Station Rota, Spain	Expand (1)	1,080	No	31	—		

Table B.4—Continued

Service	Region	Base	Changes	Operational Personnel to/ from CONUS	Close Base?	Personnel ($ millions)	Base ($ millions)	Personnel to/from OCONUS Locations	Inactivate (personnel)
Navy	EUCOM	TBD, Poland	Unchanged	35	No	1	—		
Navy	PACOM	Misawa AB, Japan	Close	–379	No	–8	—		
Navy	PACOM	Changi Naval Base, Singapore	Expand rotational forces	300	No	8	—		
Navy	PACOM	Subic Bay/Cubi Pt, Philippines	Add small permanent support	200	No	6	—		

NOTE: All personnel are operational personnel. AB = Air Base.

Table B.5
Estimated Costs of Major Contingency Posture Changes

Service	Region	Base	Changes	Operational Personnel to/ from CONUS	Close Base?	Personnel ($ millions)	Base ($ millions)	Personnel to/from OCONUS Locations	Inactivate (personnel)
Army	EUCOM	Ansbach/ Illesheim, Germany	12 AVN to Kuwait		Yes	—	–115	–2,759	
Army	EUCOM	Bamberg, Germany	173 elements to Vicenza	–866	Yes	–48	–115		–2,298
Army	EUCOM	Baumholder, Germany	170 inactivates, 92 MP to Wiesbaden rest to U.S.	–644	Yes	–67	–115	–388	–3,758
Army	EUCOM	Grafenwoehr, Germany	172 inactivates, rest to U.S.	–785	Yes	–28	–308		–1,073
Army	EUCOM	Heidelberg, Germany	Units to U.S.	–2,243	Yes	–34	–115		
Army	EUCOM	Hohenfels, Germany	Units to U.S.	–917	Yes	–14	–115		
Army	EUCOM	MK, Romania	Add 2 CAV 30 MED, 21 TSC, 1 SF BN from CONUS	300	No	5	—	5,590	
Army	EUCOM	Schweinfurt, Germany	173 elements to Vicenza	–2,217	Yes	–70	–115	–70	–2,384
Army	EUCOM	Vicenza, Italy			No	—	—	936	
Army	EUCOM	Vilseck, Germany	2 CAV to MK Romania		Yes	—	–115		
Army	EUCOM	Wiesbaden, Germany	V HQ to CONUS,		No	—	—	388	
Army	PACOM	Camp Casey, South Korea	Move units to CH, except Fires, add 2nd Fires Brigade	928	No	15	—	–4,254	

Table B.5—Continued

Service	Region	Base	Changes	Operational Personnel to/ from CONUS	Close Base?	Personnel ($ millions)	Base ($ millions)	Personnel to/from OCONUS Locations	Inactivate (personnel)
Army	PACOM	Camp Humphreys, South Korea	ABCT, AVN,		No	—	—	9,455	
Army	PACOM	Camp Red Cloud, South Korea	Continue move of 2 ID, add CAB Bde		No	—	—	–1,465	
Army	PACOM	Yongsan, South Korea	Units to Humphreys		Yes	—	–79	–3,736	
Air Force	EUCOM	Alconbury, UK		–30	No	–1	—		
Air Force	EUCOM	Lakenheath, UK	Move 48 FW to Andersen	–2,646	Yes	–103	–211		
Air Force	EUCOM	Lask, Poland	Move DET to CONUS	–10	Yes	–0.4	—		
Air Force	EUCOM	Spangdahlem, Germany	Add additional Figther Squadron from CONUS	700	No	27	—		
Air Force	PACOM	Andersen, U.S.	Add 3 FS from EUCOM	2,646	No	80	—		
Air Force	PACOM	Cocos Islands, Australia	Station RQ-9	100	No	3	—		
Navy	PACOM	Changi Naval Base, Singapore	Unknown	100	No	3	—		
Navy	EUCOM	Deveselu AB, Romania	N/A	35	No	1	—		
Navy	PACOM	Hickham AFB, Hawaii	Unknown	5,560	No	157	—		

Table B.5—Continued

Service	Region	Base	Changes	Operational Personnel to/ from CONUS	Close Base?	Personnel ($ millions)	Base ($ millions)	Personnel to/from OCONUS Locations	Inactivate (personnel)
Navy	EUCOM	Larissa, Greece	Unknown	–100	Yes	–3	—		
Navy	EUCOM	Naval Station Rota, Spain	349	1,080	No	31	—		
Navy	PACOM	Perth, Australia	N/A	200	No	6	—		
Navy	PACOM	Port Blair, India	N/A	100	No	3	—		
Navy	PACOM	Sattahip Port, Thailand	Unknown	100	No	3	—		
Navy	PACOM	Subic Bay/Cubi Pt, Philippines	N/A	1,540	No	44	—		
Navy	EUCOM	TBD, Poland	N/A	35	No	1	—		
Marine Corps	PACOM	Guam (Navy)	Add MEU	2,200	No	22	—		

NOTE: All personnel are operational personnel. AB = Air Base; AFB = Air Force base.

APPENDIX C

Security Cooperation Cost Differential Between Forward-Based and U.S.-Based Forces

DoD's primary tool for collecting security cooperation data is the Theater Security Cooperation Management Information System. Not all security cooperation events are captured in this database, while those that are may reflect only a portion of the associated forces and costs. Comparisons between COCOMs are especially problematic, as each COCOM has its own approach to managing its data. Nevertheless, a review of selected data from EUCOM and AFRICOM can at least give some sense of the various types of security cooperation and the forces and costs involved in executing these activities.

Table C.1 shows a breakdown of FY 2012 EUCOM security cooperation activities by category. The descending order of the categories roughly corresponds to the forces required to conduct the engagement (e.g., SABER STRIKE and ANATOLIAN FALCON required over 500 U.S. personnel, while the Fusion Cell Development event required fewer than 15 U.S. personnel). The total cost listed for each category is the total represented in the COCOMs' databases, which typically includes airlift and sealift costs.

Activities at the top of the table are generally more complex and have greater operational influence. For example, SABER STRIKE was used to conduct NATO interoperability training with the Baltic States, focusing on current International Security Assistance Force operations in Afghanistan. The costs of these events can be assumed to be higher not only because of the transport costs, but also because of the planning, conferences, coordination, and facilities required to conduct them. While these are the most expensive security cooperation activities listed by EUCOM, they often provide the most intensive training for U.S. and partner forces and can be necessary to fulfill predeployment and other NATO training requirements.

A reduction in permanent forces in the EUCOM AOR would disproportionately affect the categories higher in the table. Considering the examples for each category, "Combined exercises" generally employ the large units based in Europe, on the rationale that these forces are closer and bring benefits in terms of their expertise working in the region and with multinational partners. "Training" and "Education" events use

Table C.1
Summary of EUCOM Security Cooperation in FY 2012

Activity Type	Example	Number of Activities	Total Cost (millions)
Combined/multinational exercise	SABER STRIKE/ANATOLIAN FALCON	35	$11.7
Combined/multinational training	• Operation mentoring liaison team training • Combat Lifesaver Course • Night-vision goggle training and certification • Leadership development courses	124	$37.7
Combined/multinational education	• MRAP Operator/Driver Course • Basic C-IED Course • Instructor training course • Unit Movement Officer Course	129	$1.8
Defense and military contacts	• ROTC cadet exchanges • Command and Staff Engagement • U.S. Navy port visits • Medical conferences • Flag visits	218	$3.9
Info sharing/intelligence course	• Fusion cell development • Intelligence targeting group training	42	$0.136

NOTE: MRAP = mine-resistant ambush protected vehicle; C-IED = counter–improvised explosive device; ROTC = Reserve Officer Training Corps.

smaller units but are also generally conducted by forces based in Europe for similar reasons. The "Defense and military contacts" and "Info sharing/intelligence cooperation" events, however, are often facilitated by rotational forces due to their nature or the expertise desired.

Because few U.S. forces are based in Africa, AFRICOM uses far more rotational forces than EUCOM for all these categories of events. While the advantages and disadvantages of using rotational forces are discussed in Chapter Two, our analysis of the COCOMs' databases provided insights about relative costs of conducting security cooperation with forces based in the United States compared with those based overseas. As Table C.2 shows, AFRICOM conducts fewer activities in the top three categories, which are more complex and force intensive. AFRICOM was more active than EUCOM in "Defense and Military Contacts," which involve fewer forces, and "Humanitarian Assistance," which EUCOM did not break out as a separate category.

In Table C.3, we examined two multinational exercises in an attempt to draw comparisons. Both SABER STRIKE and AFRICAN LION were conducted over similar time periods. Our assumption was that SABER STRIKE, a EUCOM-sponsored

Table C.2
Summary of AFRICOM Security Cooperation in FY 2012

Activity Type	Example	Number of Activities	Total Cost (millions)
Combined/multinational exercise	AFRICAN LION	8	$8.2 ($6.6 for AFRICAN LION)
Combined/multinational training	• ADAPT II logistics exercise • UN Peacekeeping Support Training • Military mentorship field training	53	$3.2
Combined/multinational education	• Basic instructor training • Familiarization visits to U.S. training facilities	8	$1.3
Defense and military contacts	• ROTC cadet language training • Professional development workshops • Senior Leader Engagement	315	$9.7
Info sharing/intelligence course	• Fusion cell development • Intelligence analysis training	17	$3.9 (primarily counter-terrorism events)
Humanitarian assistance	• Mine action courses • ROTC cadet work with non-profit organizations	23	$2

NOTE: ROTC = Reserve Officer Training Corps.

Table C.3
Cost and Purpose of U.S. Participation in Major Exercises

Exercise	Location (duration)	U.S. Forces Required	Total Cost (millions)	Lift Cost (millions)	Notes
SABER STRIKE (2012)	Estonia, Latvia (11 days)	About 750 (USARER, MARFOREUR, USAFE)	$5.7	$3.8	Improve Baltic nations' interoperability with Coalition Forces (ISAF and ongoing Afghanistan operations)
AFRICAN LION (2012)	Morocco (9 days)	About 1,100 (MARFORAF, 24th MEU, Iwo Jima ARG)	$6.6	Airlift: $2.7 Sealift: $2 Commercial: $0.25	Bulk of U.S. force part of ARG/MEU on regular deployment (en route C6F AOR)

NOTE: ISAF = International Security Assistance Force.

event conducted at a regional training center in the Baltics, would incur lower lift costs than AFRICAN LION, which involved a rotational MEU and ARG from the United States. While this appeared to be true, the difference was not substantial, especially considering the larger number of forces involved in the African exercise. It is difficult to draw any larger conclusions from this or other comparisons we attempted, since the COCOMs' databases do not provide sufficient detail on what is included in the cost calculations of various events in different COCOMs. Nevertheless, the comparison is a useful illustration of how COCOMs can use permanent or rotational forces for similar events.

While the Theater Security Cooperation Management Information System data provided only limited opportunities for analysis, RAND has produced additional analysis relevant to this topic in the past.

A 2012 RAND study for the Air Force concluded that, provided the United States maintains the current posture in Europe, USAFE security cooperation efforts are cost-effective.[1] Generating USAFE's current security cooperation activities from the United States could greatly increase the marginal cost of providing security cooperation. The study estimated that, if USAFE's current security cooperation activities were replicated from the United States, the marginal cost to provide security cooperation could increase fourfold, from $59 million per year to over $250 million per year.

If USAFE forces were moved to the United States, some security cooperation activities would need to be significantly curtailed to be cost neutral with regard to direct security cooperation costs. We found that even when replicating only about half of USAFE's status quo security cooperation activities, both the marginal security cooperation costs and the total operating costs would be more for U.S.-based forces. While the marginal costs to provide security cooperation in USAFE are very sensitive to whether forces are located in Europe, as opposed to the United States, these changes still have a small overall budget impact relative to total USAFE operating costs.

Though we did not conduct a similar analysis for the Army, Chapter Eight does address the costs involved in moving Army battalions from the United States to Germany to conduct training with partners. Most multinational ground exercises take place in Europe at USAREUR's JMTC, which conducted 25 multinational exercises in 2011 alone. If the Army reduces one or both of its remaining combat brigades in Europe yet still desires to continue supporting multinational training, it will sometimes need to transport forces from U.S. bases. As discussed in Chapter Eight, this would reduce the cost savings resulting from those reductions.

---

[1] Moroney et al., 2012.

APPENDIX D

U.S. Military Overseas Prepositioned Equipment

This appendix provides a summary of major U.S. military overseas prepositioned assets.

Army Prepositioned Equipment

Army prepositioned stocks are an important part of the Army's strategic mobility triad (prepositioning, airlift, and sealift) and help ensure the rapid buildup of theater reception capability prior to the arrival of heavy units from CONUS, provide a capability to mitigate enemy anti-access strategies, and provide a flexible deterrent option usable in a short-notice crisis.[1] There are four categories of Army prepositioned stocks: prepositioned unit sets, operational project stocks, Army war reserve sustainment stocks, and war reserve stocks for allies. Prepositioned units sets consist of the equipment and ammunition necessary to rapidly deploy an Army unit. Operational project stocks are materiel in addition to what is normally provided to a unit that are critical to the Army's ability to conduct force-projection operations and that support Army operations, plans, and contingencies. Army war reserve sustainment stocks are stockpiles of materiel intended to meet initial combat demands until wartime production and supply lines can be established. War reserve stocks for allies consist of supplies and equipment owned and controlled by the United States that can be released to multinational forces in a crisis.[2] The Army currently has four sets of prepositioned stocks overseas: three land-based and one afloat. Details of these stocks are provided in the Table D.1.

Army Prepositioned Stocks-3 is carried on board eight Military Sealift Command–controlled vessels: six ships carrying unit equipment and sustainment stocks and two ships carrying ammunition.

---

[1] Headquarters, Department of the Army, *Army Prepositioned Operations*, FM 3-35.1, July 2008, pp. 1-1 to 1-2.

[2] Headquarters, Department of the Army, 2008, pp. 1-2 to 1-3.

Table D.1
Army Overseas Prepositioned Stocks

Name	Location	Composition
APS-2 (EUCOM)	Livorno, Italy	Mine-resistant ambush protected vehicles, force provider modules, special forces
APS-3 (Afloat)	Diego Garcia	Sustainment brigade, war reserve sustainment stocks, munitions
	Guam	IBCT, IBCT enablers, theater opening/port opening package, war reserve sustainment stocks
APS-4 (PACOM)	Camp Carroll, South Korea	ABCT, ABCT enablers, sustainment brigade (–), operational project stocks, theater sustainment stocks, war reserve sustainment stocks
	Pusan, South Korea	Operational project stocks
	Sagami General Depot, Japan	Sustainment brigade (–), operational project stocks, war reserve sustainment
	Yokohama North Dock, Japan	Sustainment brigade (–), Army watercraft
	Okinawa, Japan	Operational project stocks
APS-5 (CENTCOM)	Camp Arifjan, Kuwait	ABCT, ABCT enablers, IBCT, IBCT enablers, sustainment brigade, theater sustainment stocks
	Camp As Sayliyah, Qatar	Sustainment brigade, fires brigade, operational project stocks, war reserve sustainment stocks
	Bagram Air Base, Afghanistan	Infantry Battalion Task Force

SOURCE: Data provided to RAND by the U.S. Army.

NOTE: APS = Army prepositioned stocks; (–) = less than full authorization.

Marine Corps Prepositioned Equipment

The U.S. Marine Corps prepositioning program has two primary components: the Maritime Prepositioning Force and the Marine Corps Prepositioning Program–Norway .

A Maritime Prepositioning Force squadron consists of the equipment and supplies required to support the deployment and employment of an MEB consisting of some 16,000 marines and sailors. An MEB has a ground combat element built around a marine infantry regiment, an aviation combat element built around a fixed-wing aircraft group and a rotary-wing aircraft group, a logistics combat element, and a naval construction force. Each MEB also deploys with 30 days of supplies. The MEB is carried aboard a maritime prepositioning ships squadron (MPSRON) consisting of between four and five ships. The USMC currently has two afloat MEBs, one on MPSRON-2 operating out of Diego Garcia and the other on MPSRON-3 operating out of Guam and Saipan.[3] Upon deployment, the MEB's personnel and additional required equipment, the fly-in echelon, will be flown into the operational theater by some 330 strategic airlift sorties.

The Marine Corps Prepositioning Program–Norway consists of supplies and equipment for a notional MEB stored in the Trondheim region of Norway. The ground equipment and supplies are stored in six climate-controlled caves, and the aviation support equipment is stored in two dehumidified storage buildings.[4]

Air Force Prepositioned Equipment

USAF prepositions equipment overseas under the war reserve materiel (WRM) program to support contingency deployment and operation of combat aircraft from the United States. The WRM program consists primarily of basic expeditionary airfield resources (BEAR); vehicles; munitions; fuel equipment; medical supplies; tanks, racks, adapters, and pylons (TRAP); aerospace ground equipment (AGE); and air base operability equipment (i.e., C-Wire, sandbags, etc.). In FY 2012, the USAF had 22 major WRM storage sites, 13 for PACAF, six for USAFE, and three for Allied Forces Central Europe. Major WRM sites are intended to support forward operating locations that can handle deployments of up to 72 tactical fighter aircraft and a base population of 3,300 personnel. The support packages include fuel support, aircraft tanks, pylons, racks, adapters, vehicles, aircraft generation equipment, rations, and other direct mission support equipment. The USAF also has 21 overseas minor WRM storage sites at MOBs that are intended to support additive forces. Nine of these sites are with PACAF, five with USAFE, and seven with Allied Forces Central Europe. Additional equipment sets are located at 31 storage sites for fuels and operational readiness capability equipment (FORCE) and 184 AMC en route support locations.[5] Not all of the AMC sites are located overseas.

The USAF has prepositioned WRM at 15 overseas locations, primarily in northeast Asia and in Persian Gulf countries, to allow operations from austere facilities that do not have a regular USAF presence.[6] The core of this capability is the USAF's BEAR, which provide expeditionary basing assets for use at austere airfields. BEAR supports

---

[3] Information provided to RAND by the USMC on August 8, 2012; Headquarters Marine Corps, *Prepositioning Programs Handbook*, 2nd Edition, January 2009.

[4] Headquarters Marine Corps, 2009, pp. 23–24.

[5] FORCE includes fuel trucks, bladders and pumps. AMC en route support sites store material handling equipment and Aerial Port Squadron assets to support strategic lift operations. Department of the Air Force, *Fiscal Year (FY) 2013 Budget Estimates: Operations and Maintenance, Air Force*, Volume I, February 2012, p. 294.

[6] Ronald G. McGarvey, Robert S. Tripp, Rachel Rue, Thomas Lang, Jerry M. Sollinger, Whitney A. Conner, and Louis Luangkesorn, *Global Combat Support Basing: Robust Prepositioning Strategies for Air Force War Reserve Materiel*, Santa Monica, Calif.: RAND Corporation, MG-902, 2010.

the Agile Combat Support Force structure, infrastructure, and flightline for 143,000 deployed personnel at 64 locations and is capable of bedding down 2,239 aircraft. BEAR consists of a variety of systems to support operations such as billeting, messing, hygiene, power, water, environmental, aircraft shelters, flightline equipment, and industrial equipment.[7] The USAF also prepositions vehicle sets intended to support expeditionary operations at austere airfields. These sets are usually, but not always, collocated with BEAR sets.[8] Table D.2 provides a list of sites where BEAR and WRM vehicle sets, other WRM, and munitions are prepositioned.

The Military Sealift Command operates two long-term chartered container ships that carry USAF munitions.[9] The USAF also has stocks of munitions stored at a number of overseas bases.

Navy Prepositioned Equipment

The U.S. Navy prepositions equipment primarily to support the USMC's maritime prepositioning force. This support has three components: the Naval Support Element, the Naval Construction Force, and an expeditionary medical facility. The Naval Support Element provides the personnel and material required to offload and backload maritime prepositioning force ships either in-stream or at pier-side. It consists of a naval beach group, an amphibious construction battalion, an assault craft unit, a beachmaster unit, and a naval cargo handling battalion. The personnel and additional equipment required to complete a naval support element can be deployed in approximately 23 strategic airlift sorties. A Naval Construction Force unit is attached to an MEB and provides direct horizontal, vertical, and general engineering support. It consists of a naval mobile construction battalion of 813 personnel and a naval construction regiment (NCR) of 116 personnel. It has a fly-in echelon requiring some eight strategic airlift sorties. Both MSPRON-2 and MSPRON-3 have the equipment required for a naval support element and a Naval Construction Force unit.[10] THE USN also prepositions a number of expeditionary medical facilities that are scalable theater hospital assets that can provide comprehensive level III surgical and medical support to meet USMC or other COCOM requirements. The USN currently has prepositioned a

---

[7] Headquarters Air Force, *USAF Supply Manual: Volume 2*, AFMAN 23-110, 2009, pp. 26-48, 26-69; Headquarters Air Force, *Planning and Design of Expeditionary Airbases*, AFPAM 10-219, Vol. 6, February 2006 (Certified Current March 2012), pp. 20, 52; AF/A4LX.

[8] McGarvey et al., 2010, pp. 57–58.

[9] Military Sealift Command, "Prepositioning," U.S. Navy, undated b.

[10] Information provided to RAND by the USMC and the USN on August 16, 2012; Headquarters Marines Corps, 2009, p. 19; Headquarters Marine Corps, *Maritime Prepositioning Force (MPF) Force Lists (F/L)*, Department of the Navy, MCBuL 3501, April 14, 2010; Naval Facilities Engineer Command, *Navy Equipment Sets in Support of Maritime Prepositioning Force*, 2012.

Table D.2
USAF Overseas WRM Storage Sites

Location	Composition
Shaikh Isa, Bahrain	BEAR, WRM vehicles
Andersen AFB, Guam	BEAR, WRM vehicles, munitions
Diego Garcia	WRM vehicles
Ramstein AB, Germany	Munitions
Camp Darby, Italy	Munitions
Kadena AB, Japan	WRM vehicles, munitions
Misawa AB, Japan	BEAR, WRM vehicles, munitions
Al Jabar AB, Kuwait	BEAR, munitions
Sanem, Luxembourg	BEAR, WRM vehicles
Sola, Norway	WRM
Masirah, Oman	BEAR, munitions
Salalah Port, Oman	BEAR, munitions
Thumrait, Oman	BEAR, WRM vehicles, munitions
Al Udeid, Qatar	BEAR, WRM vehicles, munitions
Kimhae AB, South Korea	BEAR, WRM vehicles, munitions
Kunsan AB, South Korea	WRM, munitions
Kwang Ju AB, South Korea	BEAR, WRM vehicles, munitions
Osan AB, South Korea	WRM, munitions
Suwon AB, South Korea	BEAR, WRM vehicles, munitions
Taegu, South Korea	BEAR, WRM vehicles, munitions
Al Dhafra AB, UAE	BEAR, munitions
RAF Fairford, UK	WRM vehicles, munitions

SOURCE: McGarvey et al.,2010, pp. 25, 57–58; Headquarters Air Combat Command, *Basic Expeditionary Airfield Resources (BEAR), Mission Brief*, November 2, 2011; CENTCOM.

NOTE: This list is not exhaustive as open sources do not list all WRM sites.

150-bed expeditionary medical facility with MPSRON-2, two 250-bed expeditionary medical facilities in South Korea, and three 150-bed expeditionary medical facilities in Okinawa.[11]

---

[11] In the past, MSPRON-3 has had a 150-bed expeditionary medical facility. There is an additional 10-bed medical set prepositioned in Korea. Two 150-bed expeditionary medical facilities and several smaller medical sets are stored at Perry Point, Md. Information provided to RAND by the USMC and the USN on August 16, 2012.

APPENDIX E

Deployment Analysis Scenario APOD and APOE Details

Tables E.1 and E.2 show the APODs and APOEs (respectively) for the 24 different scenarios used in our analysis.

Each point of embarkation was not used in every scenario, as scenario points of embarkation were chosen based on geographic proximity. The exceptions to this are Joint Base Lewis-McChord, Washington, and Little Rock Air Force Base, Arizona, which were used in virtually every scenario. Joint Base Lewis-McChord was used frequently because it has all of the U.S. Army's active CONUS-based SBCTs, and Little Rock Air Force Base has the USAF active CONUS-based C-130 squadrons.

Results for the humanitarian relief scenarios are shown in Figure E.1. The results are similar to the more sortie-intense FID cases shown in Chapter Two, but the aircraft numbers required are naturally smaller, as the task force size is smaller by 225 sorties. Historically, many relief operations have been responded to by naval and marine forces. But we have included two cases of landlocked African countries, Mali and Burundi, for this airlift-based comparison.

Table E.1
Deployment Scenario APODs

Country	Type	Primary APOD	ICAO Code
South Korea	Deterrent	Osan AB	RKSO
Estonia	Deterrent	Tallinn	EETN
Vietnam	Deterrent	Noibai IAP	VVNB
Kuwait	Deterrent	Ali Al Salem AB	OKAS
Kosovo	Deterrent	Pristina IAP	BKPR
Georgia	Deterrent	Tbilisi	UGTB
Colombia	Deterrent	Palonegro IAP*	SKBG
Syria	PKO/WMD-E	Aleppo IAP*	OSAP
Zimbabwe	PKO	Harare IAP	FVHA
Myanmar	PKO	Naypyitaw IAP	VYEL
Nigeria	FID	Port Harcourt IAP	DNPO
Yemen	FID	Aden IAP	OYAA
Indonesia	FID	Polonia IAP	WIMM
Bolivia	FID	Viru Viru IAP	SLVR
Libya	FID	Tripoli IAP	HLLT
Philippines	FID	Clark IAP	RPLC
Tajikistan	FID	Dushanbe	UTDD
Saudi Arabia	FID	King Abdulaziz AB	OEDR
Bangladesh (tsunami)	Humanitarian	Zia IAP	VGZR
Mali (drought & famine)	Humanitarian	Tombouctou	GATB
Burundi (flooding)	Humanitarian	Bujumbura IAP	HBBA
Sri Lanka (tsunami)	Humanitarian	Bandaranaike IAP	VCBI
Peru (earthquake)	Humanitarian	Velazco Astete*	SPZO
Pakistan (earthquake)	Humanitarian	Masroor	OPMR

SOURCE: RAND analysis.

NOTE: APOD = aerial port of debarkation; AB = Air Base; IAP = international airport; * = Weight limitations; WBC < 585 TRT; WMD-E = WMD-elimination.

APPENDIX F

USFJ-Related Costs Borne by Japan

Table F.1
USFJ-Related Costs Borne by Japan (nominal billions of yen, based on Japanese fiscal year estimates)

	2009	2010	2011	2012
Promotion of base measures, etc.				
(1) Expenses related to measures for local communities	115.5	117.9	118.5	118.5
(2) Cost sharing for stationing USFJ				
a) SMA				
Labor cost	116.0	114.0	113.1	113.9
Utilities	24.9	24.9	24.9	24.9
Training relocation costs[a]	0.6	0.5	0.4	0.4
Sub-subtotal for SMA	141.5	139.5	138.4	139.2
(b) Facility improvements	21.9	20.6	20.6	20.6
(c) Measures for base personnel, etc.	29.3	27.9	26.8	26.9
Subtotal for cost sharing for stationing USFJ (a)+(b)+(c)	192.8	188.1	185.8	186.7
(3) Facility rentals, compensation expenses, etc.	131.6	130.5	129.3	136.6
Total promotion of base measures, etc. (1)+(2)+(3)	439.9	436.5	433.7	441.8
Total SACO-related expenses	11.2	16.9	10.1	8.6
Total U.S. Forces realignment-related expenses	60.2	132.0	123.0	70.7
Grand total	511.3	585.3	566.7	521.1

SOURCE: Government of Japan, Ministry of Defense, 2010–2012.

NOTES: All years refer to Japanese fiscal years.

[a] MOFA attributes another Y5.1 billion to spending on training relocation under the U.S.-Japan SMA, of which Y1.1 billion is SACO-related and Y4.0 billion is realignment-related. These costs appear to be included in the Ministry of Defense totals for each category but are not identified as "SMA."

Table F.2
Addendum to Table F.1 (in nominal billions of yen, based on Japanese fiscal year estimates)

	2009	2010	2011	2012
MOFA-reported "stationing of USFJ-related costs"[a]				
(1) Costs that might be covered under MOD budget report, including measures to improve surrounding living environments, rent for facilities, relocation, and compensation for fisheries	N/A	N/A	N/A	182.2
(2) Costs additional to those covered under MOD budget report				
(a) Estimated cost of government-owned land provided for use as USFJ facilities	N/A	N/A	N/A	165.8
(b) Expenditures borne by other Ministries (base subsidy, etc)	N/A	N/A	N/A	38.1
Subtotal of additional costs (a)+(b)	N/A	N/A	N/A	203.9
Total MOFA-reported "stationing of USFJ-related costs" (1)+(2)	N/A	N/A	N/A	386.1

SOURCE: Government of Japan, Ministry of Foreign Affairs, undated.

NOTES: All years refer to Japanese fiscal years.

[a] MOFA also reports costs incurred by non–Ministry of Defense ministries and estimated (non-budgetary) costs

Detailed Estimates of Host Nation Contributions from Japan, South Korea, and Germany

This appendix provides more detailed estimates of the contributions made by Japan, South Korea, and Germany to the U.S. bases that they host. It focuses specifically on the trajectories of Japan's contributions to labor costs, utility costs, and facility improvements and of South Korea's contributions to construction. The appendix also provides line-item detail on the "various other costs" that Germany covers.

Table H.1 quantitatively summarizes the information presented in the U.S.-Japan SMA; the text of the "Security Consultative Committee Document," issued jointly by the U.S. Secretaries of State and Defense and the Japanese Ministers for Foreign Affairs and Defense in June 2011[1] (a "2+2" statement); and discussions of the content of those documents found on the MOFA website and in government-to-government exchanges. On the basis of that information, we provide yen and U.S. dollar estimates for 2010 and the SMA period (i.e., 2011–2015). We draw directly from the figures in Table 7.6 for the 2011 and 2012 estimates and, using data on wage growth, utility cost growth, and exchange rates to make projections, we extrapolate from those estimates for the next three years. Because the SMA and 2+2 statement refer to 2010 as a benchmark, we include estimates for that year too.

As addressed in Chapter Seven, in the U.S.-South Korea SMA provisions on construction, South Korea committed to a shift from cash to in-kind contributions to be implemented fully by 2011, except for expenses associated with "design and construction oversight of facilities"; in 2009, the cash component of South Korea–funded construction was 70 percent and by 2011 that share was expected to drop to 12 percent. With that trajectory in mind, Table H.2 provides estimates of South Korea–funded construction, cash and in-kind, for 2009–2012.

Table H.3 is a more detailed version of Table 7.10, which is a compilation of the available quantitative data on Germany's contributions to the United States in 2009; most if not all of the figures contained therein refer to direct support.

[1] U.S.-Japan Security Consultative Committee, 2011.

Table H.1
Japanese Labor, Utilities, and FIP Contributions (based on Japanese fiscal year estimates)

	2010	2011	2012	2013p	2014p	2015p
Labor cost sharing						
"Upper limit of the number of workers"	23,055	23,055	23,055	22,947	22,840	22,625
Change in number from prior year	N/A	0	0	–108	–107	–215
Percentage change from prior year	N/A	0.00%	0.00%	0.47%	0.47%	0.94%
Contributions in billions of Yen[a]	114	113.1	113.9	114.7	115.5	115.8
Contributions in millions of U.S. Dollars	1,299	1,417	1,429	1,439	1,449	1,453
Utility cost sharing						
Percentage of predicted utility expenditures	76%	76%	75%	74%	73%	72%
"Upper limit of the expenditure" in billions of yen	24.9	24.9	24.9	24.9	24.9	24.9
Contributions in billions of Yen[b]	24.9	24.9	24.9	24.9	24.9	24.9
Contributions in millions of U.S. dollars	284	312	312	312	312	312
Facilities improvement program						
Minimum funding commitment	20.6	20.6	20.6	20.6	20.6	20.6
Contributions in billions of yen	20.6	20.6	20.6	20.6	20.6	20.6
Contributions in millions of U.S. dollars	234.7	258.1	258.4	258.4	258.4	258.4
Total labor, utility, and facilities improvement program contributions						
In billions of yen[c]	159.5	158.6	159.4	160.2	161.0	161.3
In millions of U.S. dollars[c]	1,817	1,987	1,999	2,010	2,020	2,023

SOURCES: For the yen-denominated estimates, Government of Japan, Ministry of Defense, "Defense Programs and Budget of Japan," 2010–2012, and authors' projections (see notes below) for 2013–2015; the U.S.-dollar-denominated figures are based on annual exchange rates from OECD, 2012a, and quarterly exchange rates from OECD, 2012b, downloaded on October 21–23, 2012, for 2010–2012 and carried forward from 2012.

NOTE: p = projection.

[a] Projections assume changes in cost of labor commensurate with OECD-reported estimates of changes in hourly earnings, in OECD, 2012c.

[b] Projections assume that usage is non-declining and upper limit is binding, owing to rising utility costs (OECD-reported estimates of changes in energy costs, in OECD, 2012c) that would more than offset declining shares.

[c] Projections assume that the 188.1 billion yen commitment to cost sharing, described in Chapter Seven, is not a binding maximum for "cost sharing for stationing USFJ." If it were, either other forms of cost sharing (specifically, training relocation or measures for base personnel) would need to decline slightly to accommodate the projected increase in labor cost sharing or spending for utilities, labor, and facilities improvement program contributions would, in combination, need to be limited to 159.5 billion yen, which was their aggregate level in 2010.

Table H.2
South Korea–Funded Construction (based on calendar year estimates and measured in nominal currency)

	2009	2010	2011	2012
Cash				
Proportion	0.70	0.40	0.12	0.12
Billions of won	205	126	40	44
Millions of U.S. dollars	160	109	36	39
In-kind				
Proportion	0.30	0.60	0.88	0.88
Billions of won	88	189	293	326
Millions of U.S. dollars	69	164	265	285
Total South Korea–funded construction				
Billions of won	292	316	333	370
Millions of U.S. dollars	229	273	301	324

SOURCES: Author's estimates of millions of U.S. dollars calculated on the basis of data on proportional allocations and billions of Won, provided to RAND by USFK on September 25, 2012. Calculations used annual exchange rates from OECD, 2012a, and quarterly exchange rates from OECD, 2012b, downloaded on October 21–23, 2012.

Table H.3
Detailed Breakouts of the Compilation of Data on Germany's Contributions to the United States in 2009

Type of Contribution	Millions of Euros	Millions of U.S. Dollars
Construction		
Estimated value of construction work	450	625.1
Construction-related "reimbursements"[a]	70	97.2
Subtotal construction	520	722.3
Payments to third parties for accommodations leased for the U.S. forces	51.1	71.0
Various other costs		
Benefits for former U.S. forces employees	8.2	11.4
Associated administrative costs	0.2	0.3
Property, building, and room management	5	6.9
Leases and rents	0.7	1.0
Maintaining properties and structures	0.5	0.7
Court and similar fees	0	0
Payments made in connection with consigning objects	0.2	0.3
Reimbursing federal states for personnel and material expenses	0.2	0.3
Settling damages caused in connection with the stationing of foreign forces	9.5	13.2
Compensating occupation damages	0.3	0.4
Procuring property	0.2	0.3
Expenses for traffic, communications, supply, and disposal facilities	0	0
Financial support for development [measures]	0	0
Expenses for completing the reinforcement of roads, bridges, etc.	0.3	0.4
Residual-value compensations	1.5	2.1
Subtotal various other costs	26.8	37.2
Total	597.9	830.6

SOURCES: Poss, 2010; Schlaufmann, 2010; exchange rate for 2009 from OECD, 2012a.

NOTES: Most if not all of these contributions appear to be direct contributions.

[a] As worded in the source document, but not constituting a cash payment.

APPENDIX I

Summary Tables of Illustrative Postures

This appendix presents summary tables of the illustrative postures for EUCOM, PACOM, CENTCOM, and AFRICOM.

EUCOM

Table I.1
EUCOM Bases Open or Closed in Illustrative Postures

		CRP	GREP	MCP
USAF	Retain/ no changes	Ramstein (Germany); Croughton, Fylingdales (UK); Papa (Hungary); Lask (Poland)	Ramstein (Germany); Welford, Fairford, Mildenhall, All intel/command and control bases (UK); Papa (Hungary); Lask (Poland)	Ramstein (Germany); Aviano (Italy); Mildenhall, Fairford, Welford, All intel/ command and control bases (UK); Lajes (Portugal); Moron (Spain); Incirlik (Turkey); Papa (Hungary)
	Increase presence	Move 31st FW from Avia to Spangdahlem (Germany)	Move 52nd FW from Spangdahlem to Aviano (Italy); rotations to Bulgaria/ Romania	Add 1 F-16 squadron from CONUS to Spangdahlem (Germany); ABW Romania and fighter squadron rotations Bulgaria/ Romania
	Reduce presence	RAF Mildenhall (remove 100th ARW) shifts to expansible base (UK); reduce ABW Incirlik, (Turkey)	Reduce to ABW at Spangdahlem (Germany), RAF Lakenheath shifts to expansible base (remove 48th FW), reduce size of ABWs at Lajes (Portugal) and Incirlik (Turkey)	(Action not taken in posture)
	Close/give up access	Lajes (Portugal); Fairford, Lakenheath (48th FW), Upwood, Molesworth, Welford, Alconbury, Menwith Hill (UK); Aviano (31st FW) (Italy); Akrotiri (Cyprus); Moron (Spain); Graf Ignatievo (Bulgaria)	48th FW, Upwood (UK)	RAF Lakenheath (48th FW), Upwood (UK); Lask (Poland)

References

"Agreement Between the United States of America and Japan Concerning New Special Measures Relating to Article XXIV of the Agreement Under Article VI of the Treaty of Mutual Cooperation and Security Between the United States of America and Japan, Regarding Facilities and Areas and the Status of United States Armed Forces in Japan," Tokyo, January 21, 2011.

"Agreement Between the United States of America and the Republic of Korea Concerning Special Measures Relating to Article V of the Agreement Under Article IV of the Mutual Defense Treaty Between the Republic of Korea and the United States of America Regarding Facilities and Areas and the Status of United States Armed Forces in the Republic of Korea," Seoul, January 15, 2009.

Air Force Pamphlet 10-1403, *Air Mobility Planning Factors*, December 18, 2003.

Air Force Pamphlet 10-1403, *Air Mobility Planning Factors*, December 12, 2011.

Air Mobility Command, *Air Mobility Command Global En Route Strategy White Paper*, version 7.2.1, July 14, 2010.

Albion, Robert G., "Distant Stations," *Proceedings*, Vol. 80, March 1954.

Assistant Secretary of the Army for Financial Management and Comptroller, "Financial Management," 2010. As of November 11, 2012:
http://asafm.army.mil/offices/CE/ForcesInfo.aspx?OfficeCode=1400

Barnett, Michael N., *Dialogues in Arab Politics*, New York: Columbia University Press, 1998.

Berchtold, David, email to authors, Department of Defense Office of Acquisition, Technology, and Logistics, June 28, 2012.

Berejikian, Jeffrey D., "A Cognitive Theory of Deterrence," *Journal of Peace Research*, Vol. 39, No. 2, 2002.

Berteau, David J., Michael J. Green, Gregory Kiley, and Nicholas Szechenyi, *U.S. Force Posture Strategy in the Asia Pacific Region: An Independent Assessment*, Washington, D.C.: Center for Strategic and International Studies, August 2012.

Binnendijk, Hans, "Testimony Before the Senate Foreign Relations Committee: Defense Issues for the NATO Summit," May 10, 2012.

Blaker, James R., *United States Overseas Basing: An Anatomy of the Dilemma*, New York: Praeger, 1990.

Bloomfield, Lincoln P., Jr., "Politics and Diplomacy of the Global Defense Posture Review," in Carnes Lord, ed., *Reposturing the Force: U.S. Overseas Presence in the Twenty-First Century*, Newport, R.I.: Naval War College Press, 2006.

Bonn International Center for Conversion, *Restructuring the US Military Presence in Germany: Scope, Impacts, and Opportunities*, June 1995.

Bowie, Christopher J., *The Anti-Access Threat and Theater Air Bases*, Washington, D.C.: Center for Strategic and Budgetary Assessments, 2002.

Bowie, Christopher J., Suzanne M. Holroyd, John Lund, Richard E. Stanton, James R. Hewitt, Clyde B. East, Tim Webb, and Milton Kamins, *Basing Uncertainties in the NATO Theater*, Santa Monica, Calif.: RAND Corporation, 1989, not available to the general public.

Boyne, Walter J., "El Dorado Canyon," *Air Force Magazine*, Vol. 82, No. 3, March 1999.

———, *The Two O'Clock War: The 1973 Yom Kippur Conflict and the Airlift That Saved Israel*, New York, N.Y.: Thomas Dunne Books, 2002.

Buddin, Richard J., Brian P. Gill, and Ron W. Zimmer, *Impact Aid and the Education of Military Children*, Santa Monica, Calif.: RAND Corporation, MR-1272-OSD, 2001. As of January 24, 2013: http://www.rand.org/pubs/monograph_reports/MR1272.html

Bueno de Mesquita, Bruce, *The War Trap*, New Haven, Conn.: Yale University Press, 1981.

Burns, Robert, "Clinton: U.S. Will Consider Keeping Manas," *Air Force Times*, December 2, 2010.

Calder, Kent E., *Embattled Garrisons: Comparative Base Politics and American Globalism*, Princeton, N.J.: Princeton University Press, 2007.

Carney, Stephen A., *Allied Participation in Operation Iraqi Freedom*, Washington, D.C.: Center of Military History, United States Army, 2011.

Carter, Ashton B., "The U.S. Strategic Rebalance to Asia: A Defense Perspective," speech delivered in New York City, N.Y., August 1, 2012.

Cavas, Christopher P., "Undersecretary: 8 LCS Could Be Based in Gulf," *Navy Times*, May 21, 2012.

Chanlett-Avery, Emma, and Ian E. Rinehart, *The U.S. Military Presence in Okinawa and the Futenma Base Controversy*, Washington, D.C.: Congressional Research Service, August 3, 2012.

Clarke, Duncan L., and Daniel O'Connor, "U.S. Base Rights Payments After the Cold War," *Orbis*, Vol. 37, No. 3, Summer 1993.

Coletta, Paolo E., and K. Jack Bauer, *United States Navy and Marine Corps Bases Overseas*, Westport, Conn.: Greenwood Press, 1985.

Commission on Review of Overseas Military Facility Structures of the United States, *Report of the Commission on Review of Overseas Military Facility Structures of the United States*, H-8-H10, May 2005.

Congressional Budget Office, *Options for Changing the Army's Overseas Basing*, May 2004.

———, *Options for Strategic Military Transportation Systems*, September 2005.

Converse, Elliott V., III, *Circling the Earth: United States Military Plans for a Postwar Overseas Military Base System, 1942–1948*, Maxwell Air Force Base, Ala.: Air University Press, 2005.

Cooley, Alexander, *Base Politics: Democratic Change and the U.S. Military Overseas*, Ithaca, N.Y.: Cornell University Press, 2008.

Cooley, Alexander, and Hendrik Spruyt, *Contracting States: Sovereign Transfers in International Relations*, Princeton, N.J.: Princeton University Press, 2009.

Cordesman, Anthony H., *U.S. Forces in the Middle East: Resources and Capabilities*, Boulder, Colo.: Westview Press, 1997.

———, *Saudi Arabia, the US, and the Structure of Gulf Alliances*, Washington, D.C.: Center for Strategic and International Studies, 1999.

Daalder, Ivo H., "A New Shield Over Europe," *New York Times*, June 6, 2012.

Davis, Lynn E., Stacie L. Pettyjohn, Melanie W. Sisson, Stephen M. Worman, and Michael J. McNerney, "U.S. Overseas Military Presence: What Are the Strategic Choices?" Santa Monica, Calif.: RAND Corporation, MG-1211-AF, 2012. As of February 28, 2013:
http://www.rand.org/pubs/monographs/MG1211.html

Davis, Paul K., *Analytic Architecture for Capabilities-Based Planning, Mission-System Analysis, and Transformation*, Santa Monica, Calif.: RAND Corporation, MR-1513-OSD, 2002. As of January 17, 2013:
http://www.rand.org/pubs/monograph_reports/MR1513.html

Defense Manpower Data Center, "Military Personnel Statistics: Active Duty Military Personnel by Service by Region/Country," various dates. As of March 12, 2013:
http://siadapp.dmdc.osd.mil/personnel/MILITARY/miltop.htm

———, "Active Duty Military Personnel Strengths by Regional Area and by Country," December 31, 2011. As of January 17, 2013:
http://siadapp.dmdc.osd.mil/personnel/MILITARY/history/hst1112.pdf

Defense Travel Management Office "Overseas COLA Calculator," online, undated. As of January 19, 2013:
http://www.defensetravel.dod.mil/site/colaCalc.cfm

———, "Per Diem Rates Query," online, August 26, 2011. As of January 17, 2013:
http://www.defensetravel.dod.mil/site/perdiemCalc.cfm

Department of the Air Force, *Fiscal Year (FY) 2012 Budget Estimates: Active MILPERS Appropriation*, February 2011. As of January 17, 2013:
http://www.saffm.hq.af.mil/shared/media/document/AFD-110204-046.pdf

———, *Fiscal Year (FY) 2013 Budget Estimates: Military Personnel Appropriation*, February 2012a. As of January 19, 2013:
http://www.saffm.hq.af.mil/shared/media/document/AFD-120206-029.pdf

———, *Fiscal Year (FY) 2013 Budget Estimates: Operations and Maintenance, Air Force*, Volume I, February 2012b. As of January 17, 2013:
http://www.saffm.hq.af.mil/shared/media/document/AFD-120206-061.pdf

Department of the Air Force, Headquarters Air Force Civil Engineer Support Agency, *Engineering Technical Letter (ETL) 09-1: Airfield Planning and Design Criteria for Unmanned Aircraft Systems (UAS)*, Tyndall AFB, Fl., September 28, 2009.

Department of the Army, *Fiscal Year (FY) 2013 Budget Estimate: Military Personnel, Army, Justification Book*, February 2012. As of January 19, 2013:
http://asafm.army.mil/Documents/OfficeDocuments/Budget/BudgetMaterials/FY13/milpers/mpa.pdf

Department of Defense, *Allied Contributions to the Common Defense*, various dates. As of January 17, 2013:
http://www.defense.gov/pubs/allied.aspx

———, "Military Construction Program Budget: North Atlantic Treaty Organization Security Investment Program, Justification Data Submitted to Congress," various dates.

———, *Quadrennial Defense Review Report*, September 30, 2001.

———, *Facilities Recapitalization Front-End Assessment*, August 2002.

———, *Strengthening U.S. Global Defense Posture: Report to Congress*, September 2004a.

———, *2004 Statistical Compendium on Allied Contributions to the Common Defense*, 2004b. As of January 17, 2013:
http://www.defense.gov/pubs/allied_contrib2004/allied2004.pdf

———, "Deterrence Operations Joint Operating Concept," Version 2.0, December 2006.

———, *Quadrennial Defense Review Report*, February 2010a.

———, "DoD Real Property Inventory Data Element Dictionary," RPIM, Version 4.0, April 22, 2010b.

———, "Unclassified Report on the Military Power of Iran," Washington, D.C., April 2010c. As of January 17, 2013:
http://www.fas.org/man/eprint/dod_iran_2010.pdf

———, *Nuclear Posture Review Report*, April 2010d.

———, *Sustaining U.S. Global Leadership: Priorities for 21st Century Defense*, Washington, D.C., January 2012a.

———, Financial Management Regulation, DoD 7000.14-R, Volume 12, Chapter 24, October 12, 2012b.

Department of Defense, Office of the Assistant Secretary of Defense (Public Affairs), "Joint Statement of the Security Consultative Committee," April 26, 2012. As of January 17, 2013:
http://www.defense.gov/releases/release.aspx?releaseid=15220

Department of Defense, Office of the Under Secretary of Defense, Acquisition, Technology, and Logistics, "Facility Quality Rating Guidance," Memorandum, September 5, 2007.

Department of Defense, Under Secretary of Defense, Comptroller, *National Defense Budget Estimates*, Washington, D.C., various years.

———, "Fiscal Year (FY) 2005 Burdensharing Contribution Report," January 16, 2006.

———, "Fiscal Year (FY) 2006 Burdensharing Contribution Report," February 20, 2007.

———, "Fiscal Year (FY) 2007 Burdensharing Contribution Report," transmittal date not available.

———, "Fiscal Year (FY) 2008 Burdensharing Contribution Report," November 4, 2008.

———, "Fiscal Year (FY) 2009 Burdensharing Contribution Report, November 21, 2009.

———, "Fiscal Year 2010 Burden Sharing Contribution Report," November 24, 2010.

———,"Fiscal Year 2011 Burden Sharing Contribution Report," December 28, 2011.

———, *Construction Programs (C-1): Department of Defense Budget Fiscal Year 2013*, February 2012.

Department of Education, Office of Elementary and Secondary Education, "About Impact Aid: Impact Aid Programs," August 27, 2008. As of January 19, 2013:
http://www2.ed.gov/about/offices/list/oese/impactaid/whatisia.html

Department of the Navy, *Fiscal Year (FY) 2013 Budget Estimates: Justification of Estimates, Military Personnel, Marine Corps*, February 2012a. As of January 19, 2013:
http://www.finance.hq.navy.mil/FMB/13pres/MPMC_Book.pdf

———, *Fiscal Year (FY) 2013 Budget Estimates: Justification of Estimates, Military Personnel, Navy*, February 2012b. As of January 19, 2013:
http://www.finance.hq.navy.mil/FMB/13pres/MPN_Book.pdf

———, *Fleet Response Plan*, OPNAV Instruction 3000.15, August 31, 2006.

———, *Personnel Tempo of Operations Program*, OPNAV Instruction 3000.13C, January 16, 2007a.

———, *Surface Force Training Manual*, COMNAVSURFOR Instruction 3502.1D, July 1, 2007b.

Department of the Navy, Assistant Secretary of the Navy (Energy, Installations, and Environment), *Report to Congress on Camp Lemonnier, Djibouti, Master Plan*, Washington, D.C., August 12, 2012.

Department of the Navy, Headquarters United States Marine Corps, *Manpower Unit Deployment Program Standing Operating Procedures*, Marine Corps Order P3000.15B, October 11, 2001.

Deputy Secretary of Defense Gordon England, *DoD Policy and Responsibilities Relating to Security Cooperation*, Department of Defense Directive Number 5132.03, October 24, 2008.

Directorate of Manpower, Air Force Personnel Center, *USAF Manpower Standards*, undated, not available to the general public.

DMDC—*See* Defense Manpower Data Center

Dobbins, James, Seth G. Jones, Keith Crane, and Beth Cole DeGrasse, *The Beginner's Guide to Nation-Building*, Santa Monica, Calif: RAND Corporation, MG-557-SRF, 2007. As of January 17, 2013:
http://www.rand.org/pubs/monographs/MG557.html

Dobbins, James, Seth G. Jones, Keith Crane, Andrew Rathmell, Brett Steele, Richard Teltschik, and Anga R. Timilsina, *The UN's Role in Nation-Building: From the Congo to Iraq*, Santa Monica, Calif.: RAND Corporation, MG-304-RC, 2005. As of January 17, 2013:
http://www.rand.org/pubs/monographs/MG304.html

Dobbins, James, John G. McGinn, Keith Crane, Seth G. Jones, Rollie Lal, Andrew Rathmell, Rachel M. Swanger, and Anga R. Timilsina, *America's Role in Nation-Building: From Germany to Iraq*, Santa Monica, Calif.: RAND Corporation, MR-1753-RC, 2003. As of January 17, 2013:
http://www.rand.org/pubs/monograph_reports/MR1753.html

DoD—*See* Department of Defense

Duke, Simon, *United States Military Forces and Installations in Europe*, Oxford: Oxford University Press, 1989.

Ek, Carl, *NATO Common Funds Burdensharing: Background and Current Issues*, Washington, D.C.: Congressional Research Service, RL30150, February 15, 2012.

Erickson, Andrew S., Walter C. Ladwig III, and Justin D. Mikolay, "Diego Garcia and the United States' Emerging Indian Ocean Strategy," *Asian Security*, Vol. 6, No. 3, 2010.

Fidler, Stephen, "U.S. to Keep Troops Longer in Europe," *Wall Street Journal*, April 8, 2011.

Fields, Todd W., "Eastward Bound: The Strategy and Politics of Repositioning U.S. Military Bases in Europe," *Journal of Public and International Affairs*, Vol. 15, Spring 2004.

Flournoy, Michele, and Janine Davidson, "Obama's New Global Posture: The Logic of U.S. Foreign Deployments," *Foreign Affairs*, Vol. 91, No. 4, July/August 2012.

Fontenot, Gregory, E. J. Degen, and David Tohn, *On Point: The United States Army in Operation Iraqi Freedom*, Annapolis, Md.: Naval Institute Press, 2004.

Freedom House, *Freedom in the World 2012: The Arab Uprisings and Their Global Repercussions*, 2012. As of January 17, 2013:
http://www.freedomhouse.org/sites/default/files/FIW%202012%20Booklet_0.pdf

Frickenstein, Scott G., "Kicked Out of K2," *Air Force Magazine*, Vol. 93, No. 9, September 2010.

Gaffney, H. H., and Robert Benbow, Jr., *Employment of Amphibious MEUs in National Responses to Situations*, Alexandria, Va.: The CNA Corporation, 2006.

GAO—*See* Government Accountability Office

Gause, Gregory, III, *The International Relations of the Persian Gulf*, Cambridge: Cambridge University Press, 2010.

Gillem, Mark L., *America Town: Building the Outposts of Empire*, Minneapolis, Minn.: University of Minnesota Press, 2007.

Gilmore, Gerry J., "Ships, Planes Deliver Stryker Brigade to Afghanistan," American Forces Press Service, Washington, D.C., August 21, 2009.

Gisick, Michael, "U.S. Base Projects Continue in Iraq Despite Plans to Leave," *Stars and Stripes*, June 1, 2010. As of January 17, 2013:
http://www.stripes.com/news/u-s-base-projects-continue-in-iraq-despite-plans-to-leave-1.105237

Government Accountability Office, *Overseas Basing: Withdrawal of U.S. Forces from Thailand: Ways to Improve Future Withdrawal Operations*, LCD-77-446, November 1977.

———, *Overseas Basing: Costs of Relocating the 401st Tactical Fighter Wing*, GAO/NSIAD-89-225, September 1989.

———, *European Drawdown: Status of Residual Value Negotiations in Germany*, Washington, D.C., June 1994.

———, *Defense Infrastructure: Factors Affecting U.S. Infrastructure Costs Overseas and the Development of Comprehensive Master Plans*, GAO-04-609, July 2004.

———, *Overseas Master Plans Are Improving, but DOD Needs to Provide Congress Additional Information About the Military Buildup on Guam*, GAO-07-1015, 2007.

———, *Defense Transportation: Additional Information Is Needed for DoD's Mobility Capabilities and Requirements Study 2016 to Fully Address All of Its Study Objectives*, GAO-11-82R, December 8, 2010.

———, *Comprehensive Cost Information and Analysis of Alternatives Needed to Assess Military Posture in Asia*, GAO-11-316, May 2011.

Government of Japan, Ministry of Defense, "Defense Programs and Budget of Japan," budget documents for FYs 2010–2012. As of January 17, 2013:
http://www.mod.go.jp/e/d_budget/index.html

Government of Japan, Ministry of Foreign Affairs, "U.S. Forces in Japan-Related Costs Borne by Japan (JFY 2012 Budget)," undated. As of January 17, 2013:
http://www.mofa.go.jp/region/n-america/us/security/pdfs/arrange_ref6.pdf

———, "The Japan-U.S. Security Arrangements," July 2012. As of January 17, 2013:
http://www.mofa.go.jp/region/n-america/us/security/arrange.html

Government of Japan, Ministry of Internal Affairs and Communications, Statistics Bureau, *Report on the Consumer Price Index, Historical Data*, various years. As of January 18, 2013:
http://www.stat.go.jp/english/data/cpi/1588.htm#his

Grimmett, Richard F., *U.S. Military Installations in NATO's Southern Region*, report prepared for the U.S. Congress, Washington, D.C.: U.S. Government Printing Office, 1986.

Han, Zhang, and Huang Jingjing, "New Missile 'Ready by 2015': Global Times," People's Daily Online, February 18, 2011. As of January 18, 2013:
http://english.peopledaily.com.cn/90001/90776/90786/7292006.html

Hardy, James, "US Confirms USS *Freedom* Deployment to Singapore," *Jane's Navy International*, May 11, 2012.

Harkavy, Robert E., *Bases Abroad: The Global Foreign Military Presence*, New York: Oxford University Press, 1989.

———, "Thinking About Basing," in Carnes Lord, ed., *Reposturing the Force: U.S. Overseas Presence in the Twenty-First Century*, Newport, R.I.: Naval War College Press, 2006.

Headquarters Air Combat Command, *Basic Expeditionary Airfield Resources (BEAR), Mission Brief*, November 2, 2011.

Headquarters Air Force, *Planning and Design of Expeditionary Airbases*, AFPAM 10-219, Vol. 6, February 2006.

———, *USAF Supply Manual*, Volume 2, AFMAN 23-110, April 2009.

Headquarters, Department of the Army, *Army Prepositioned Operations*, FM 3-35.1, July 2008.

Headquarters Marine Corps, *Prepositioning Programs Handbook*, 2nd Edition, January 2009. As of January 17, 2013:
http://www.marines.mil/Portals/59/Publications/Prepositioning%20Programs%20Handbook%20 2d%20Edition.pdf

———, *Maritime Prepositioning Force (MPF) Force Lists (F/L)*, Department of the Navy, MCBuL 3501, April 14, 2010.

Headquarters, U.S. Army Europe, "Strong Soldiers, Strong Teams," briefing, Heidelberg, Germany, August 29, 2012.

Henry, Ryan, "Transforming the U.S. Global Defense Posture," in Carnes Lord, ed., *Reposturing the Force: U.S. Overseas Presence in the Twenty-First Century*, Newport, R.I.: Naval War College Press, 2006.

Huth, Paul K., "Extended Deterrence and the Outbreak of War," *American Political Science Review*, Vol. 82, No. 2, June 1988.

———, "Deterrence and International Conflict," *Annual Review of Political Science*, Vol. 2, 1999.

IISS—*See* International Institute for Strategic Studies

Ingimundarson, Valur, "Relations on Ice Over U.S. Jets," *New York Times*, July 12, 2003.

International Institute for Strategic Studies, *The Military Balance*, Vol. 110, No. 1, London: Routledge, 2010.

Jaffe, Greg, "2 Army Brigades to Leave Europe in Cost-Cutting Move," *Washington Post*, January 12, 2012.

Jervis, Robert, "Cooperation Under the Security Dilemma," *World Politics*, Vol. 30, No. 2, January 1978.

———, "The Confrontation Between Iraq and the US: Implications for the Theory and Practice of Deterrence," *European Journal of International Relations*, Vol. 9, No. 2, June 2003.

Jervis, Robert, Richard Ned Lebow, and Janice Gross Stein, *Psychology and Deterrence*, Baltimore, Md.: Johns Hopkins University Press, 1985.

Joint Federal Travel Regulations, Volume 1, *Uniformed Service Members*, Alexandria, Va.: The Per Diem, Travel and Transportation Allowance Committee, October 1, 2012. As of March 13, 2013:
http://www.defensetravel.dod.mil/Docs/perdiem/JFTR(Ch1-10).pdf

Joint Publication 1-02, *Department of Defense Dictionary of Military and Associated Terms*, Joint Chiefs of Staff, November 15, 2012. As of January 17, 2013:
http://www.dtic.mil/doctrine/new_pubs/jp1_02.pdf

Kerr, Julian, and James Hardy, "First Tranche of US Marines Arrives in Darwin," *Jane's Defense Weekly*, April 5, 2012.

Krepinevich, Andrew, and Robert Work, *A New Global Defense Posture for the Second Transoceanic Era*, Washington, D.C.: Center for Strategic and Budgetary Assessment, 2007.

Kuffner, Stephan, "Ecuador Targets a U.S. Air Base," *Time*, May 14, 2008.

Lebow, Richard Ned, "Deterrence: Critical Analysis and Research Questions," in Myriam Dunn and Victor Mauer, eds., *The Routledge Companion to Security Studies*, London: Routledge, August 2011.

Leffler, Melvyn P., "The American Conception of National Security and the Beginnings of the Cold War, 1945–48," *American Historical Review*, Vol. 89, No. 2, April 1984.

———, *A Preponderance of Power: National Security, the Truman Administration, and the Cold War*, Stanford, Calif.: Stanford University Press, 1993.

Livermore, Seward W., "American Naval-Base Policy in the Far East 1850–1914," *Pacific Historical Review*, Vol. 13, No. 2, June 1944.

Lum, Thomas, *The Republic of the Philippines and U.S. Interests*, Washington, D.C.: Congressional Research Service, RL33233, April 5, 2012

Lutz, Catherine, "Introduction: Bases, Empire, and Global Response," in Catherine Lutz, ed., *The Bases of Empire: The Global Struggle Against U.S. Military Posts*, New York: New York University Press, 2009.

Lynch, Marc, "Paint by Numbers," *The National*, May 29, 2009. As of January 17, 2013:
http://www.thenational.ae/news/world/middle-east/paint-by-numbers#full

Maclean, William, "Djibouti: Western Bases Pose Manageable Risk," *Chicago Tribune*, July 11, 2012. As of January 17, 2013:
http://articles.chicagotribune.com/2012-07-11/news/sns-rt-us-djibouti-securitybre86a1bv-20120711_1_djiboutians-djibouti-economy-shabaab

Maehara, Seiji, Minister of Foreign Affairs of Japan, letter (translated) to His Excellency, Mr. John V. Roos, Ambassador Extraordinary and Plenipotentiary of the United States of America, referring to the U.S.-Japan Special Measures Agreement, Tokyo, January 21, 2011.

Mako, William P., *U.S. Ground Forces and the Defense of Central Europe*, Washington, D.C.: Brookings Institution, 1983.

Manyin, Mark E., Emma Chanlett-Avery, Mary Beth Nikitin, and Mi Ae Taylor, *U.S.–South Korea Relations*, Washington, D.C.: Congressional Research Service, R41481, November 3, 2010.

McGarvey, Ronald G., Robert S. Tripp, Rachel Rue, Thomas Lang, Jerry M. Sollinger, Whitney A. Conner, and Louis Luangkesorn, *Global Combat Support Basing: Robust Prepositioning Strategies for Air Force War Reserve Materiel*, Santa Monica, Calif.: RAND Corporation, MG-902-AF, 2010. As of January 17, 2013:
http://www.rand.org/pubs/monographs/MG902.html

McLeary, Paul, "U.S. Army's Uncertain Future: Generals Study Battalion, Vehicle, Equipment Mix," *Defense News*, October 21, 2012. As of January 17, 2013:
http://www.defensenews.com/apps/pbcs.dll/article?AID=2012310210007

McNerney, Michael, and Thomas Szayna, *Assessing Security Cooperation as a Preventive Tool,* Santa Monica, Calif.: RAND Corporation, forthcoming.

Mearsheimer, John, *Conventional Deterrence*, Ithaca, N.Y.: Cornell University Press, 1983.

Meernik, James, *U.S. Foreign Policy and Regime Instability*, Carlisle, Pa.: United States Army War College Strategic Studies Institute, May 2008. As of November 27, 2012: http://www.strategicstudiesinstitute.army.mil/pdffiles/pub845.pdf

Military Impacted Schools Association, "DoD Impact Aid Funding for Military Children," undated. As of January 24, 2013: http://militarystudent.whhive.com/Content/Media/File/Funding/appropriations_dod_2010.pdf

Military Sealift Command, "MSC Ship Inventory," U.S. Navy, undated a. As of January 17, 2013: http://www.msc.navy.mil/inventory/

———, "Prepositioning," U.S. Navy, undated b. As of January 17, 2013: http://www.msc.navy.mil/pm3/

———, *2012 Ships of the U.S. Navy's Military Sealift Command*, U.S. Navy, undated c. As of January 17, 2013: http://www.msc.navy.mil/latestnews/posters/MSC_USNavyShips.pdf

Miller, Edward S., *War Plan Orange: The U.S. Strategy to Defeat Japan, 1897–1945*, Annapolis, Md.: Naval Institute Press, 1991.

Millett, Allan R., and Peter Maslowski, *For the Common Defense: A Military History of the United States of America*, New York: The Free Press, 1994.

Mills, Patrick, Adam Grissom, Jennifer Kavanagh, Leila Mahnad, and Steven Worman, *The Costs of Commitment: Cost Analysis of Overseas Basing*, Santa Monica, Calif.: RAND Corporation, RR-150-AF, forthcoming.

Moroney, Jennifer D. P., Beth Grill, Joe Hogler, Lianne Kennedy-Boudali, and Christopher Paul, *How Successful Are U.S. Efforts to Build Capacity in Developing Countries?* Santa Monica, Calif.: RAND Corporation, TR-1121-OSD, 2011. As of January 17, 2013: http://www.rand.org/pubs/technical_reports/TR1121.html

Moroney, Jennifer D. P., Patrick Mills, David T. Orletsky, and David E. Thaler, *Working with Allies and Partners: A Cost-Based Analysis of U.S. Air Forces in Europe*, Santa Monica, Calif.: RAND Corporation, TR-1241-AF, 2012. As of January 17, 2013: http://www.rand.org/pubs/technical_reports/TR1241.html

Morral, Andrew R., and Brian A. Jackson, *Understanding the Role of Deterrence in Counterterrorism Security*, Santa Monica, Calif.: RAND Corporation, OP-281-RC, 2009. As of January 17, 2013: http://www.rand.org/pubs/occasional_papers/OP281.html

Mueller, Karl P., Jasen J. Castillo, Forrest E. Morgan, Negeen Pegahi, and Brian Rosen, *Striking First: Preemptive and Preventive Attack in U.S. National Security Policy,* Santa Monica, Calif.: RAND Corporation, MG-403-AF, 2006. As of January 17, 2013: http://www.rand.org/pubs/monographs/MG403.html

National Air and Space Intelligence Center, *Ballistic and Cruise Missile Threat*, Wright-Patterson Air Force Base, Ohio, NASIC-1031-0985-09, April 2009.

National Geospatial-Intelligence Agency, *Automated Air Facilities Intelligence File (AAFIF)*, St. Louis, Mo.: U.S. Government, 2010, not available to the general public.

NATO—*See* North Atlantic Treaty Organization

Naval Facilities Engineer Command, *Navy Equipment Sets in Support of Maritime Prepositioning Force*, 2012.

Nichol, Jim, *Central Asia: Regional Developments and Implications for U.S. Interest*, Washington, D.C.: Congressional Research Service, RL33458, September 19, 2012.

North Atlantic Treaty Organization, "Paying for NATO," undated. As of January 17, 2013:
http://www.nato.int/cps/en/natolive/topics_67655.htm?selectedLocale=en

———, "Agreement Between the Parties to the North Atlantic Treaty Regarding the Status of Their Forces," Washington, D.C., April 4, 1949.

———, *Active Engagement, Modern Defence, North Atlantic Treaty Organization Strategic Concept*, 2010.

———, "International Security Force Assistance (ISAF), Key Facts and Figures," December 3, 2012. As of January 17, 2013:
http://www.nato.int/isaf/docu/epub/pdf/placemat.pdf

OECD—*See* Organisation for Economic Co-operation and Development

Office of the Deputy Under Secretary of Defense (Installations and Environment), Business Enterprise Integration, "RPAD 2009–2011," July 31, 2012, not available to the general public.

Office of the President of the Philippines, "The 1987 Constitution of the Republic of the Philippines—Article XVIII." As of January 17, 2013:
http://www.gov.ph/the-philippine-constitutions/the-1987-constitution-of-the-republic-of-the-philippines/the-1987-constitution-of-the-republic-of-the-philippines-article-xviii/

Office of the Secretary of Defense, *Annual Report to Congress: Military Power of the People's Republic of China*, 2009.

———, *Annual Report to Congress: Military and Security Developments Involving the People's Republic of China*, 2011.

O'Hanlon, Michael, *Unfinished Business: U.S. Overseas Military Presence in the 21st Century*, Washington, D.C.: Center for New American Security, 2008.

Organisation for Economic Co-operation and Development, "Prices: Consumer Prices," Main Economic Indicators, database, 2012a.

———, *Monthly Statistics of International Trade,* database, 2012b.

———, *Main Economic Indicators,* Vol. 2012, No. 10, October 2012c.

Paul, Christopher, Colin P. Clarke, Beth Grill, Stephanie Young, Jennifer D. P. Moroney, Joe Hogler, and Christine Leah, *What Works Best When Building Partner Capacity and Under What Circumstances?* Santa Monica, Calif.: RAND Corporation, MG-1253/1-OSD, 2013. As of March 18, 2013:
http://www.rand.org/pubs/monographs/MG1253z1.html

Payne, Keith B., "The Fallacies of Cold War Deterrence and a New Direction," *Comparative Strategy*, Vol. 22, No. 5, 2003.

Pellerin, Cheryl, "Panetta Calls Kuwait Important U.S. Partner," American Forces Press Service, December 11, 2012. As of January 18, 2013:
http://www.defense.gov/news/newsarticle.aspx?id=118754

Pettyjohn, Stacie L., *U.S. Global Defense Posture, 1783–2011*, Santa Monica, Calif.: RAND Corporation, MG-1244-AF, 2012. As of February 28, 2013:
http://www.rand.org/pubs/monographs/MG1244.html

Poss, Ralf, "Subject: German Financial Contribution to U.S. Forces Construction Work in Germany," letter (as translated) to Glendon Pitts, November 8, 2010.

Public Law 112-81, National Defense Authorization Act for Fiscal Year 2012, December 31, 2011.

Putnam, Robert D., "Diplomacy and Domestic Politics: The Logic of Two-Level Games," *International Organization*, Vol. 42, No. 3, Summer 1998.

Quintal, Paul, "Italian Contributions to Forward Deployed U.S. Forces," memorandum provided to the authors by OSD point of contact, June 5, 2012.

Rabiroff, Jon, "US Artillery Unit May Not Realign South of Seoul," *Stars and Stripes*, June 18, 2012.

Rasmussen, Anders Fogh, "Building Security in an Age of Austerity," keynote speech delivered at Munich Security Conference, February 4, 2011. As of January 18, 2013:
http://www.nato.int/cps/en/natolive/opinions_70400.htm

Reiff, Greg H., "Korea Relocation Plan Construction Update," briefing slides, U.S. Army Corps of Engineers, June 15, 2011.

Reilly, James, *Strong Society, Smart State: The Rise of Public Opinion in China's Japan Policy*, New York: Columbia University Press, 2011.

Richardson, Doug, "China Plans 4,000 km-Range Conventional Ballistic Missile," *Jane's Missiles & Rockets*, March 1, 2011. Accessed on January 11, 2013.

Ross, Robert S., "The 1995–96 Taiwan Strait Confrontation" *International Security*, Vol. 25, No. 4, 2000.

Rowland, Ashley, "Fewer Bases, Same Number of Troops in South Korea, US Ambassador Says," *Stars and Stripes*, February 15, 2012.

Rumsfeld, Donald H., U.S. Secretary of Defense, *The Global Posture Review of United States Military Forces Stationed Overseas: Hearing Before the Committee on Armed Services*, United States Senate, One Hundred Eighth Congress, Second Session, September 23, 2004, Washington D.C.: U.S. Government Printing Office, 2005.

Sandars, Christopher, *America's Overseas Garrisons: The Leasehold Empire*, Oxford: Oxford University Press, 2000.

Schelling, Thomas, *Arms and Influence*, New Haven, Conn.: Yale University Press, 1966.

Scher, Robert, and David F. Helvey, "U.S. Force Posture in the Pacific Command Area of Responsibility: Joint Statement before the House Armed Services Subcommittee on Readiness," August 1, 2012.

Schlaufmann, Michael, "U.S. Forces Stationed in Germany; Direct Support," letter (as translated) to Glendon Pitts, November 22, 2010.

Schnabel, James F., *The Joint Chiefs of Staff and National Policy, 1945–1947*, Washington, D.C.: Office of Joint History, 1996.

Seelke, Clare Ribando, *Ecuador: Political and Economic Situation and U.S. Relations*, Washington D.C.: Congressional Research Service, RS21687, May 21, 2008.

Shlapak, David A., David T. Orletsky, Toy I. Reid, Murray Scot Taner, and Barry A. Wilson, *A Question of Balance: Political Context and Military Aspects of the Cross-Strait Dispute*, Santa Monica, Calif.: RAND Corporation, MG-888-SRF, 2009. As of January 18, 2013:
http://www.rand.org/pubs/monographs/MG888.html

Shlapak, David A., John Stillion, Olga Oliker, and Tanya Charlick-Paley, *A Global Access Strategy for the U.S. Air Force*, Santa Monica, Calif.: RAND Corporation, MR-1216-AF, 2002. As of January 18, 2013:
http://www.rand.org/pubs/monograph_reports/MR1216.html

Siegel, Adam B., *Basing and Other Constraints on Ground-Based Aviation Contributions to U.S. Contingency Operations*, Washington, D.C.: Center for Naval Analysis, March 1995.

Snyder, Don, and Patrick H. Mills, *Supporting Air and Space Expeditionary Forces: A Methodology for Determining Air Force Deployment Requirements*, Santa Monica, Calif.: RAND Corporation, MG-176-AF, 2004. As of January 18, 2013:
http://www.rand.org/pubs/monographs/MG176.html

Snyder, Thomas S., and Daniel F. Harrington, *Historical Highlights: United States Air Forces in Europe 1942–1997*, Ramstein Air Base, Germany: USAFE Office of History, March 14, 1997.

Sortor, Ronald E., *Army Active/Reserve Mix: Force Planning for Major Regional Contingencies*, Santa Monica, Calif.: RAND Corporation, MR-545-A, 1995. As of January 18, 2013:
http://www.rand.org/pubs/monograph_reports/MR545.html

Stillion, John, and David T. Orletsky, *Airbase Vulnerability to Conventional Cruise-Missile and Ballistic-Missile Attacks: Technology, Scenarios, and U.S. Air Force Responses*, Santa Monica, Calif.: RAND Corporation, MR-1028-AF, 1999. As of January 18, 2013:
http://www.rand.org/pubs/monograph_reports/MR1028.html

Surface Deployment and Distribution Command Transportation Engineering Agency, *Deployment Planning Guide: Transportation Assets Required for Deployment*, SDDCTEA Pamphlet 700-5, January 2012.

Swartz, Karl L., Great Circle Mapper, website, 2013. As of January 18, 2013:
www.gcmap.com

Swartz, Peter M., *Sea Changes: Transforming U.S. Navy Deployment Strategy: 1775–2002*, Alexandria, Va.: Center for Naval Analysis, July 31, 2002.

Tan, Michelle, "2 Europe-Based BCTs Pack to Move Out," *Army Times*, October 13, 2012. As of January 18, 2013:
http://www.armytimes.com/news/2012/10/army-2-europe-based-bcts-pack-to-move-out-101312/

Trachtenberg, David J., "US Extended Deterrence: How Much Strategic Force is Too Little?" *Strategic Studies Quarterly*, Vol. 6, No. 2, Summer 2012. As of January 18, 2013:
http://www.au.af.mil/au/ssq/2012/summer/trachtenberg.pdf

United States of America and the Kyrgyz Republic, *Agreement Between the Government of the United States of America and the Government of the Kyrgyz Republic Regarding the Transit Center at Manas International Airport and Any Related Facilities/Real Estate*, Bishkek, May 13, 2009.

United States of America and the Republic of Djibouti, *Arrangement in Implementation of the Agreement Between the Government of the United States of America and the Government of the Republic of Djibouti on Access to and Use of Facilities in the Republic of Djibouti of February 19, 2003, Concerning the Use of Camp Lemonier and Other Facilities and Areas in the Republic of Djibouti*, May 11, 2006.

United States of America and the Republic of Korea, *Implementing Arrangement for Special Measures Agreement*, Seoul, March 24, 2009.

United States–Japan Security Consultative Committee Document, "United States–Japan Roadmap for Realignment Implementation," May 1, 2006. As of January 18, 2013:
http://www.mofa.go.jp/region/n-america/us/security/scc/doc0605.html

U.S. Air Force, "U.S. Air Force Fact Sheet: KC-10 Extender," December 2011a. As of January 18, 2013:
http://www.af.mil/information/factsheets/factsheet.asp?id=109

———, "U.S. Air Force Fact Sheet: KC-135 Stratotanker," December 2011b. As of January 18, 2013:
http://www.af.mil/information/factsheets/factsheet.asp?id=110

U.S. Air Forces in Europe, "Base Conditions and Facility Costs," Section 3, slides 20–24, August 30, 2012a.

———, "Host Nation Support, Financial Arrangements, and Effects on Costs," Chapter 5 of briefing book, provided to RAND staff during site visit, August 2012b.

"US and Japan Revise US Forces Realignment Plans," *Jane's Intelligence Weekly*, May 1, 2012.

U.S. Army Corps of Engineers, DoD Area Cost Factors (ACF): PAX Newsletter No 3.2.1 Revision-1, dated 21 Mar 12 to PAX Newsletter 3.2.1, 19 Jan 12, TABLE – B, PART I and II (US and Overseas Locations), 2012. As of February 26, 2013:
http://www.usace.army.mil/Portals/2/docs/2012PAX321ACFTable%20FinalMar2012.pdf

U.S. Army Europe, "RAND Overseas Basing Study," briefing, August 29, 2012.

U.S. Code, Title 10, Subtitle A, Part 4, Chapter 138, Subchapter II, Sections 2350j and 2350k. University of North Texas, *BRAC Commission—Legal Documents*, UNT Digital Library, 2013. As of January 18, 2013:
http://digital.library.unt.edu/ark:/67531/metadc24569/

U.S. European Command, "EUCOM Releases Command Statement on Force Posture," February 16, 2012. As of January 18, 2013:
http://www.eucom.mil/article/23125/useucom-releases-command-statement-on-force-posture

U.S. Forces Japan, "Special Measures Agreement Overview," briefing, June 27, 2012.

U.S. Forces Korea (USFK), Regulation 37-5, "Special Measures Agreement Process," Headquarters, June 23, 2010.

U.S. House of Representatives, *Conference Report on H.R 1540, National Defense Authorization Act for Fiscal Year 2012,* Report 112-239, Washington, D.C.: U.S. Government Printing Office, December 12, 2011.

U.S.-Japan Security Consultative Committee, "Security Consultative Committee Document Host Nation Support," issued by Secretary of State Clinton, Secretary of Defense Gates, Minister for Foreign Affairs Matsumoto, and Minister of Defense Kitazawa, June 21, 2011.

U.S. Marine Corps, *Organization of Marine Corps Forces*, MCRP 512-D, October 1998.

———, *Prepositioning Programs Handbook*, 2nd edition, January 2009.

———, *The United States Marine Corps: America's Expeditionary Force in Readiness*, July 2012. As of January 18, 2013:
http://www.mca-marines.org/news/marine-corps-connection-americas-expeditionary-force-readiness-38

U.S. Marine Corps, Budget and Execution Division, Programs and Resources Department, correspondence to the author, December 14, 2012.

U.S. Marine Corps, Pacific Division, Plans, Policies and Operations, "U.S. Marine Corps DPRI (Distributed Laydown) Update Brief," briefing, December 2012a.

———, correspondence to the author, December 29, 2012b, not available to the general public.

U.S. Senate, *The Gulf Security Architecture: Partnership with the Gulf Cooperation Council*, a majority staff report prepared for the use of the Committee on Foreign Relations, United States Senate, One Hundred Twelfth Congress, Second Session, Washington, D.C.: U.S. Government Printing Office, June 19, 2012.

U.S. Transportation Command, Joint Distribution Process Analysis Center, *USAFRICOM Case Study*, February 2009, slide 29.

Vick, Alan J., and Jacob L. Heim, *Assessing U.S. Air Force Basing Options in East Asia*, Santa Monica, Calif.: RAND Corporation, 2013, not available to the general public.

Walt, Stephen, *The Origins of Alliances*, Ithaca, NY: Cornell University Press, 1987.

Wasserbly, Daniel, "Panetta: Pacific to Host 60 Per Cent of US Naval Power by 2020," *Jane's Defense Weekly*, June 4, 2012.

Watman, Ken, Dean Wilkening, Brian Nichiporuk, and John Arquillla, *U.S. Regional Deterrence Strategy*, Santa Monica, Calif.: RAND, MR-490-A/AF, 1995. As of January 18, 2013: http://www.rand.org/pubs/monograph_reports/MR490.html

Whitlock, Craig, "Remote U.S. Base at Core of Secret Operations," *Washington Post*, October 25, 2012.

Willard, Robert F., USN, Commander, U.S. Pacific Command, "Testimony Before the Senate Armed Services Committee on U.S. Pacific Command Posture," February 28, 2012.

Yeo, Andrew, "U.S. Military Base Realignment in South Korea," *Peace Review: A Journal of Social Justice*, Vol. 22, No. 2, May 2010a.

———, "Ideas and Institutions in Contentious Politics: Anti-U.S. Base Movements in Ecuador and Italy," *Comparative Politics*, Vol. 42, No. 4, July 2010b.

———, *Activists, Alliances, and Anti-U.S. Base Protests*, Cambridge: Cambridge University Press, 2011.

Yost, Davis S., "Assurance and US Extended Deterrence in NATO," *International Affairs*, Vol. 85, No. 4, 2009.